POLYMER SCIENCE AND TECHNOLOGY
Volume 2

WATER-SOLUBLE POLYMERS

POLYMER SCIENCE AND TECHNOLOGY

A Continuation Order Plan is available for this series. A continuation order will bring delivery of each new volume immediately upon publication. Volumes are billed only upon actual shipment. For further information please contact the publisher.

POLYMER SCIENCE AND TECHNOLOGY
Volume 2

WATER-SOLUBLE POLYMERS

Proceedings of a Symposium held by the American Chemical Society, Division of Organic Coatings and Plastics Chemistry, in New York City on August 30-31, 1972

Edited by
N. M. Bikales

Rutgers University
Ralph G. Wright Chemistry Laboratory
New Brunswick, New Jersey

PLENUM PRESS • NEW YORK - LONDON • 1973

Library of Congress Catalog Card Number 73-79431
ISBN 0-306-36402-6

© 1973 Plenum Press, New York
A Division of Plenum Publishing Corporation
227 West 17th Street, New York, N.Y. 10011

United Kingdom edition published by Plenum Press, London
A Division of Plenum Publishing Company, Ltd.
Davis House (4th Floor), 8 Scrubs Lane, Harlesden, London, NW10 6SE, England

Printed in the United States of America

To the memory of

AHARON KATCHALSKY

and

HERBERT K. LIVINGSTON

two outstanding polymer chemists who met a similar tragic
death—one at Lod Airport, the other in Detroit, Michigan

PREFACE

Water-soluble polymers have been attracting increasing attention because of their utility in industrial applications of great
current concern. Perhaps preeminent among these is their ability
to flocculate suspended solids, e.g., wastes in municipal sew-
age-treatment plants or pulp in papermaking. Other important applications are to aid in so-called secondary recovery of petroleum,
to reduce turbulent friction of water, and as components of
water-based finishes developed in response to environmental constraints. Some water-soluble polymers have shown interesting biologic activity, which is being investigated further.

This book is based on papers presented at a symposium held by
the American Chemical Society, Division of Organic Coatings and
Plastics Chemistry, in New York City on 30-31 August 1972. The
large attendance and the favorable response of the audience confirmed not only our view of the importance of the field but also
the need to bring these topics together. The chapters in this
book are generally enlarged and more detailed, with more complete
bibliographies, than the papers presented at the Symposium. They
include not only the important applications described above, but
also descriptions of new syntheses and characterization methods.

Current industrial manufacture of polyacrylamide, commercially
the most important synthetic water-soluble polymer, is described
in detail - probably for the first time - and new polymers and
synthetic methods are discussed. Characterization of water-soluble
polymers by new instrumental techniques, such as exclusion chromatography and sedimentation equilibrium, as well as viscoelastic
properties and detailed characterization of some individual classes
are described.

We hope this book will help to fill an obvious void in the
literature of polymers.

7 January 1973   Norbert M. Bikales
                                        Livingston, New Jersey

CONTENTS

III. CHARACTERIZATION

CONTRIBUTORS

G. L. BEYER, Research Laboratories, Eastman Kodak Company, Rochester, New York 14650.

NORBERT M. BIKALES, Consulting Chemist, 8 Trafalgar Drive, Livingston, New Jersey 07039.

GEORGE E. F. BREWER, Ford Motor Company, Detroit, Michigan; present address: Coating Consultant, 11065 East Grand River Road, Brighton, Michigan 48116.

J. L. BRONSON, Research Laboratories, Eastman Kodak Company, Rochester, New York 14650.

ROBERT C. BURR, Northern Regional Research Laboratory, U. S. Department of Agriculture, Peoria, Illinois 61604.

GEORGE B. BUTLER, Department of Chemistry and Center for Macro-molecular Science, University of Florida, Gainesville, Florida 32601.

D. CASSON, Jet Propulsion Laboratory, California Institute of Technology, 4800 Oak Grove Drive, Pasadena, California 91103.

W. M. DOANE, Northern Regional Research Laboratory, U. S. Department of Agriculture, Peoria, Illinois 61604.

A. EISENBERG, Department of Chemistry, McGill University, Montreal, Quebec, Canada.

E. J. ELLIS, Department of Chemistry, Lowell Technological Institute, Lowell, Massachusetts 01854.

GEORGE F. FANTA, Northern Regional Research Laboratory, U. S. Department of Agriculture, Peoria, Illinois 61604.

HOWARD G. FLOCK, Research and Development, Calgon Corporation, Pittsburgh, Pennsylvania 15230.

WILLIAM A. FOSTER, Designed Products, Dow Chemical U.S.A.,
    2040 Dow Center, Midland, Michigan 48640.

ROBERT H. FRIEDMAN, Getty Oil Company, 3903 Stoney Brook,
    Houston, Texas 77042.

SUSAN P. GASPER, Research Laboratories, Eastman Kodak Company,
    Rochester, New York 14650.

J. W. HOYT, Naval Undersea Research and Development Center,
    3202 E. Foothill Boulevard, Pasadena, California 91107.

NORIO ISE, Department of Polymer Chemistry, Kyoto University,
    Kyoto, Japan.

ALLENE JEANES, Northern Regional Research Laboratory, U. S.
    Department of Agriculture, Peoria, Illinois 61604.

G. D. JONES, Physical Research, Dow Chemical U.S.A.,
    Midland, Michigan 48640.

OH-KIL KIM, Chemistry Division, Naval Research Laboratory,
    Washington, D. C. 20390.

M. KING, Department of Chemistry, McGill University,
    Montreal, Quebec, Canada.

CAROLE KLEEMAN, Physical Research, Dow Chemical U.S.A.,
    Midland, Michigan 48640.

E. D. KLUG, Hercules Research Center, Hercules Incorporated,
    Wilmington, Delaware 19899.

C. A. LEWIS, Hercules Research Center, Hercules Incorporated,
    Wilmington, Delaware 19899.

D. C. MACWILLIAMS, Dow Chemical U.S.A., 2800 Mitchell Drive,
    Walnut Creek, California 94598.

THOMAS J. MIRANDA, Elisha Gray II Research and Engineering Center,
    Whirlpool Corporation, Monte Road, Benton Harbor,
    Michigan 49022.

TSUNEO OKUBO, Department of Polymer Chemistry, Kyoto University,
    Kyoto, Japan.

EMERSON G. RAUSCH, Calgon Corporation, Pittsburgh, Pennsylvania
    15230.

WILLIAM REGELSON, Division of Medical Oncology, Medical College
    of Virginia, Richmond, Virginia 23219.

ALAN REMBAUM, Jet Propulsion Laboratory, California Institute of
    Technology, 4800 Oak Grove Drive, Pasadena, California 91103.

J. H. ROGERS, Dow Chemical U.S.A., 2800 Mitchell Drive, Walnut
    Creek, California 94598.

C. R. RUSSELL, Northern Regional Research Laboratory, U. S.
    Department of Agriculture, Peoria, Illinois 61604.

J. C. SALAMONE, Department of Chemistry, Lowell Technological
    Institute, Lowell, Massachusetts 01854.

MAURICE SCHMIR, Jet Propulsion Laboratory, California Institute
    of Technology, 4800 Oak Grove Drive, Pasadena, California
    91103.

A. L. SPATORICO, Research Laboratories, Eastman Kodak Company,
    Rochester, New York 14650.

DONALD L. SUSSMAN, Department of Civil and Environmental
    Engineering, University of Rhode Island, Kingston,
    Rhode Island 02881.

JULIA S. TAN, Research Laboratories, Eastman Kodak Company,
    Rochester, New York 14650.

ROBERT Y. TING, Chemistry Division, Naval Research Laboratory,
    Washington, D. C. 20390.

DOREEN VILLANI-PRICE, Physical Research, Dow Chemical U.S.A.,
    Midland, Michigan 48640; present address: G. D. Searle &
    Co., Skokie, Illinois.

ROBERT H. WADE, Naval Undersea Research and Development Center,
    3202 E. Foothill Blvd., Pasadena, California 91107.

EDWARD CHUN-CHIN WANG, Department of Civil and Environmental
    Engineering, University of Rhode Island, Kingston,
    Rhode Island 02881.

T. J. WEST, Dow Chemical U.S.A., 2800 Mitchell Drive, Walnut
    Creek, California 94598.

ROY L. WHISTLER, Department of Biochemistry, Purdue University,
    Lafayette, Indiana 47907.

D. P. WINQUIST, Hercules Research Center, Hercules Incorporated, Wilmington, Delaware 19899.

CHESTER WU, Department of Chemistry and Center for Macromolecular Science, University of Florida, Gainesville, Florida 32601.

S. P. S. YEN, Jet Propulsion Laboratory, California Institute of Technology, 4800 Oak Grove Drive, Pasadena, California 91103.

T. YOKOYAMA, Department of Chemistry, McGill University, Montreal, Quebec, Canada; present address: Engineering Department, Nagasaki University, Nagasaki, Japan.

MONICA A. YORKE, Environmental Research Group, Calgon Corporation, Pittsburgh, Pennsylvania 15230.

I. APPLICATIONS

WATER-SOLUBLE POLYMERS AS FLOCCULANTS IN PAPERMAKING

William A. Foster

Dow Chemical U.S.A.

Midland, Michigan

The papermaking process involves three basic operations. First, a wet web of fibers is formed from a fiber - water slurry and drained on a continuously moving wire. Second, additional water is removed by pressing the web between felts. Third, the web is dried on a series of steam-heated drier drums. The water systems surrounding this basic process will be described in more detail below, as will the complete papermaking system. Figure 1 shows a Fourdrinier paper machine (the most common type), looking from the continuously moving wire toward the driers.

Papermakers use water-soluble polymer flocculants in papermaking for two main reasons:

1. To improve retention, within the sheet, of fiber fines, inorganic fillers, and other small particulate matter.

2. To improve liquid-water removal, or drainage, during the papermaking operation.

Historically, improved retention of titanium dioxide in the sheet has been the primary reason for using filler retention aids (1,2). This is because of the high cost of titanium dioxide (20 to 22¢/lb.) compared to other components in the system. The expanding use of other relatively expensive pigments has increased the need for use of a filler retention aid. Also, recent increases in pollution control measures have made the papermaker more interested in improved retention of the cheaper materials such as clay and fiber fines. Papermakers also have become more aware recently of the potential economics in improved retention of fiber fines.

3

Fig. 1.   Fourdrinier machine in operation.

Even though most of the water drained from the wire during
papermaking is recirculated to the wire, optimum operating standards
require the "first pass" retention to be as high as possible. When
inadvertant spills or leaks occur, or when the machine is shut down
and drained, the loss of expensive filler and fiber is directly related to the concentration of these materials in the recirculating
"white water" (so called because the pigment build-up gives it a
white color). Typical papermaking practice calls for frequent
change in the grade of paper being produced. Often a grade of paper
which requires a high level of titanium dioxide is followed by a
grade which does not require TiO_2. In such case, the TiO_2 buildup
from the first grade will be bled into the second grade until a new
equilibrium is built up. In this case the price obtained for the
second grade would probably not pay for the TiO_2 being used in it,
so the papermaker would operate at a reduced profit. On the other
hand, if the grade without TiO_2 was made first, the system would
have to be "slugged" with large amounts of TiO_2 so that the sheet
following would conform to specifications. This would also cost
the papermaker money. Use of a retention aid would minimize TiO_2
buildup in the system and avoid these losses.

The use of flocculants to improve drainage, or water removal, in the papermaking process is a relatively new development compared with filler retention. Papermakers are primarily interested in drainage aids because they improve machine speeds. Increased machine speed gives increased production for a given amount of equipment which, in turn, makes more money for the papermaker. It is common for a papermaker to get a production increase of ten percent or more. In a typical case, a production increase of ten percent resulted in a net profit increase to the mill of about $250,000 per year (3).

Increased drainage rates also improve sheet formation and increase sheet strength. The improved formation (distribution of fibers throughout the sheet) results from increase dilution of the fiber slurry going to the papermaking process and from removal of excess water by the improved drainage rate. The strength improvement is obtained by increasing the mechanical refining on the fiber itself, which improves strength but decreases drainage. The loss in drainage is compensated for by the increased drainage rate made possible by the polymer. Action of the polymer does not reduce this strength improvement.

THE PAPERMAKING SYSTEM

A typical papermaking system is shown in Figure 2. The stock input system is comprised of the virgin fiber used and the broke (or scrap) from the papermaking operation. It also includes many other ingredients such as fillers, chemical additives, dyes, etc., which are used in the sheet. These materials are all run through a stock-proportioning device and stored in the machine chest to be drawn on by the paper machine.

The primary system we are interested in within the papermaking process is the wet-end recirculation system. We are particularly concerned here with the tray water (white water) which is removed from the early stages of the Fourdrinier wire and recirculated around the system. To this material is added the thick stock from the machine chest. The combination then goes through a fan pump, some cleaning and screening devices and into the headbox. From here the diluted stock, at about 0.5% concentration, flows into the Fourdrinier wire. Most of the water is again drained off in the tray section and recirculated.

The wire in the tray section is supported by a series of rolls (called table rolls) or foils which function as drainage devices. As the wire and table roll surfaces separate on the down-stream side of the roll, the internal tension in the water draws more water from the web until the gap between the two surfaces becomes so large that cavitation occurs. This phenomonon occurs with each roll as long

Paper Machine Wet End System

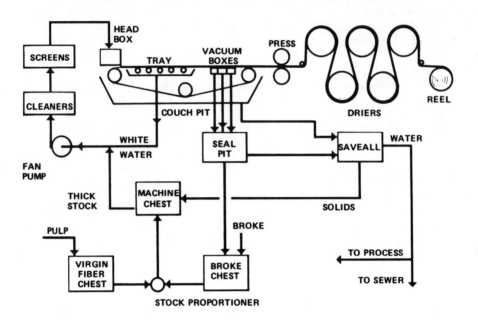

Fig. 2.   Schematic diagram of a typical papermaking system.

as enough water remains in the web to form a liquid continuum at
the roll-wire nip. When there is less water in the web than the
required amount, further water removal must be accomplished by vacuum devices. Foils operate on the same principle, but in a more
gentle fashion (4). Foils are basically stationary flat surfaces
which are slanted a few degrees from the plane of the wire. With
table rolls a pressure buildup on the upstream side of the nip causes
disruptions in the web structure. This pressure buildup does not
occur with foils. However, some lesser amount of web disruption
occurs even with foils (5).

Water removal at the table roll or foil section of the wire
accounts for 60 to 75% of the total drainage on the wire. An additional 20 to 25% is accounted for by the vacuum boxes. The remainder of the water is removed from the wire by hydrostatic pressure
from the weight of the water, and by inertial pressure provided by
the inpingement of the jet from the slice (the outlet from the headbox) into the wire, and by pressure from the rolls riding on the
upper surface of the sheet, such as the dandy roll (6). A dandy
roll is a light wire roll used to give the top or felt side of the
sheet a surface similar to the bottom or wire side.

The next system we are interested in is the vacuum drainage - seal pit - saveall system. A vacuum of about 19 inches Hg is used here. Under a pressure drop of this sort, the web structure compresses, giving a reduced pore size which in turn decreases web permeability. The seal pit collects the water from the vacuum drainage devices on the Fourdrinier wire. The seal pit water (which is somewhat lower in solids concentration than the tray water) is used to feed the saveall. However some of the water from the seal pit can perform other functions, such as reslurrying the broke which comes back from the paper machine.

The saveall is a solids-liquid separator used to remove additional solids from the process water before it leaves the system. The recovered solids are returned to the system in the case of the higher value papers. These generally contain expensive pigments and/or fibers. With lower value papers, the solids are discarded. The trim from the edges of the sheet coming off the wire, and the total sheet, during periods of down time, are dropped into a couch pit below the machine. This material is usually pumped to the saveall for solids recovery, along with the seal pit water. However, in some cases the material from the couch pit is sent directly to the machine chest for immediate reuse. Some of the water leaving the saveall finds its way back to process, but much of it is sent directly to a sewer. Hence, the efficiency of separation in the saveall has a direct bearing on mill cost savings as well as mill pollution control.

After the sheet of paper leaves the Fourdrinier wire (at about 20% solids) it goes through a press section. Here the paper is sandwiched between two felts before it is run through mechanical presses. More water is pressed out of the sheet at this point and is picked up by the felts. These felts then run through felt washers, which remove any fines or chemicals which are picked up from the sheet.

Finally, the pressed sheet (at about 38% solids) runs over a series of steam-heated drier drums to evaporate the remaining water before the sheet is wound up at the reel. The moisture content at the reel is about 6%.

We have dealt here only with the Fourdrinier paper machine, since it is the one most commonly used. Another type of machine used in papermaking is called the cylinder machine (7). On this equipment a number of plys are wet-laminated together to form heavy-weight boxboard. Each ply is formed by a wire roll turning in a vat of stock.

The papermaking operation is most commonly carried out at a pH of from 4.5 to 5.0. However, the range of papermaking process pH values runs from about 3.5 to over 9.0. While there is a gradual shift to making paper at neutral or slightly alkaline pH values, the bulk of papermaking is still done on the acid side. A further description of papermaking can be found in Reference 21.

POLYMERS USED FOR DRAINAGE AND RETENTION IMPROVEMENT

Many different types of polyelectrolytes are used by the paper industry for drainage and retention improvement. They are of both the anionic and cationic type. Molecular weights range from intermediate to high. Dry products are sold, as well as liquid products ranging from 5% solids to about 35% solids. The most commonly used polymers may be classified into the following types:

1. <u>Polyacrylamides</u> (anionic and cationic liquid and dry products available).

2. <u>Polyamides</u> (often referred to as polyamide-amines, cationic, liquid products).

3. <u>Polyamines</u> (cationic, liquid products).

4. <u>Cationic Starches</u> (dry products derived from various sources, e.g., corn, potatoes, etc.).

Representative structures for the above product types are shown in Figure 3. It should be noted that these structures are not necessarily those of specific products on the market today. In addition to the polymer types listed above, combinations of anionic and cationic polymers are often used as drainage and retention systems. In this case, the intermediate- rather than high-molecular-weight products are generally used to avoid overflocculation. Britt describes a combination polymer system where a cationic polymer is used on the thick stock and an anionic polymer on the diluted stock (8). This system gives a rather strong flocculation effect, so some turbulence is needed to break down the larger flocs before the sheets is formed. Britt's experimental machine results with this system gave first-pass-retention values of around 80%, which are quite high. It is noteworthy that no loss in formation was observed. In this case, the anionic polymer was added before the fan pump, which probably resulted in a breakdown of the larger flocs due to the shearing section of the fan pump.

Addition of Drainage and Retention Aids to Papermaking Systems

The main guidelines for adding flocculants to the papermaking system are:

Polyacrylamides

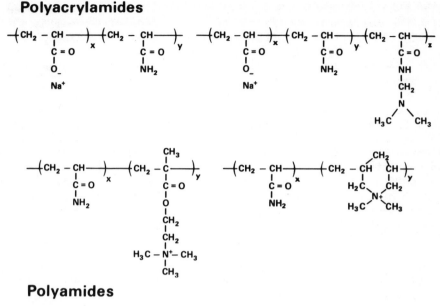

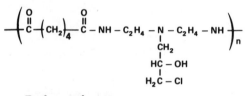

Polyamides

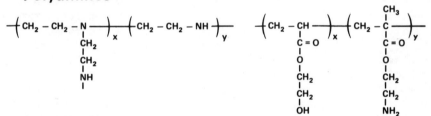

Polyamines

Cationic Starch

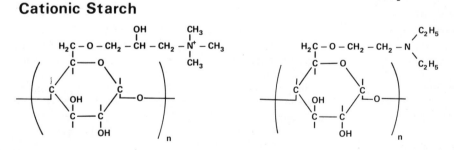

Fig. 3. Structures of water-soluble polymers used for drainage and retention improvement.

1. That they be added to the system after the thick stock has been
added to the recirculating tray water. The tray water contains a
large part of the fines and fillers which have to be retained.

2. That they be added to the system after points of high shear,
but far enough ahead of the headbox to give a good mixing.

Since flocs formed by different polymers have different sensi-
tivities to shear (8), the common practice when testing a flocculant
on a paper machine is to try different points of addition and see
which works best. An addition point in current favor is the outlet
of the centrifugal screens. In this case, however, it is necessary
to have sufficient mixing between the screen and the headbox. If
sufficient mixing is not allowed, "stratification," or localized
over-concentrations will occur, causing streaks in the sheet. The
immediate inlet side of the fan pump also can be used as an addition
point. Here the fan pump is the in-line mixer. The flocs probably
do not form until the stock and polymer have passed through the
pump. When a fan pump addition point is used, however, flocculants
should be used which give flocs less susceptible to breakdown by
the shear forces in the cleaners and screens. When two or more
polymers are used in combination, the last polymer is added as described above. The preceding polymer(s) is added at a point further
back in the system. The first polymer often is added to the thick
stock (8). In such cases the first polymer often functions as a
conditioner for the second polymer.

Generally, it has been observed that for any given polymer
larger amounts need to be added for drainage improvement than are
necessary for improved retention. With the synthetic polyelectrolytes, addition levels for retention are usually in the range of
0.01% to 0.05%, based on dry paper solids and dry polymer solids.
When used as drainage aids, synthetic polyelectrolytes are used at
levels ranging from approximately 0.03% to about 0.20%. Cationic
starch, when used as a combination strength - retention aid, is
used at levels from approximately 0.25% to 0.75%. Cationic starch
is used at higher levels than the synthetics because polymer usage
for strength improvement requires higher levels of addition than
for drainage and retention improvement. Drainage and retention
aids are added to the papermaking system as very dilute solutions.
Because of their inherently very high viscosity, extreme dilution
is required to obtain rapid and thorough mixing with the stock.
Otherwise, localized areas of over-flocculation can occur, causing
poor formation and decreases in optical efficiency of the pigments
used. The synthetic polymers are usually added at concentrations
of 0.05% to 0.5% solids. Cationic starch is added at levels of
about 1% solids. Higher solids are required with cationic starch
because of the need to add larger amounts.

Flocculants are often used at the saveall, when improved retention is the objective. When used in this manner, the flocculants are normally mixed with the incoming water as it enters the saveall. Often, however, the use of a flocculant on the paper machine itself gives a sufficiently improved saveall operation. In some cases, a small fraction of the polymer used on the machine is added immediately before the saveall, and is in addition to that used on the machine. A polymer of opposite charge to that used on the paper machine also may be used on the saveall. There is no set rule on how to best use a flocculant at the saveall. Each mill has its own point of view. Unfortunately, the saveall operation is still governed largely by guesswork. Furthermore, it often receives a disproportionately low share of attention from the machine technical crews.

Drainage and Retention Mechanisms

Drainage and retention aids can alter the structure of the web in three primary ways:

1. They can flocculate or agglomerate the small particulate matter to the larger whole fibers. This is why the flocculant is added to the system after the thick stock is mixed with the fines and filler in the recirculating white water. Figure 4 shows two photomicrographs of a fiber - fines system - one with a flocculant (in this case, polyethylenimine), one without (3). The agglomeration of fines onto fiber is readily visible. Table I shows fiber-length-classification data on the same two pulp systems (with and without polyethylenimine) used to produce the photomicrographs (9). Again, the depletion of fines with the flocculant caused by the agglomeration of fines to whole fibers can readily be seen. Date and Schute (2) show a similar agglomeration of TiO_2 particles to fibers using anionic polyacrylamide and alum.

2. Drainage and retention aids can redistribute the small particulate matter within the web structure. Without the use of a flocculant, fines (and fillers) tend to be retained in the sheet via a filtration mechanism, whereby they tend to plug the pores in the structure and decrease permeability. On the other hand, when a flocculant is used to agglomerate the fines to the larger fibers, the fines are not free to move with the water stream. If allowed to freely move with the water stream, they will eventually be stopped by a pore restriction or pass from the web. If they are stopped by a pore restriction, they will reduce the wet and dry permeability of the web. This concept of fines redistribution is reinforced by the observation that the dry porosity of handsheets and many commercial papers is increased by the use of drainage and retention aids. The formation (distribution of whole fibers) with these sheets remains the same with or without the flocculant.

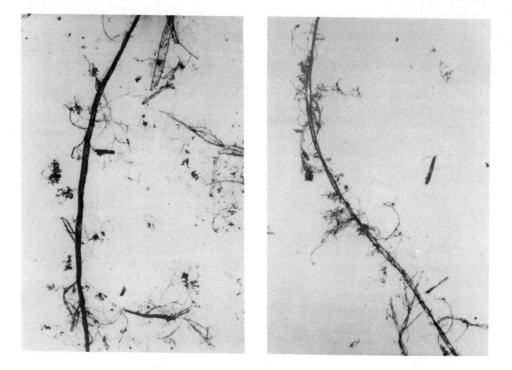

Fig. 4. Effect of polyethylenimine flocculant on fiber - fines
system; left, no flocculant; right, with flocculant.

TABLE I. Fiber Length Distribution
With and Without Drainage Aids

Screen No.	Screen Opening	Weight of Pulp on Screen	
		Control	Drainage Aid
14	1.19 mm	2.58 g	3.03 g
28	0.595	1.45	1.65
48	0.297	1.34	1.50
100	0.149	1.46	1.58
100		1.27	0.34
		8.10	8.10

However the effect of flocculants on the distribution of fillers, and probably of fines, across the sheet from the felt to wire side (two sidedness) is somewhat uncertain. It has been observed with handsheets and some commercial sheets containing colored pigments that flocculants cause less two-sidedness. This reinforces the fines - filler redistribution concept.

By contrast, Groen (10), Truman (11), and Schwable (12) state they have observed higher than normal filler content on the felt side, but the same filler content on the wire side (i.e., worse two-sidedness) when filler agglomeration occurs. However, in their particular observation, large amounts of filler were used (20-30%) and the role of the flocculant was not clearly defined.

3. <u>Drainage and retention aids can reduce or collapse the hydration shell on the fibers and fines.</u> The cause of this is believed to be flocculation of the surface fibrillation, developed on the fibers and larger fines during refining of the stock (13). Collapse or reduction of the swollen hemicellulose on the surface of fibers and fines and reduction of the electrical double layer on the fibers and fines also are effects which the polyelectrolyte could cause which would reduce the hydration shell on the fibers and fines. Any of these effects would increase the <u>wet</u> permeability of the web.

Flocculation is undoubtedly the mechanism by which retention aids perform their designed function. Agglomeration of the fines and filler onto the fibers may make them less susceptible to removel by the liquid-water stream. On the other hand, agglomeration of the fines and filler to themselves may cause larger particles which filter out more easily in the top and center portions of the sheet. It is unclear which of these mechanisms predominates, although both probably occur to some degree. The nature of the interaction between polymer and fiber or fines and the dynamics of the machine itself probably dictate which of the mechanisms functions most dominantly in the process. In any event, retention aids are being used successfully on all types of commercial machines. Table II shows some typical machine results using a retention aid (in this case a cationic acrylamide polymer). The ash retention in Table II is defined as follows:

$$\frac{(\% \text{ Ash in paper})}{(\% \text{ Ash in headbox solids})} \quad \text{x } 100 = \% \text{ ash retention}$$

It is essentially equivalent to first-pass retention.

The mechanism by which drainage aids improve drainage is primarily fines redistribution and, to some extent, hydration-shell collapse, resulting in an increased wet-web permeability. Increased

TABLE II.   Effect of Cationic Acrylamide Polymer Retention Aid[a]

| | Headbox | | White Water | | Paper | % Ash |
	% Solids	% Ash	% Solids	% Ash	Ash	Retention
Without Retention Aid	0.668	50.0	0.360	73.4	17.9	35.8
Without Retention Aid	0.728	52.5	0.418	73.2	19.2	36.6
2 Hours After Addition 0.3 lb/ton	0.536	41.8	0.238	68.1	19.7	47.4
$3\frac{1}{2}$ Hours After Addition 0.5 lb/ton	0.478	34.3	0.174	62.1	18.9	55.1

[a]Pulp: 50/50 bleached HW/SW kraft

permeability allows faster water removal at the table rolls or
foils, the vacuum section, and in the presses. This will allow the
papermaker to run his machine faster if he is limited by wire drainage capacity or press capacity. Figure 5 shows a typical increase
in drainage, achieved through the use of a drainage aid. In this
case the polymer is polyethylenimine and the pulp system is the same
as that used for the photomicrographs and fiber classification data.

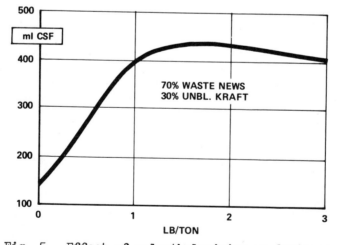

Fig. 5.   Effect of polyethylenimine on drainage.

A more important result of increased wet-web permeability, is that it will allow the sheet to get into the drier at a slightly higher percent solids. Papermakers have a rule-of-thumb which says that a 1% increase in solids going into the drier (e.g., 38% to 39%) is equivalent to a 5% increase in drier capacity. Since paper-machine speeds are commonly limited by lack of dr er capacity, a small increase in water removal before the drier can result in sub-stantial machine-speed increases. In fact, drainage aids are often referred to as drainage and drying improvers. Limited data on a machine making hardboard has shown that a machine-speed increase of approximately ten percent resulted from a two percent increase in solids ahead of the drier.

In addition to increasing solids through improved permeability, fines redistribution and the resultant increase in sheet porosity will probably allow steam to escape from the sheet more rapidly in the drier. This would further enhance the capability of the drain-age aid to improve drying time and drying efficiency. Machine speed increase resulting from flocculants is almost certainly a result of increased web permeability. Less certain are the relative import-ance of improved steam release, better water removal at the presses and increased water removal on the wire. The relative importance of each of these factors probably will vary widely with different furnishes and different machines. Very little work has been pub-lished in this area.

Other Factors Affecting Performance of Drainage and Retention Aids

Papermaking fibers have an anionic surface charge. This is largely due to the uronic acid (carboxyl) groups of the hemicellu-lose portion of the fiber. These hemicelluloses exist as a coating around the cellulose fibrils which comprise the fiber. Therefore, the fiber surfaces which the flocculants see are mostly hemicellu-lose. Anionic polyelectrolytes would act as dispersants rather than flocculants for the anionic fibers if they were not used in conjunction with some multivalent metal ion. In the papermaking process this cationic charge is supplied by alum. In the case of the cationic polyelectrolytes, direct interaction with the anionic fibers is possible.

In many cases, the anionic charge on the fibers can be dimin-ished or even reversed by the effect of preceding chemicals (13-16). The best drainage and retention performance generally occurs near the point of zero net surface charge. At this point the ionic re-pulsion forces are at a minimum. This maximizes the inherent floc-culation tendency. It also minimizes the effect of ionic repulsion forces reinforcing the shear forces which tend to break down the flocs that have already formed. In addition, a low surface charge

will result in a lower tendency for repeptization of the fines and
fillers which are detached from the fibers by the shear forces.

When the charge on the fiber - fines system is sufficiently
diminished, but still anionic, a weaker cationic charge on the polymer is desirable, since more polymer can be absorbed on the system
without completely reversing the charge. If a strongly anionic
fiber - fines system is used and little neutralization has occurred,
a strongly cationic polymer is more desirable. If the charge has
been completely reversed, an anionic will probably work best. Unfortunately, because of the complexity of the system, it is usually
difficult to predict whether the surface charge on the system will
be anionic or cationic on any given papermaking machine. A practical
approach to deciding which polymer(s) should be tested on the machine is to evaluate a sample of headbox stock in the laboratory
for drainage and retention values using a series of different types
of polymers.

Moore has shown that when the charge on the fiber has been
greatly diminished or reversed in a system containing alum, the
performance of the cationic polymer can be increased by the addition
of more alum (17). This is contrary to what might be expected.
Moore postulates that polyvalent anions, such as sulfate, can bridge
the gap between the cationic alum groups complexed with the fiber
surface and the cationic polymer sites. The mechanism is probably
not a simple one, however, since the presence of additional aluminum
and sulfate ion are both required for improved performance.

Other types of materials also can have a detrimental effect on
the performance of a drainage and retention aid. For example, it
has been known for some time that oxidized starches have a detrimental effect on filler retention (18). They adsorb on the pigment
surfaces and form a protective colloid, making the pigments difficult to flocculate. With unbleached kraft paper and board grades,
the anionic waste pulping residues interfere with the action of the
cationic polymers. Cationic polymers will work if added in sufficient quantities, but the amounts required are uneconomical. This
same observation holds for the unbleached neutral sulfite semichemi-
cal (NSSC) pulps used for making corrugating medium. Here, however,
even larger amounts of polymer are required for good performance.
Anionic polymers do not seem to do well in unbleached kraft systems
either, even though alum is present. Alum cannot be used in the
NSSC systems because it interferes with absorbency. Consequently,
anionics cannot be expected to work in NSSC systems.

The use of drainage and retention aids on boxboard furnishes
has been notoriously erratic. Boxboard is made from a mixture of
secondary fibers from many different sources. As a result, the
system is an extremely complex mixture of ionic species. When poly-

mers do not work in boxboard systems, performance can usually be attained by replacing the soluble fraction of the pulp with pure water. While this is not a practical approach in the mill, it does show that the problem often lies in the soluble portion of the system.

Machine dynamics is another factor affecting the performance of a drainage and retention aid. Shear forces in the diluted stock system preceding the wire, and shear forces in the stock on the wire itself can both adversely affect the distribution of particles in the web structure. These forces have been discussed to some degree above. Britt (8) has developed a laboratory technique for measuring drainage and retention under shear which correlates well with observed mill results. As discussed earlier, he has shown that combinations of anionic and cationic polymers give flocs that are less sensitive to shear than flocs formed by single polymers, either anionic or cationic. Single polymer systems are currently being used successfully, but further increases in machine speed will give approaches such as Britt's greater importance.

Problems and Side Effects When Using Drainage and Retention Aids

Overflocculation is probably the biggest problem associated with the use of drainage and retention aids. When a retention aid is used to retain titanium dioxide, for example, some loss in optical efficiency will result. This is caused by the TiO_2 particles being agglomerated with a resultant loss in optical scattering surfaces. Increased retention of TiO_2 will usually offset this slight loss in optical efficiency. At higher levels of retention aid, however, the loss of efficiency will surpass the incremental increase in retention and the opacity will decrease. Different polymers or polymer systems can be expected to have different ratios of retention to loss of optical efficiency. Pummer (19) has given a good illustration of the technique of Brecht and co-workers (20) for relating the various optical properties of paper. They plot scattering power, absorption power, opacity, and brightness on a chart which they call the "plane of optical status." The scattering and absorption power of the sheet are determined using the well-known Kubelka-Monk theory. Becht's technique is an excellent one for showing the relevant optical properties of a sheet of paper on a single graph. Pummer evaluated different types of polymers for their effect on the optical properties of the sheet. He concluded that the lower-molecular-weight polymers have less effect on optical efficiency of fillers than do polymers of high molecular weight. However, lower-molecular-weight polymers usually require substantially higher addition levels to obtain good retention.

Some polyelectrolytes (polyamines, for example) have a delete-
rius effect on optical brighteners (19). They presumably complex
with the optical brighteners and cause them to lose efficiency in
absorbing and re-emitting light.

SUMMARY AND CONCLUSIONS

Drainage and retention aids can function in three ways:

1. to agglomerate fines and fillers to whole fibers;

2. to redistribute fines and fillers within the web;

3. to collapse or reduce the hydration shell on the fibers.

Improved retention is achieved by agglomeration of fines and fillers
to the whole fibers. Redistribution of fines and fillers and collapse of the hydration shell on the fibers and fines gives better
web permeability. This can increase drainage through improved
water removal at the wet end, and improve steam escape on the dryers.
Drainage and retention aids work best at the point of zero net surface charge on the fibers. Combination anionic and cationic polymer systems are less shear-sensitive than single polymer systems.
Evaluation of any retention aid system should include an analysis
of optical properties.

A key point to remember when considering flocculants in papermaking is that these water-soluble polymers are tools that the
papermaker can use to give him more flexibility in operating his
system. He can achieve many advantages by efficiently using these
materials, but they generally involve readjustment of his operation
to take full advantage of them. This is an aspect of the use of
polymers that papermakers often do not fully realize. This implies
a need on the part of the supplier industry for rather large amounts
of technical service to accompany introduction and profitable use
of water-soluble polymer flocculants.

BIBLIOGRAPHY

1. J. F. Reynolds and R. F. Ryan, Tappi, 40 (1), 918 (1957).

2. J. D. Date and J. M. Schute, Tappi, 42 (10), 824 (1959).

3. W. H. Ellis and W. A. Foster, Paper Trade J., 152 (23), 34
   (June 3, 1968).

4. T. O. P. Speidel and N. Beauchemin, Paper Trade J., 155 (11),
   68 (March 15, 1971).

5. P. E. Wrist, in F. Bolam, Ed., Formation and Structure of Paper, Vol. II, Tech. Sect., Brit. Paper and Board Makers Assoc., Inc., London, 1962, p. 839.

6. W. H. Kennedy and P. E. Wrist, in C. E. Libby, Ed., Pulp and Paper Sci. and Tech., Vol. II, McGraw-Hill Book Co., Inc., New York, 1962, p. 163.

7. J. C. Nutter, in R. G. MacDonald, Ed., Pulp and Paper Mfg., Vol. III, McGraw-Hill Book Co., Inc., New York, 1970, p. 297.

8. K. W. Britt, Preprint TAPPI Papermakers Conf., Atlanta, June, 1972, p. 249.

9. Chemical 26, 2 (7), 23 (1966).

10. L. J. Groen, in F. Bolam, Ed., Formation and Structure of Paper, Vol. II, Tech. Sect., Brit. Paper and Board Makers Assoc., Inc., London, 1962, p. 697.

11. A. B. Truman, in F. Bolam, Ed., Consolidation of the Web, Vol. II, Tech. Sect., Brit. Paper and Board Makers Assoc., Inc., London, 1966, p. 739.

12. H. C. Schwalbe, ibid, p. 740.

13. M. Y. Chang and A. A. Robertson, Pulp and Paper Mg. Can., 68 (9), T-438 (1967).

14. D. L. Kenaga, W. A. Kindler, and F. J. Meyer, Tappi, 50 (7), 381 (1967).

15. E. Strazdins, Tappi, 53 (1), 80 (1970).

16. W. F. Linke, Tappi, 51 (11), 59A (1968).

17. E. E. Moore, Preprint TAPPI Papermakers Conf., Atlanta, June, 1972, p. 261.

18. H. C. Brill and F. L. Hecklau, Tappi, 43 (4), 229A (1960).

19. H. Pummer, Preprint TAPPI Papermakers Conf., Atlanta, June, 1972, p. 31.

20. W. Brecht, E. Maier, and W. Volk, Das Papier, 18 (7), 298 (1964).

21. R. P. Whitney et al, in N. M. Bikales, Ed., Encyclopedia of Polymer Science and Technology, Interscience Publishers, New York, Vol. 9, 1968, pp. 714-747.

APPLICATION OF POLYELECTROLYTES IN MUNICIPAL WASTE TREATMENT

Howard G. Flock and Emerson G. Rausch

Calgon Corporation

Pittsburgh, Pennsylvania

WASTE TREATMENT

Raw municipal waste water consists of domestic sewage alone or a combination of domestic sewage, industrial waste and/or storm sewer run-off that contains five major pollutants: suspended solids, organic matter, phosphate compounds, nitrogen compounds, and pathogenic organisms.

Primary Treatment. Initially, conventional primary municipal waste-treatment plants were designed to remove 30-50 percent of the organic matter (measured as biochemical oxygen demand - BOD) from raw waste waters utilizing liquid - solids separation processes, and to disinfect the treated waste water by chlorination prior to being discharged to a receiving stream. In municipal primary waste treatment (Figure 1), raw waste water enters the plant, receives preliminary screening and degritting, and in some plants preaeration, to remove large or coarse materials and grease before it flows to the conventional primary treatment process that consists of the primary sedimentation basin. Primary sedimentation is the only treatment that the raw waste water receives, and it removes 30-50 percent of the suspended solids and 30-40 percent of the BOD by gravity settling. The effluent from the primary sedimentation basis is chlorinated and discharged to a receiving stream.

The solids collected in the sedimentation basin are called primary sludge. This primary sludge (2-10 percent solids) is then processed to 20+ percent solids by dewatering. The two most common dewatering methods employed are vacuum filtration or centrifugation. The primary sludge also can be processed through a digester, where

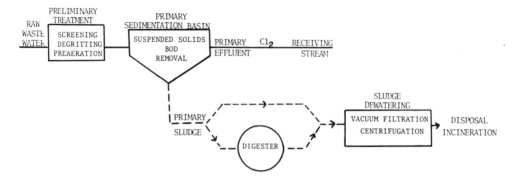

Fig. 1.   Conventional primary municipal waste-treatment plant.

anerobic (bacterial) digestion reduces the volume and renders the
sludge inert prior to dewatering by vacuum filtration or centrifugation and disposal by incineration, landfill, etc.

Primary sedimentation and sludge dewatering are the two liquid - solids separation processes utilized in conventional primary
waste treatment plants.

Secondary Treatment. Initially, conventional secondary municipal waste-treatment plants were designed to remove 85-95 percent
of the suspended solids and organic matter (BOD) from raw waste
waters utilizing liquid - solids separation and bacteriological
processes, and to disinfect the treated waste water by chlorination
prior to being discharged to a receiving stream.

In municipal secondary waste treatment (Figure 2), the raw
waste water enters the plant and receives conventional primary
treatment as previously described. Effluent from the primary sedimentation basin, which still contains suspended solids and BOD,
then flows to a conventional secondary treatment process which consists of biological oxidation (bacteria as well as aeration) followed by a secondary sedimentation basin. The two types of biological processes used by waste-treatment plants are activated sludge
and trickling filter.

In the activated sludge process, the bacterial growths convert
70-90 percent of the remaining raw waste water BOD, suspended or
dissolved organics, into biological suspended solids. In the trickling filter process, the bacterial growths convert 60-80 percent of
this remaining BOD into biological slimes. Effluent from the

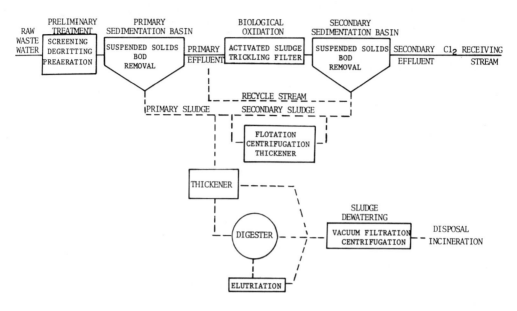

Fig. 2. Conventional secondary municipal waste-treatment plant.

activated sludge or trickling filter process then flows to the
secondary sedimentation basin where 70-90 percent of the remaining
suspended solids are removed by gravity settling. Effluent from
the secondary sedimentation basin is chlorinated and discharged to
a receiving stream.

The solids captured in the primary and secondary sedimentation
basins are called primary sludge and secondary sludge, respectively.
The secondary sludge, waste activated or trickling filter humus, is
usually blended with the primary sludge. However, in some instances
the activated sludge can be processed through the sludge concentration steps of gravity thickening (thickener), air flotation or centrifugation, before it is blended with the primary sludge. This
primary-secondary sludge blend can then be processed through the
sludge handling steps of 1) thickener or sludge-holding tank; 2)
thickener and digester; or, 3) thickener, digester, and elutriation
before it is dewatered by vacuum filtration or centifugation and
disposed of by incineration, landfill, etc.

This brief description of primary and secondary sedimentation,
sludge concentration, and sludge dewatering encompasses the four
liquid - solids separation processes utilized in conventional secondary waste-treatment plants.

Methods of Upgrading Removals. Due to the increase in popu-
lation and industrial growth, many municipal primary and secondary
waste treatment plants have not been able to maintain their designed
removals for suspended solids and BOD for the following reasons:
1) increased raw waste water flows (hydraulic overloading); or,
2) suspended solids and BOD levels increased (solids and BOD overloading) in raw waste water.

Recently, stream discharge standards have required municipal
primary and secondary waste-treatment plants to upgrade their existing facilities for higher suspended solids and BOD removals. Most
primary municipal waste-treatment plants are now required to include
secondary treatment processes and some municipal waste-treatment
plants are being required to remove over 80 percent of the phosphate
from raw waste water. To date, the most effective and economical
method of upgrading removals in waste treatment plants, including
phosphate removal, have been accomplished by increasing the solids
removal efficiency of liquid - solids separation processed with
chemical treatment.

Chemicals have been used in municipal waste treatment since
the beginning of the century. Around 1900, the Worcester, Massachusetts, and Providence, Rhode Island, sewage plants added lime
to their raw waste water to remove solids in their primary sedimentation basins with the captured sludge solids being dewatered
by filter presses. For the next 50 years, inorganic chemical flocculants consisting mainly of lime, aluminum salts, and iron salts,
because of their ability to agglomerate solids by the mechanism of
chemical precipitation, charge neutralization, or adsorption, were
used to improve solids removal efficiency in primary and secondary
sedimentation and in concentrating and dewatering sludge solids.
However, due to high chemical costs and material-handling problems,
the inorganic chemical flocculants were only employed in concentrating and dewatering sludge solids.

During the late 1950s, water-soluble organic polyelectrolytes
emerged as promising replacements for inorganic flocculants to increase the removal efficiency of municipal waste-treatment liquid - solids separation processes. This was due principally to
the higher molecular weight of the synthetic organic polyelectrolytes and their ability to flocculate (agglomerate) solids by the
mechanisms of adsorption, charge neutralization, and interparticle
bridging via hydrogen bonding. Since, then, synthetic organic polyelectrolytes have been produced from a wide variety of complex monomers with the major differences occurring in molecular weight and
type and degree of charge on the polymer. Improvements in the
structure of polyelectrolytes (i.e., degree of branching, higher
molecular weight and charge density), as well as lower costs and
reduced material-handling problems, have made it possible to re-

place inorganic chemicals in most of the municipal waste-treatment
liquid - solids separation processes.

Polyelectrolytes are classified into three main groups (anionic, nonionic, and cationic) depending on the residual charge of the
polymer in an aqueous solution. The major structural classes of
these three polymer types and some examples of each are listed below. Nonionic polymers, although they are not polyelectrolytes in
the strict sense of the word, are included because of their similarity to the anionic and cationic polymers.

Anionic Polyelectrolytes

Carboxylic
Examples - Hydrolyzed polyacrylic esters, amides, and nitriles

Sulfonic
Examples - Polystyrenesulfonate and polyethylenesulfonate

Phosphonic
Examples - Polyvinylphosphonate and organic polyphosphate
            esters

Nonionic Polymers

Polyols
Example - Poly(vinyl alcohol)

Polyethers
Example - Poly(ethylene oxide)

Polyamides
Example - Polyacrylamide

Poly(N-Vinyl Heterocyclics)
Examples - Poly(N-vinyl -4-methyl-2-oxazolidinone) and
            poly(vinylpyrrolidinone)

Cationic Polyelectrolytes

Ammonium
    Primary, Secondary, and Tertiary Amines (Protonated)
    Example - Poly(ethylenimine hydrochloride)

    Quaternary
    Example - Poly(2-methacryloyloxyethyltrimethylammonium
                chloride)

Sulfonium
Example - Poly(2-acryloxyethyldimethylsulfonium chloride)

Phosphonium
Example - Poly(glycidyltributylphosphonium chloride)

Recently, Gutcho (1) published a patent literature review on the type of polyelectrolytes that have been investigated as flocculants in waste treatment. Also, Hoover (2) conducted a literature review of cationic quaternary polyelectrolytes and mentioned the types of quaternary polymers that have been investigated as flocculants in waste treatment. From these reviews, the major structural classes of polyelectrolytes that have been investigated as flocculants in waste treatment are:

Anionic polyelectrolytes: carboxylic and sulfonic
Nonionic polymers: polyols, polyethers and polyamides
Cationic polyelectrolytes: ammonium (amines and quaternaries) and sulfonium

This chapter reviews the types of synthetic organic polyelectrolytes (anionic, nonionic, or cationic), their cost - performance relationships and application technology required to achieve improved liquid - solids separation in the following application areas of municipal waste treatment: conventional primary treatment, conventional secondary treatment, sludge concentration, and sludge dewatering. See also the chapter by D. L. Sussman and E. C. C. Wang.

CONVENTIONAL PRIMARY TREATMENT

To Increase Removal of Suspended Solids and BOD

Raw waste waters contain 150 mg/l to 400 mg/l of suspended solids and organic matter (BOD) with the average being 200 mg/l for suspended solids and 200 mg/l for organic matter (BOD). In municipal primary and secondary waste-treatment plants, about 40% of the suspended solids and 35% of the suspended BOD are removed from the raw waste water in the gravity settling primary sedimentation basin without chemical treatment. To increase the removal efficiency of suspended solids and BOD in the primary sedimentation basin (Figure 3), a dilute polymer feed solution containing about 0.05 to 0.1% active polymer is applied to the raw waste water at the grit chamber or preaeration unit. This addition point provides sufficient time for the polymer to flocculate (agglomerate) the suspended solids and/or suspended BOD before it reaches the primary sedimentation basin. The types of anionic, nonionic, and cationic

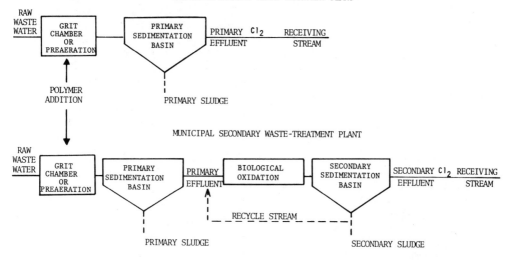

Fig. 3. Conventional primary treatment; polyelectrolyte addition
for suspended solids and BOD removal in municipal waste-
treatment plant.

polyelectrolytes that have been reported in the literature as flocculants in this application area are as follows.

Anionic Polyelectrolytes. This class of polyelectrolytes has
found the greatest utility as flocculants for removing suspended
solids and BOD from raw waste water.

Laboratory studies to determine the effect of anionic polymers
as raw waste water flocculants have been reported by Chamont et al
(3), Harrison et al (4), Glavis et al (5), Priesing et al (6),
Eck et al (7), and Eck (8). A summary of their studies shows that
hydrolyzed polyacrylamide, blends of hydrolyzed polyacrylamide and
cationic polyelectrolytes, sodium acrylate - acrylamide copolymers,
sodium acrylate - vinyl alcohol copolymers, poly(sodium styrenesulfonate), blends of poly(sodium styrenesulfonate) and cationic
polyelectrolytes, wax dispersions or latex dispersions are effective
flocculants.

The application of hydrolyzed polyacrylamide, sodium acrylate-
acrylamide copolymer, poly(sodium styrenesulfonate) (also known as

"sodium polystyrene sulfonate"), or a blend of poly(sodium styrene-sulfonate) and polyamine as raw waste water flocculants in municipal primary and secondary waste-treatment plants have been reported by Crowe (9), Freese et al (10), Harrison et al (4), Mogelnicki (11), Rizzo et al (12), Wirts (13), and others (14). The affect of an anionic polymer on the performance of the primary sedimentation basin in a municipal primary and secondary waste treatment plant is illustrated in Table I. For comparative purposes, the performance of the primary sedimentation basin before and after treatment is shown. In the primary waste-treatment plant, the suspended solids and BOD removals without treatment were 34% and 18%, respectively. With the addition of 0.25 mg/l hydrolyzed polyacrylamide, suspended solids and BOD removal efficiencies increased to 66% and 28%, respectively. In the secondary waste-treatment plant, the suspended solids and BOD removals without treatment were 43% and 23%, respectively. With the addition of 1.0 mg/l poly(sodium styrenesulfonate), suspended solids and BOD removal efficiencies increased to 76% and 33%, respectively. Table II illustrates the effect of anionic polymers on the performance of the primary sedimentation basin in several municipal primary and secondary waste-treatment plants. For comparative purposes, the performance of the primary sedimentation basins before and after the addition of the anionic polymers is shown. As indicated, without polymer treatment, the average values for suspended solids and BOD removals were 40% and 26%, respectively. As a result of polymer treatment, suspended solids and BOD removal efficiencies increased to 63% and 41%, respectively.

Generally, when 0.2 mg/l to 1.0 mg/l of the above anionic polymers are applied in conventional primary treatment, they can increase suspended solids removal efficiency by 20 to 40% and BOD by 8 to 26%.

Crowe (9) reported the comparison of a ferric sulfate + lime treatment vs a 0.85 mg/l poly(sodium styrenesulfonate) treatment to increase the removal of suspended solids and BOD in primary sedimentation. His results showed that the ferric sulfate and lime treatment removed 62% of the suspended solids and 38% of the BOD, while the poly(sodium styrenesulfonate) removed 65% of the suspended solids and 58% of the BOD. The cost of the treatment with ferric sulfate and lime was $3.13 per million gallons of waste water treated, while the poly(sodium styrenesulfonate) was $6.50 per million gallons of waste water treated. Although the poly(sodium styrenesulfonate) treatment costs more, it was noted by Crowe that the poly(sodium styrenesulfonate) was more economical since it would replace both inorganic chemicals, and reduce the plant personnel required to maintain the chemical feed equipment.

Mogelnicki (11) and others (14,15) have reported that the application of anionic polymers to the primary sedimentation basin

TABLE I

CONVENTIONAL PRIMARY TREATMENT

EFFECT OF ANIONIC POLYELECTROLYTES ON PRIMARY SEDIMENTATION BASIN PERFORMANCE

MUNICIPAL PRIMARY WASTE-TREATMENT PLANT

TREATMENT	SUSPENDED SOLIDS mg/1		PERCENT REMOVAL	BOD mg/1		PERCENT REMOVAL
	INFLUENT	EFFLUENT		INFLUENT	EFFLUENT	
NONE	201	133	34	147	121	18
HPAM (0.25 mg/1)	172	58	66	158	114	28

MUNICIPAL SECONDARY WASTE-TREATMENT PLANT

TREATMENT	SUSPENDED SOLIDS mg/1		PERCENT REMOVAL	BOD mg/1		PERCENT REMOVAL
	INFLUENT	EFFLUENT		INFLUENT	EFFLUENT	
NONE	365	208	43	350	268	23
SPSS (1.0 mg/1)	370	89	76	381	254	33

HPAM – Hydrolyzed Polyacrylamide;   SPSS – Sodium Polystyrene Sulfonate

TABLE II

CONVENTIONAL PRIMARY TREATMENT

MUNICIPAL PRIMARY AND SECONDARY WASTE-TREATMENT PLANTS

EFFECT OF ANIONIC POLYELECTROLYTES ON PRIMARY SEDIMENTATION BASIN PERFORMANCE

TYPE AND AMOUNT OF ANIONIC POLYMER	PERFORMANCE WITHOUT POLYMER TREATMENT				PERFORMANCE WITH POLYMER TREATMENT				REFERENCE
	S.S. REMOVAL mg/l	PERCENT	BOD REMOVAL mg/l	PERCENT	S.S. REMOVAL mg/l	PERCENT	BOD REMOVAL mg/l	PERCENT	
1. HPAM (0.15 mg/l)	153	58	--	--	229	78	--	--	MOGELNICKI
2. HPAM (0.25 mg/l)	52	31	47	31	80	51	58	46	WIRTS
3. HPAM (0.33 mg/l)	126	39	68	24	217	64	61	25	ANON
4. SA/AM (0.10 mg/l)	111	30	--	--	271	76	--	--	HARRISON
5. SPSS (0.74 mg/l)	--	50	--	--	--	63	--	--	FREESE
6. SPSS (0.75 mg/l)	26	18	16	17	70	50	25	27	ANON
7. SPSS (1.0 mg/l)	113	53	62	35	170	73	102	51	CROWE
8. SPSS + P.A. (1.14 mg/l)	--	43	--	--	--	63	--	--	FREESE
MEAN		40		26		64		37	

HPAM – Hydrolyzed Polyacrylamide, SA/AM – Sodium Acrylate/Acrylamide Copolymer, P.A. – Polyamine
SPSS – Sodium Polystyrene Sulfonate

of municipal secondary waste-treatment plants resulted in a higher
secondary treatment removal of suspended solids and BOD. The effect
of feeding 1.0 mg/l poly(sodium styrenesulfonate) to the primary
sedimentation basin of a secondary, trickling filter, waste-treatment
plant is illustrated in Table III. As indicated, without polymer
treatment the secondary treatment plant removals were 72% for suspended solids and 79% for BOD. As a result of polymer treatment,
suspended solids and BOD removal efficiencies increased to 85% and
85%, respectively. Table IV illustrates the effect of feeding
0.3 mg/l hydrolyzed polyacrylamide to the primary sedimentation
basin of a secondary, activated sludge, waste-treatment plant. As
indicated, without polymer treatment, the secondary treatment plant
removals were 79% for suspended solids and 82% for BOD. As a result
of polymer treatment, suspended solids and BOD removal efficiencies
increased to 87% and 91%, respectively.

From the above reported studies, it can be concluded that
anionic polyelectrolytes are effective additives to increase or
upgrade the removal efficiency of suspended solids and BOD in over-
loaded (hydraulic and/or solids) municipal primary and secondary
waste-treatment plants.

When anionic polymers are applied to conventional primary
treatment, they are usually employed at treatment dosages of 0.25
mg/l to 0.50 mg/l. Therefore, a 10 million gallon per day waste
treatment plant that applied a 0.50 mg/l dosage would use 43 pounds
of polymer per day. At a typical anionic polymer cost of $1.20/lb,
the cost of increasing the removal of suspended solids and BOD would
be $52 per day.

Nonionic Polymers. This class of polymers has found virtually
no utility as flocculants for increasing the removal of suspended
solids or BOD from raw waste waters.

Laboratory studies to determine the effect of nonionic polymers
as raw waste water flocculants have been reported by Akyel et al
(16), Larson et al (17), and the Environmental Protection Agency
(EPA) (18). Poly(vinyl alcohol), poly(ethylene oxide) and polyacrylamide were more effective flocculants than inorganic salts or
natural organic polyelectrolytes. Larson et al (17) and the EPA
(18) found that polyacrylamide exhibited no significant activity
as a flocculant. However, in plant applications, Asendorf (19) reported that with a ferric sulfate and lime treatment polyacrylamide
was an effective flocculant.

Since nonionic polymers have no charge on the chain of the
molecule for adsorption or charge neutralization, it has generally
been accepted that nonionic polymers have little or no utility as
flocculants for removing suspended solids or BOD from waste waters
when compared to the performance of anionic polyelectrolytes.

TABLE III

MUNICIPAL SECONDARY WASTE-TREATMENT PLANT – ACTIVATED SLUDGE

EFFECT OF ANIONIC POLYELECTROLYTE ADDITION TO CONVENTIONAL PRIMARY TREATMENT

TREATMENT	SUSPENDED SOLIDS			BOD		
	mg/1		PERCENT	mg/1		PERCENT
	INFLUENT	EFFLUENT	REMOVAL	INFLUENT	EFFLUENT	REMOVAL
NONE	324	68	79	283	40	82
HPAM (0.3 mg/1)	340	45	87	247	26	91

HPAM – Hydrolyzed Polyacrylamide

TABLE IV

MUNICIPAL SECONDARY WASTE-TREATMENT PLANT – TRICKLING FILTER

EFFECT OF ANIONIC POLYELECTROLYTE ADDITION TO CONVENTIONAL PRIMARY TREATMENT

TREATMENT	SUSPENDED SOLIDS			BOD		
	mg/1		PERCENT REMOVAL	mg/1		PERCENT REMOVAL
	INFLUENT	EFFLUENT		INFLUENT	EFFLUENT	
NONE	365	103	72	350	75	79
SPSS (1.0 mg/1)	370	58	84	381	55	85

SPSS – Sodium Polystyrene Sulfonate

Cationic Polyelectrolytes. This class of polyelectrolytes has found limited utility as flocculants for removing suspended solids and BOD from raw waste waters.

Laboratory studies to determine the effect of cationic polymers as raw waste water flocculants have been reported by Priesing et al (6), Nagy (20), Thurman et al (21), Suen et al (22), Scanley et al (23), Hoke (24), Rogers et al (25), Lees (26), Jones (27), Larson et al (17), and the Environmental Protection Agency (18). A summary of their studies shows that polyamines, polyvinylimidazoline, poly[N-(3-dimethylaminopropyl)acrylamide], poly[N-(1,1-dimethyl-3-dimethylaminopropyl)acrylamide], poly(dimethyldiallyammonium chloride) and copolymers of diethylaminomethylacrylamide - acrylamide, dimethylaminopropylacrylate - acrylamide, 2-vinylpyridine - acrylamide, dimethyldiallylammonium chloride - acrylamide, 3-dimethylamino-propylacrylamide - acrylamide, N-(1,1-dimethyl-3-dimethylaminopropyl)-acrylamide - acrylamide, acrylamide - β-methacryloyloxyethyltri-methylammonium methylsulfate are effective flocculants. Although the above studies showed significant reductions in suspended solids and/or BOD, dosages from 1 mg/l to 50 mg/l were required.

The plant application of a polyamine and acrylamide - β-meth-acryloyloxyethyltrimethylammonium (AM/MTMMS) copolymer as raw waste water flocculants in municipal primary and secondary waste-treatment plants has been reported by Odom et al (28), Freese et al (10), and Lees (26).

Odom et al (28) mentioned that 20 mg/l of a polyamine would increase the total solids removal of the primary sedimentation basin by 9% in a municipal primary waste-treatment plant. Lees (26) claimed that 0.54 mg/l of an AM/MTMMS copolymer would increase the suspended solids or BOD removal in primary sedimentation from 32% to 63%.

These studies show that although cationic polyelectrolytes can increase the removal efficiency of suspended solids and BOD in conventional primary treatment, they generally require higher treatment dosages than anionic polymers and when compared on a cost - performance basis to anionic polymers, they are considered uneconomical.

It can be concluded that anionic polyelectrolytes are the most effective class of polymers that can be used to increase the removal efficiency of suspended solids and BOD in the primary sedimentation basin of municipal primary and secondary waste-treatment plants.

Phosphate Removal

Raw waste water contains about 20-30 mg/l of phosphate as ($PO_4^=$). In municipal primary and secondary waste-treatment plants only 0-15% of the phosphate is removed from the waste water via the primary sedimentation basin. The removal of 80% phosphate has been accomplished in conventional primary treatment with the application of inorganic chemicals (lime, iron salts, or aluminum salts) and synthetic organic polyelectrolytes. To obtain phosphate removal (Figure 4), the inorganic chemical is applied as a dilute solution 1-10% solids) to the raw waste water just as it enters the plant. This addition point must provide sufficient contact time (2-10 min) for the inorganic chemical to precipitate the phosphate. Meanwhile, the dilute polymer feed solution containing 0.05 to 0.1% active polymer is applied to the raw waste water containing the precipitated phosphate at the grit chamber or preaeration unit. This polymer addition point provides sufficient contact time (0.5-1.0 min) for the polymer to flocculate the chemically precipitated phosphate before it reaches the primary sedimentation basin.

The types of anionic, nonionic, and cationic polyelectrolytes that have been reported in the literature as flocculants for phosphate removal with inorganic chemicals follows:

Anionic Polyelectrolytes. This class of polyelectrolytes has found the greatest utility as flocculants to increase phosphate removals with inorganic chemicals.

Laboratory studies by Wukash et al (29,30) and Daniels et al (31) reveal that a separate two-stage addition of an inorganic chemical and an anionic polymer was an effective method for removing phosphate from raw waste water.

Wukash (29) reported that the addition of 20 mg/l of Al^{+++} (as aluminum sulfate) resulted in a reduction of phosphate from 24 mg/l to 11 mg/l. With the addition of 20 mg/l Al^{+++} and 0.5 mg/l sodium acrylate - acrylamide copolymer the phosphate was reduced from 24 mg/l to 1.0 mg/l.

Wukash et al (30) and Daniels et al (31) claimed that the addition of an anionic polymer [hydrolyzed polyacrylamide, sodium acrylate - acrylamide copolymer, sodium methacrylate - acrylamide copolymer or poly(sodium styrenesulfonate)] at a dosage between 0.1 and 1.0 mg/l would enhance the phosphate removal with inorganic chemicals.

The application of inorganic chemicals and an anionic polymer in municipal primary and secondary waste-treatment plants has been reported by Wukash (32), Brenner (33) and Flock et al (34). The

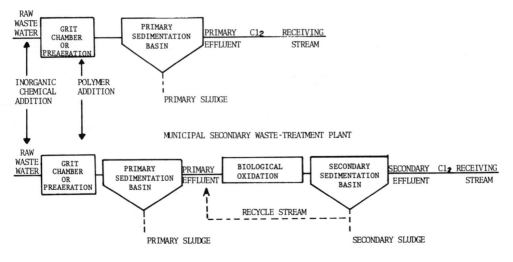

Fig. 4. Conventional primary treatment; inorganic chemical and
         polyelectrolyte addition for phosphate removal in municipal
         waste-treatment plant.

effect of inorganic chemicals and hydrolyzed polyacrylamide on the
performance of the primary sedimentation basin is illustrated briefly
in Table V. For comparative purposes, the performance of the primary sedimentation basin before and after chemical treatment is
shown. As indicated, without chemical treatment the phosphate removal varied from 0 to 20%. With chemical treatment, phosphate removal efficiency increased to 72 and 84%. In addition, this treatment increased the removal of suspended solids and BOD in the primary sedimentation basin. Without chemical treatment, the suspended
solids and BOD removals were 45% and 40%, respectively. With
chemical treatment, suspended solids and BOD removal increased to
79% and 65%, respectively.

Brenner (33) reported that the application of ferric chloride
and and anionic polymer to the primary sedimentation basin would
increase the phosphate removal through a municipal secondary waste
treatment plant. Without chemical treatment, the phosphate removal
of the secondary waste treatment plant was 35%. With the addition
of 21 mg/l of ferric chloride (as Fe^{+++}) and 0.25 mg/l of hydrolyzed
polyacrylamide to the primary sedimentation basin, phosphate removal
efficiency was increased to 91%.

TABLE V

PHOSPHATE REMOVAL IN CONVENTIONAL PRIMARY TREATMENT

EFFECT OF INORGANIC CHEMICALS AND AN ANIONIC POLYELECTROLYTE ON PRIMARY SEDIMENTATION BASIN PERFORMANCE

	REMOVAL WITHOUT CHEMICAL TREATMENT				REMOVAL WITH CHEMICAL TREATMENT				REFERENCE
	PO_4		S.S.	BOD	PO_4		S.S.	BOD	
	mg/1	PERCENT	PERCENT	PERCENT	mg/1	PERCENT	PERCENT	PERCENT	
1. FeCl$_2$ (18 mg/1) NaOH (30 mg/1) HPAM (0.4 mg/1)	0	0	50	41	11.1	72	78	58	WUKASH
2. FeCl$_2$ (43 mg/1) LIME (66 mg/1) HPAM (0.4 mg/1)	—	—	—	—	——	84	74	59	BRENNER
3. LIME (175 mg/1) HPAM (0.6 mg/1)	1	20	42	39	6.3	80	66	58	FLOCK
4. ALUM (150 mg/1) HPAM (0.5 mg/1)	1	20	42	39	10.9	81	81	83	FLOCK
MEAN	0-20		45	42		72-84	79	65	

HPAM – Hydrolyzed Polyacrylamide

Generally, when inorganic chemical and anionic polymer treatment is employed for phosphate removal, the anionic polymer dosage is between 0.25 mg/l and 0.5 mg/l. The cost to obtain 80% phosphate removal with the inorganic chemical + anionic polymer treatment is between $25 and $45 per million gallons.

Nonionic Polymers. Generally, this class of water-soluble polymers has found limited utility as flocculants to increase the removal of phosphate with inorganic chemicals.

A laboratory study by Flock and Rausch (35) showed that phosphate removals of greater than 80% could be obtained with a blend of ferric chloride and polyacrylamide. Also, Wukash et al (30) claimed that polyacrylamide could be used as a flocculant in conjunction with a ferric chloride and sodium hydroxide treatment for phosphate removal. Although polyacrylamide can flocculate the inorganic chemical precipitates of phosphate at dosages of 0.5 mg/l to 1.5 mg/l, its utility as a flocculant has been limited due to the improved performance obtained with anionic polyelectrolytes.

Cationic Polyelectrolytes. Cationics generally have had limited utility as flocculants to increase the removal of phosphate with inorganic chemicals.

Laboratory studies by Eck and Zegel (38) revealed that a treatment of 40 mg/l alum and 10 mg/l of a cationic latex would remove 80% of the phosphate from raw waste water. Brenner (33) mentioned that cationic polyelectrolyte dosages of 30 mg/l or greater are required to obtain 80% phosphate removal with inorganic chemicals.

In a municipal primary waste-treatment plant, Odom et al (28) reported that the application of a polyamine at 5 mg/l (no inorganic chemicals) increased the phosphate removal efficiency of the primary sedimentation basin to 79.5%.

Although cationic polyelectrolytes can flocculate the inorganic chemical precipitates of phosphate, they generally require treatment dosages above 5 mg/l, which is uneconomical. From the above reported studies, it can be concluded that anionic polyelectrolytes are the most effective class of polymers to increase the removal efficiency of phosphate with inorganic chemicals in the primary sedimentation basin of municipal primary and secondary waste-treatment plants. In addition to obtaining 80% phosphate removal, suspended solids and BOD removals also are increased with this treatment.

CONVENTIONAL SECONDARY TREATMENT

There is little published information available on the application of synthetic organic polyelectrolytes to improve removal of suspended solids and BOD from waste water in secondary sedimentation basins in municipal secondary waste-treatment plants. This is probably due to the fact that the aerobic bacteria in the biological oxidation process (activated sludge or trickling filter) produce a polysaccharide capable of flocculating other bacteria and suspended solids. This allows for the solids to settle readily in the secondary sedimentation basin, providing normal biological growth conditions are maintained in the biological oxidation process. However, upsets in the performance of secondary sedimentation basins can occur as hydraulic and/or solids loadings are increased or as filamentous bulking (nonsettable) sludge develops. When one or more of these conditions exist, the application of synthetic organic polyelectrolytes has been successful in increasing suspended solids and BOD removal in the secondary sedimentation basin so that a satisfactory secondary effluent can be obtained. To increase the removal efficiency of suspended solids and BOD in the secondary sedimentation basin (Figure 5), a dilute polymer feed solution containing about 0.05 to 0.1% active polymer is applied in the activated sludge (biological oxidation) basin or to the trickling filter effluent stream. These polymer addition points provide sufficient contact time for the polymer to flocculate (agglomerate) the biological suspended solids before the flow enters the secondary sedimentation basin.

To Increase Removal of Suspended Solids and BOD

Anionic Polyelectrolytes. Anionics have found limited utility as flocculants to increase the removal of suspended solids and BOD in the secondary sedimentation basin.

In the laboratory, Singer et al (37) and Woldman et al (38) studied the effect of poly(sodium styrenesulfonate) to improve settling characteristics of bulking activated sludge. From their studies, they concluded that poly(sodium styrenesulfonate) was not an effective flocculant for bulking activated sludge. However, a laboratory study by Busch and Strumm (39) revealed that hydrolyzed polyacrylamide and poly(sodium styrenesulfonate) were effective flocculants for normal (nonbulking) activated sludge.

The application of poly(sodium acrylate) to increase the removal efficiency of trickling filter suspended solids and BOD in the secondary sedimentation basin has been reported by Reis (40). Without treatment, suspended solids and BOD removals through the secondary sedimentation basin averaged 7% for suspended solids and 15% for BOD. With the addition of 8 mg/l poly(sodium acrylate), the average removals were increased to 41% for suspended solids and 50% for BOD.

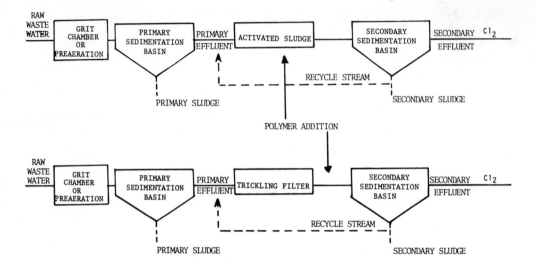

Fig. 5.   Conventional secondary treatment; polyelectrolyte addition
          for suspended solids and BOD removal in municipal secondary
          waste-treatment plants.

Nonionic Polymers.   Nonionic water-soluble polymers have had
limited utility as flocculants to increase the removals of suspended
solids and BOD in the secondary sedimentation basin.

Laboratory studies by Walker and Dougherty (41) revealed that
polyacrylamide would flocculate normal (nonbulking) activated sludge,
but its use would be limited based on the improved results obtained
with anionic and cationic polymers.

Woldman et al (30) reported that the plant application of polyacrylamide at a treatment dosage of 5 mg/l would increase the settling rate of suspended solids in the secondary sedimentation basin.
Also, this treatment increased the BOD removal efficiency of the
secondary sedimentation basin from 95% to 99%.

Cationic Polyelectrolytes.   Cationics enjoy the greatest utility as flocculants for increasing the removal of suspended solids
and BOD in the secondary sedimentation basin.

Laboratory studies by Tenney et al (42), Hurwitz et al (43),
and Jones (27) show that polyamines, polyvinylimidazoline, and

dimethyldiallylammonium chloride cyclic homopolymer were effective flocculants for normal (nonbulking) activated sludge. Also,
laboratory studies by Singer et al (37) mentioned that a polyamine
or polyethylenimine were effective flocculants for bulking activated sludge.

The plant application of a polyamine and an acrylamide - β-meth-
acryloyloyxyethyltrimethylammonium methylsulfate (AM/MTMMS) copolymer
to flocculate suspended solids in the secondary sedimentation basin
was reported by Jordan et al (44), Goodman (45), Goodman et al (46),
and others (47).

Jordan et al (44) reported that the application of polyamine
would flocculate activated sludge and control bulking in the secondary sedimentation basin. When 5 mg/l of the polyamine was added
to one of two identical sedimentation basins, the polaymine-treated
solids remained at 2.5 feet while the control without treatment rose
to 9.5 feet. This trial was considered a success but was quite expensive at a cost of $8.90 per hour of treatment.

Goodman et al (45, 46) and others (47), on the basis of a
plant-scale study, concluded that the application of an AM/MTMMS
copolymer to the biological oxidation process at a rate of 0.1 lb/ton
of secondary effluent dried suspended solids increased the overall
BOD removal efficiency of the plant to 95% and decreased the loss
of suspended solids in secondary effluent by 99%. At the beginning
of this study (Figure 6), the secondary effluent suspended solids
were 700 mg/l. Initially, the AM/MTMMS copolymer was applied at a
treatment dosage of 5 mg/l. After 5 hours of polymer treatment,
only 25 mg/l of suspended solids were found in the secondary effluent remaining at 25 mg/l. Due to the increased capture of suspended
solids in the secondary sedimentation basin, the polymer treatment
was reduced to 1.0 mg/l and was fed for another 60 hours. At this
treatment dosage, the suspended solids in the secondary effluent
remained at 25 mg/l. When the polymer treatment study was stopped
at 100 hours, the suspended solids in the secondary effluent gradually increased to 700 mg/l in 90 hours.

From the above reported studies, it is concluded that cationic
polyelectrolytes have the greatest flexibility and can be used advantageously to improve the overall removal efficiency (upgrading
secondary sedimentation basins) of municipal secondary waste-treatment plants.

Phosphate Removal

Since raw waste water contains 20-30 mg/l of phosphate, as
($PO_4^=$), only 30-40% of the phosphate is removed by a municipal

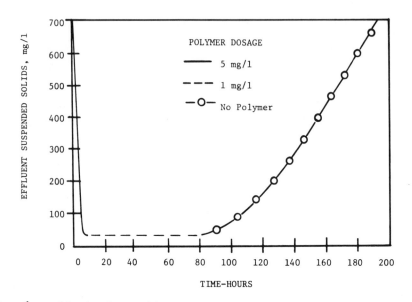

Fig. 6. Effect of a cationic polyelectrolyte on secondary sedimentation basin performance.

secondary waste-treatment plant via primary sedimentation, biological oxidation and secondary sedimentation. To increase the
phosphate removal to 80% in a secondary treatment plant via the
secondary sedimentation basin, inorganic chemicals (lime, iron salts
or aluminum salts) and a synthetic polyelectrolyte can be utilized
(Figure 7). Usually, iron or aluminum salts are applied to the
activated sludge basin (biological oxidation) while lime, iron salts,
or aluminum salts are applied to the trickling filter effluent stream
as a dilute solution containing 1-10% solids. These addition points
provide sufficient contact time for the inorganic chemical to chemically precipitate the phosphate before it reaches the secondary
sedimentation basin. A dilute polymer feed solution containing
0.05 to 0.1% active polymer is applied to activated sludge or trickling filter effluent streams. This polymer addition point provides
sufficient contact time for polymer to flocculate the chemical precipitate before it reaches the secondary sedimentation basin.

The types of anionic, nonionic, and cationic polyelectrolytes
that have been reported in the literature as flocculants to improve
phosphate removal in secondary treatment with inorganic chemicals
follow:

Anionic Polyelectrolytes. Anionics have found the greatest
utility as flocculants to increase phosphate removals with inorganic
chemicals.

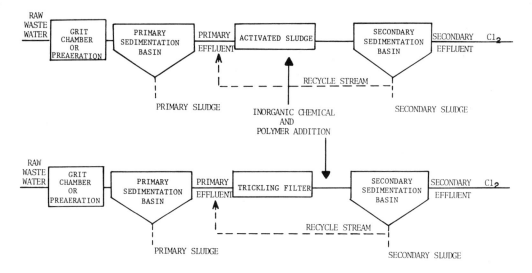

Fig. 7. Conventional secondary treatment; inorganic chemical and
polyelectrolyte addition for phosphate removal in municipal
secondary waste-treatment plants.

Flock et al (34) reported that the plant application of lime
and hydrolyzed polyacrylamide to the trickling filter effluent
would increase the removal of phosphate in the secondary sedimentation basin. Without treatment, the phosphate removal for the
secondary waste treatment plant was 40%. The addition of 165 mg/l
lime and 0.75 mg/l hydrolyzed polyacrylamide to the trickling filter
effluent increased phosphate removal efficiency to 84%. This treatment also increased suspended solids removal from 84.5% to 94.0%
and BOD removal from 95.6% to 98.0%.

Boehler and Purvis (48) claimed that the plant application of
100 mg/l aluminum sulfate, 10 mg/l sodium aluminate and 0.2 mg/l
sodium acrylate - acrylamide copolymer to the biological oxidation
process increased phosphate removals to 80% in the secondary sedimentation basin.

Davis and Love (49) reported that the addition of alum to the
activated sludge basin and the addition of 1.0 mg/l of hydrolyzed
polyacrylamide to activated sludge effluent stream would remove
more than 80% of the phosphate in the secondary sedimentation basin.

Generally, when the inorganic chemical + anionic polymer treatment is applied to municipal secondary treatment for phosphate removal, the anionic polymer dosage is between 0.25 and 1.0 mg/l.
The cost to obtain 80% phosphate removal with the inorganic chemical
+ anionic polymer treatment is between $25 and $35 per million gallons.

Nonionic Polymers. Nonionics have shown little utility as
flocculants to increase phosphate removals with inorganic chemicals.

Duff (50) reported that the plant application of 200 mg/l lime,
30 mg/l alum and a low dosage of polyacrylamide (0.003 to 0.01 mg/l)
to the effluent of the biological oxidation process would increase
phosphate removals to 96% in the secondary sedimentation basin.

Cationic Polyelectrolytes. Cationics have no commercial utility as flocculants to improve phosphate removals with inorganic
chemicals.

Brenner (33) reported that cationic polyelectrolytes can flocculate the inorganic chemical precipitates of phosphate; however,
they generally require treatment dosages above 30 mg/l, which is
considered uneconomical.

From all of the above reported studies, it can be concluded
that anionic polyelectrolytes are the most effective class of polymers to increase the removal efficiency of phosphate in secondary
sedimentation basins of municipal secondary waste-treatment plants.
In addition to obtaining 80% phosphate removal, the suspended solids
and BOD removals are also increased with the inorganic chemical +
anionic polymer treatment.

SLUDGE CONCENTRATION

Sludge concentration can be defined as a process of removing
water from solids after the initial separation of solids from waste
water. The basic objective of sludge concentration is to reduce the
volume of sludge solids to be handled in subsequent sludge-removal
processes. The four most common sludge-concentration processes used
in municipal primary and secondary treatment are gravity thickening,
dissolved-air pressure flotation, elutriation, and centrifugation.
Since centrifugation will be discussed in the sludge dewatering section, the application of synthetic organic polyelectrolytes to increase the removal efficiency in gravity thickening, dissolved air
flotation, and elutriation will be reviewed.

Gravity Thickening

This process was incorporated into municipal waste-treatment plants to increase the solids content of primary and secondary sludges prior to sludge dewatering or to increase the digester capacity for these sludges. To improve the efficiency of sludge solids capture and concentration during gravity thickening, a dilute polymer feed solution containing about 0.05% to 0.1% active polymer is applied (Figure 8) to the wash water so that the polymer can be homogeneously mixed with the sludge. This polymer addition point provides sufficient contact time for the thickener. The types of anionic, nonionic, and cationic polyelectrolytes that have been reported in the literature as flocculants in sludge thickening follow:

Anionic Polyelectrolytes. Anionics have found utility as flocculants for thickening primary sludge. Laboratory studies by Duncan (51) show that poly(sodium styrenesulfonate) was an effective flocculant for thickening primary and primary digested sludges. The plant application of hydrolyzed polyacrylamide and poly(sodium styrenesulfonate) to thicken primary sludge was reported by Hilson (52), Crowe et al (53), and Mogelnicki (11).

Hilson (52) reported on the application of hydrolyzed polyacrylamide to thicken primary sludges that contained 0.5% to 1.0% solids. In one plant, without treatment, the primary sludge was thickened to 2% solids. With the addition of 1.0 mg/l of hydrolyzed polyacrylamide, the primary sludge was thickened to 5% solids. In another plant (Figure 9), he reported that 7 mg/l of hydrolyzed polyacrylamide was required to thicken a 0.5% solids primary sludge to 5% solids while 12 mg/l of hydrolyzed polyacrylamide was required to thicken a 1.0% solids primary sludge to 5% solids.

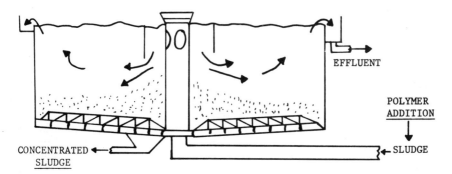

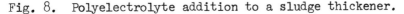

Fig. 8. Polyelectrolyte addition to a sludge thickener.

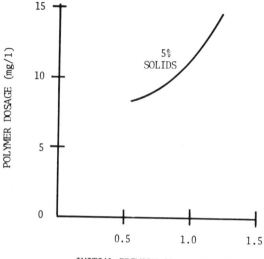

Fig. 9. Thickening primary sludge with anionic polyelectrolytes.

Crowe et al (53) stated that the application of poly(sodium styrenesulfonate) at 1.0 mg/l would capture additional suspended solids in the primary sedimentation basin and increase the concentration of the primary sludge solids from 4.3% to 8.0%, which in turn increased the effective capacity of the sludge holding tanks and digesters.

Mogelnicki (11) reported similar results to Crowe et al (53) in that 1.0 mg/l of poly(sodium styrenesulfonate) would increase the concentration of the primary sludge solids from 4.6% to 7.5% or an increase of 63%.

Nonionic Polymers. Nonionics have no known utility as flocculants to thicken sludges. Duncan (51) and Hilson (52) reported that polyacrylamide would thicken primary sludge in their laboratory and/or pilot plant studies, but its performance was below that of the anionic polymers at equal treatment dosages. From these studies, it has been generally accepted that nonionic polymers have no utility as flocculants in sludge thickening when compared to the performance of anionic or cationic polyelectrolytes.

Cationic Polyelectrolytes. This class of polyelectrolytes has found the greatest utility as flocculants for thickening waste activated sludge.

Laboratory studies with cationic polymers to thicken sludge were reported by Duncan (51) and Harrison et al (54). Duncan (51) found that a polyamine was an effective flocculant for thickening primary and primary digested sludges. Harrison et al (54) found that poly(β-methacryloyloxyethyltrimethylammonium methylsulfate) was an effective flocculant for thickening waste activated sludge.

The plant application of polyamines and an acrylamide - β-methacryloyloxyethyltrimethylammonium methylsulfate (AM/MTMMS) copolymer has been reported (44,55). Jordan et al (44) found that the application of several polyamines and an AM/MTMMS copolymer to thicken waste activated sludge that contained 0.20-0.34% solids, with the addition of 4 to 13 lb of polyamines/ton of dry solids, the waste activated sludge was thickened to between 2.3% and 2.9% solids. With the addition of 1.26 lb of the AM/MTMMS copolymer/ton of dry solids, the waste activated sludge was thickened to 2.3% solids. Ettelt et al (55) reported on the plant application of a polyamine to thicken waste activated and mixed sludges. However, a polyamine dosage of 10 lb/ton of dry solids was necessary before sludge compaction was obtained.

Dissolved-Air Pressure Flotation

This process was originally employed to concentrate waste activated sludge prior to sludge dewatering or to increase the digester capacity for this sludge. Presently, however, blends of primary and secondary sludge are being concentrated by flotation. The objective of flotation-thickening is to attach minute air bubbles to the sludge solids, which then allows the solids to separate from its waste water since bubble-particle aggregates now have a specific gravity lower than water. To improve the efficiency of sludge solids capture and concentration during flotation, a dilute polymer feed solution containing about 0.05% to 1.0% active polymer is applied (Figure 10) to the recycled wash water so that the polymer can be homogeneously mixed with the sludge before it enters the flotation unit.

The types of anionic, nonionic, and cationic polyelectrolytes that have been reported in the literature as flocculants for dissolved air flotation follow.

Anionic Polyelectrolytes. Anionics have little known utility as flocculants for the flotation of sludge. Braithwaite (56) reported the plant application of $FeCl_3$ and anionic polyelectrolytes (not identified, but possible poly(sodium styrenesulfonate) and/or hydrolyzed polyacrylamide) as flocculants for the flotation of waste activated sludge. The floated sludge solids obtained with this treatment were 7.1% compared to 4.1% without chemical treatment. In the flotation of waste activated sludge, anionic polyelectrolytes have little utility as flocculants by themselves.

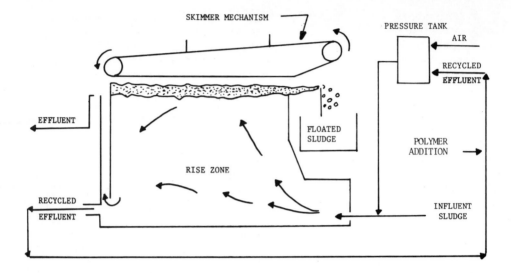

Fig. 10. Polyelectrolyte addition to a dissolved air flotation unit.

Nonionic Polymers. Nonionics have no known utility as flocculants for the flotation of sludge.

Cationic Polyelectrolytes. Cationics have found the most
utility as flocculants for the flotation of sludge.

Laboratory flotation studies by Harrison et al (54) revealed
that polyethylenimine, poly(β-methacryloyloxyethyltrimethylammonium
methylsulfate) and an acrylamide - β-methacryloyloxyethyltrimethylammonium methylsulfate copolymer were effective flocculants for the
flotation of waste activated sludge. Also, laboratory studies by
Ettelt (57) showed that a polyamine was an effective flocculant for
floating waste activated sludge.

The plant application of cationic polymers to dissolved air
flotation units has been reported by Braithwaite (56), Ettelt (57),
Crowe (9), Gatz et al (58), and Goodman (45).

Braithwaite (56) reported the application of $FeCl_3$ and polyamines as flocculants for the flotation of activated sludge. The
floated sludge solids obtained with this treatment were 6.2% compared
to 4.1% without chemical treatment.

Ettelt (57) found that the application of a polyamine as a
flocculant would increase the flotation efficiency of waste activated
sludge and blends of primary and waste activated sludge (Figure 11).

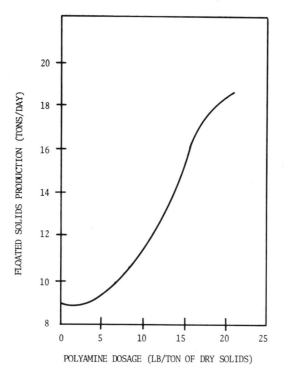

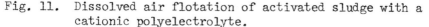

Fig. 11.   Dissolved air flotation of activated sludge with a
           cationic polyelectrolyte.

Without treatment, the floated sludge solids were 9 tons/day. With
the addition of the polyamine at 7.5 lb to 21.0 lb/ton of dry
solids, the floated solids were increased to between 10 and 19
tons/day. Also, the chemical treatment increased solids capture
from 92.5% to 99.6%, and solids concentration from 3.3% to 3.9%.

Crowe (9) claimed that the application of a polyamine at 25
lb/dry ton of solids in the flotation of waste activated sludge
increased the floated sludge solids from 4.5% to 4.6%, increased
solids capture from 61.0 to 99.8% and increased the yield from
0.63 to 3.20 lb/ft²/hr.

Gatz et al (58) stated that the application of acryloxymethyldi-
methylsulfonium methylsulfate - acrylamide copolymer on activated
and activated + primary sludge blends increased the percent solids
capture by 10-20% when compared to normal operating results at a
cost from $2.00 to $9.00/dry ton of solids.

Goodman (45) mentioned that an acrylamide - β-methacryloyloxyethyltrimethylammonium methylsulfate (AM/MTMMS) copolymer would

increase the flotation of activated sludge solids. Also, he reported that an activated sludge at 0.8% solids was floated to 2.6% solids with the application of an AM/MTMMS copolymers.

From the above reported studies, it can be concluded that cationic polyelectrolytes by themselves can significantly improve the degree of solids separation achieved by the dissolved air flotation of waste activated sludge and sludge blends containing a higher ratio of waste activated sludge to primary sludge.

Elutriation

This process was originally developed for the chemical reduction of lime and ferric chloride dosages when applied to the vacuum filtration of digested sludge. In this process, the digested sludge is mixed with plant effluent water to wash and remove excess bicarbonate ion from the digested sludge prior to vacuum filtration. Today, this process is essentially used as a digested sludge thickening technique since synthetic organic polyelectrolytes, not affected by excess bicarbonate ions, are now the products of choice for dewatering municipal sludges.

To improve the efficiency of sludge solids capture and concentration during elutriation, a dilute polymer feed solution containing about 0.05% to 0.1% active polymer is applied (Figure 12) to the wash water so that it can be homogeneously mixed with the digested sludge. This polymer addition point provides sufficient contact time for the polymer to flocculate (agglomerate) the sludge solids before it reaches the elutriation basin. The types of anionic, nonionic, and cationic polyelectrolytes which have been reported in the literature as flocculants for elutriation follow.

Anionic Polyelectrolytes. Anionics have limited utility as flocculants for elutriation. Laboratory elutriation tests by Priesing et al (59,60) showed that poly(sodium styrenesulfonate) was an effective flocculant. Also, they claimed that hydrolyzed polyacrylamide and sodium acrylate - acrylamide copolymers were effective flocculants.

The plant application of anionic polymers as flocculants in elutriation has been reported by Crowe (9) and Priesing et al (60). Crowe (9) reported that the application of poly(sodium styrenesulfonate) at 1.6 lb/dry ton of primary digested and activated sludge solids in elutriation would reduce the overflow of suspended solids from 3,835 mg/l to 365 mg/l, increase solids capture from 65.1% to 95.3% and increase the underflow solids concentration from 3.5% to 4.3%.

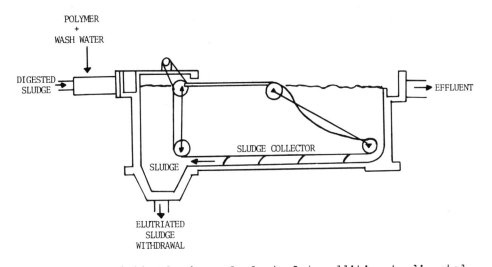

Fig. 12.   Elutriation basin; polyelectrolyte addition to digested
            sludge.

Priesing et al (60) mentioned that the application of poly(sodium styrenesulfonate) at 1.5 lb/dry ton of primary digested and activated sludge solids in elutriation would increase solids capture by 20% and increase the underflow solids concentration from 8% to 10%.

Nonionic Polymers.   Nonionics have no known utility as flocculants for elutriation.

Duncan (51) reported that polyacrylamide would thicken digested sludge in his laboratory and pilot studies, but its performance was below that of the anionic polymer [poly(sodium styrenesulfonate)] at equal treatment dosages. From this study and those of others, it has been accepted that nonionic polyelectrolytes have no utility as flocculants in elutriation compared to anionic or cationic polyelectrolytes.

Cationic Polyelectrolytes.   Cationics are the most useful flocculants for elutriation. Laboratory elutriation studies by Buhl et al (61) showed that polyethylenimine and copolymers of acrylamide - β-methacryloyloxyethyltrimethylammonium methylsulfate (AM/MTMMS) were effective flocculants.

The plant application of cationic polymers as flocculants for elutriation has been reported (61,62). Buhl (61) claimed that the

application of an acrylamide - β-methacryloyloxyethyltrimethylammonium methylsulfate copolymer at 3.4 lb/dry ton of digested sludge
solids (1700 mg/1) in elutriation would reduce the overflow of suspended solids from 2,620 mg/1 to 260 mg/1 and increase solids capture by 9.4% (98.9% vs 89.5%).

Dahl et al (62) have stated that the application of a polyamine,
acrylamide - β-methacryloyloxyethyltrimethylammonium methylsulfate
(AM/MTMMS) copolymer and dimethyldiallylammonium chloride - acrylamide
(DMDAAC/AM) copolymer in the elutriation of digested sludge increased
the capture of solids from 57% to 92%. This improvement in elutriate
clarity has enabled the plant to tolerate the recyle of this waste
liquor and has reduced the total plant effluent suspended solids and
BOD to approximately 100,000 lb/day from an average of 172,000 lb/day
for suspended solids and 134,000 lb/day for BOD.

This improvement was achieved with a polymer cost of $2.57/mil-
lion gallons of waste water treated, based on an annualized average
flow of 253 million gallons per day.

From the above studies it can be concluded that cationic polyelectrolytes have the most flexibility in improving the efficiency
of the elutriation process since the resulting sludge can be dewatered by cationic polyelectrolytes.

SLUDGE DEWATERING

The two most common methods of dewatering municipal sewage
sludges are by vacuum filtration and centrifugation.

Vacuum Filtration

Filtration has been defined by Hubbell as a process for sepa-
rating solids from a liquid by passing the liquid through a porous
medium on which the solids remain to form a cake. In rotary vacuum
filtration, the drum is continuously passed through the sludge where
it picks up solids to form a cake which is partially dewatered and
discharged. Sludge dewatering differs from thickening in that the
sludge is processed into a nonfluid form.

The nature of the sludge to be filtered is an important filtration parameter. In general, primary sludge is easier to filter
than digested sludge, and it is also easier to filter than secondary sludge. Sludge particle size, shape, electrical charge, and
density affect filterability because they affect compaction and flocculating chemical demand. Small particles exert a greater chemical
flocculant demand per unit weight than larger particles. Usually

the larger the particle size, the higher the filter rate (dry lb/ft^2/hr) and the lower the cake moisture. An increase in slimes or extremely fine particles decreases the filtration rate and increases the cake moisture.

The Federal Water Pollution Control Administration (now EPA) in their Water Pollution Control Research Series Publication WP-20-4 (On a Study of Sludge Handling and Disposal) has reported: "The success of synthetic organic polymeric flocculants represents one of the few recent major advances in the sludge handling field. These materials have captured the major portion of the municipal sludge conditioning market, exclusive of raw waste activated sludge."

To improve the efficiency of sludge dewatering by vacuum filtration with synthetic organic polyelectrolytes, a dilute polymer feed solution containing about 0.05-0.1% active polymer is applied to sludge in the sludge conditioning box (Figure 13). This application point provides sufficient mixing and contact time for the polymer to flocculate the sludge solids before they reach the filter pan and are dewatered by a rotary drum vacuum filter which contains either a cloth or coilspring porous filter media. The types of polyelectrolytes that have been used as aids to vacuum filter sludge solids are:

Anionic Polyelectrolytes. Anionics have found utility as a flocculant for dewatering primary sludge.

Laboratory studies by Priesing et al (60,63) reveal that the application of hydrolyzed polyacrylamide or poly(sodium styrenesulfonate) at treatment dosages of 0.05-1.5 lb/ton of dry solids would increase the sludge dewatering rates of cationic polymers, namely, polyamines, polyethylenimine, vinylbenzyltrimethylammonium chloride, and dimethylaminoethyl methacrylate. Also, they claim that poly(sodium styrenesulfonate) would increase the sludge dewatering rates of inorganic chemicals.

Plant applications of anionic polymers to dewater sludges by vacuum filtration have been reported by Sherbeck (64), Reilly (65), and Hopkins et al (66).

Sherbeck (64) found that the application of poly(sodium styrenesulfonate) would dewater primary sludge at a treatment dosage of 1-2 lb/ton of dry solids at an average cost of $2.75/ton. He also reported difficulty in using poly(sodium styrenesulfonate) because it had a tendency to blind the vacuum filter media due to over-treatment.

Reilly (65) stated that the application of hydrolyzed polyacrylamide would dewater primary sludge with filter cake solids of

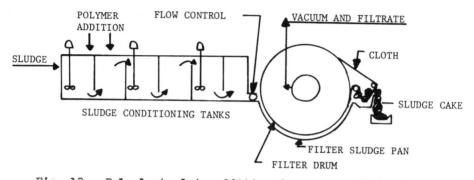

Fig. 13. Polyelectrolyte addition for vacuum filtration.

20-25% and filter production rates of 4 to 5 $lb/ft^2/hr$ at a cost
of $1.00-2.00/ton of dry solids. He also mentioned that the appli-
cation of hydrolyzed polyacrylamide would improve the performance
of a polyamine in filtering sewage sludges. With the polyamine
treatment, filter cake solids of 20-30% and filter production rates
of 4 to 5 $lb/ft^2/hr$ at cost of $4.00/ton of dry solids were obtained.
With the combined treatment of hydrolyzed polyacrylamide and polyamine, filter cake solids of 25-35% and filter production rates of
6 $lb/ft^2/hr$ at cost of $3.00-400/ton of dry solids were obtained.

Hopkins et al (66) stated the application of the above anionic
and cationic polymer treatment would dewater primary sludge at a cost
of $2.20/ton of dry solids. The two-stage treatment consists of
applying 0.4 lb/ton of hydrolyzed polyacrylamide followed by a
polyamine to the sludge.

Generally, it has been found that anionic polymers can dewater
only primary sludge. However, they have been used at low dosages
to improve the performance of polyamines in dewatering other types
of sludge.

Nonionic Polymers. Nonionics by themselves have found no
known utility as a flocculant to dewater sludge.

Laboratory filtration studies by Boothe et al (67) revealed
that polyacrylamide had no utility as a flocculant to dewater digested sludges. However, LoSasso and Rausch (68) showed that a
blend of ferric chloride and polyacrylamide would dewater sludges
at lower treatment dosages than a separate addition of $FeCl_3$ and
polyacrylamide.

Bargman et al (69) reported that the plant application of polyacrylamide would reduce the $FeCl_3$ dosage required to filter digested sludge. The $FeCl_3$ dosage required to dewater this sludge was 78 lb/ton of solids (dry) with the treatment cost being \$3.67/ ton. With the addition of polyacrylamide at 0.6 lb/ton of dry solids, the $FeCl_3$ dosage was reduced to 43 lb/ton of dry solids with the treatment cost being \$2.66/ton.

From the above studies it can be stated that nonionic poly-electrolytes, by themselves, have no utility as flocculants in de-watering sewage sludges by vacuum filtration.

Cationic Polyelectrolytes. Cationics have shown the greatest utility as a flocculant for dewatering sludge.

Laboratory filtration studies to determine the effect of cat-ionic polymers in dewatering sludges have been reported by Boothe et al (67), Bailey et al (70), Harrison et al (53), Hurwitz (43), Hoover et al (71), Jones (27), Levy et al (72), Nagy (20), Crook et al (73), Priesing et al (60), Scanley et al (23), and Schaper (74). A summary of their studies shows that polyamines, polyethyl-enimine, dimethylaminoethyl methacrylate, polyvinylimidazoline, poly(vinylbenzyltrimethylammonium chloride), poly(dimethyldial-lylammonium chloride), poly(diallylmethyl-β-propionamidoammonium chloride), poly(1,2-dimethyl-5-vinylpyridinium methylsulfate), poly(β-methacryloyloxyethyltrimethylammonium methylsulfate), quater-nized polyvinyl chloroacetate, polymers of dimethylaminomethylacryl-amide, and copolymers of dimethyldiallylammonium chloride - acryl-amide, acrylamide - β-methacryloyloxyethyltrimethylammonium meth-ylsulfate, or acrylamide - acryloxyethyl dimethylsulfonium methylsul-fate are effective flocculants.

The plant application of polyamines and polyethylenimine vs a combined ferric chloride and lime treatment to dewater primary sludge was reported by Sherbeck (64) and Crowe (9). For compara-tive purposes, the performance of vacuum filters with cationic polymers vs a combined ferric chloride and lime treatment are shown in Table VI. As indicated, the average performance parameters with cationic polymers were: 17.5 lb/dry ton, 6.3 $lb/ft^2/hr$, 31.0% cake solids and \$10.30/dry ton vs lime and ferric chloride at 392.5 lb/dry ton, 4.5 $lb/ft^2/hr$, 35% cake solids and \$11.10/dry ton.

Figure 14 illustrates the vacuum filtration of primary sludges. The ferric chloride and lime treatment dosage required to dewater primary sludge is between 3% and 20% (60-100 lb/ton) which results in filtration production rates from a low of 3 $lb/ft^2/hr$ to a high of 7 $lb/ft^2/hr$ (75,76,77). The required cationic polyelectrolyte dosage to dewater primary sludge is between 0.2% to 1/2% (4-24 lb/ton), which results in filtration production rates between 6 and 20 $lb/ft^2/hr$ (78).

TABLE VI

DEWATERING PRIMARY SLUDGE BY VACUUM FILTRATION

CATIONIC POLYELECTROLYTES VS. FERRIC CHLORIDE AND LIME

| | PERFORMANCE OF CATIONIC POLYELECTROLYTES | | | | PERFORMANCE OF $FeCl_3$ AND LIME | | | | |
| | DOSAGE | PRODUCTION | CAKE SOLIDS | COST | DOSAGE | PRODUCTION | CAKE SOLIDS | COST | REFERENCE |
TYPE	LBS/TON	LBS/FT.^{2}/HR.	PERCENT	$/TON	LBS/TON	LBS/FT.^{2}/HR.	PERCENT	$/TON	
P.A.	25.0	5.5	35	11.20	440	3.1	40	11.00	SHERBECK
P.E.I.	13.0	5.3	39	14.50	440	3.1	40	11.00	SHERBECK
P.A.	14.0	7.0	20	7.50	330	5.0	20	9.90	CROWE
P.A.	18.0	7.5	30	8.00	360	6.9	40	12.50	CROWE
MEAN	17.5	6.3	31	10.30	392.5	4.5	35	11.10	

P.A. - Polyamine, P.E.I. - Polyethylenimine

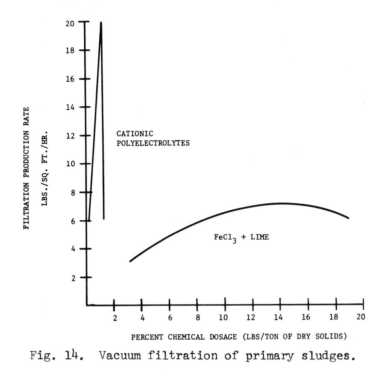

Fig. 14. Vacuum filtration of primary sludges.

The plant application of polyamines, acrylamide - β-methacrylo-yloxyethyltrimethylammonium methylsulfate and acrylamide - acrylo-xyethyl dimethylsulfonium methylsulfate copolymers vs a combined ferric chloride and lime treatment to dewater primary digested sludge was reported by Crowe (9), Sharman (79), and Conway (80). For comparative purposes, the performance of vacuum filters with cationic polymers vs a combined ferric chloride and lime treatment is shown in Table VII. As indicated, the average performance para-meters with cationic polymers were: 7.5 lb/dry ton, 12.5 lb/ft^2/hr, 32% cake solids and $5.75/dry ton vs lime and ferric chloride at 289 lb/dry ton, 7.8 lb/ft^2/hr, 35% cake solids and $7.04/dry ton.

Figure 15 illustrates the vacuum filtration of primary digested sludges. The ferric chloride and lime treatment dosage required to dewater primary digested sludge is between 4% and 20% (80-400 lb/ton), which results in filtration production rates from a lost of 2.5 lb/ft^2/hr to a high of 9.2 lb/ft^2/hr. The average ferric chloride and lime treatment required is 16%, which has a filtration production rate of 7.2 lb/ft^2/hr (75,76,77). The cationic polyelectrolyte dosage required to dewater primary digested sludges is 0.2 to 1.5% (4-30 lb/ton), which results in filtration production rates between 4 and 15 lb/ft^2/hr (78).

The plant application of polyamines, polyethylenimine, acrylam-ide - diallylmethyl-β-propionamidoammonium chloride copolymer,

TABLE VII

DEWATERING PRIMARY DIGESTED SLUDGE BY VACUUM FILTRATION

CATIONIC POLYELECTROLYTES VS. FERRIC CHLORIDE AND LIME

		PERFORMANCE OF CATIONIC POLYELECTROLYTES				PERFORMANCE OF $FeCl_3$ AND LIME			
TYPE	DOSAGE	PRODUCTION	CAKE SOLIDS	COST	DOSAGE	PRODUCTION	CAKE SOLIDS	COST	REFERENCE
	LBS/TON	LBS/FT.^{2}/HR.	PERCENT	$/TON	LBS/TON	LBS/FT.^{2}/HR.	PERCENT	$/TON	
P.A.	17.0	25.0	32	7.60	232	9.2	43	9.66	CROWE
AM/MTMMS	2.4	10.3	28	3.89	230	9.4	28	4.41	SHARMAN
AM/ADMMS	2.5	5.1	27	----	226	4.7	27	----	CONWAY
AM/ADMMS	8.0	9.7	41	----	466	8.0	41	----	CONWAY
MEAN	1.5	12.5	32	5.75	289	7.8	35	7.04	

P.A. - Polyamine, AM/MTMMS - Acrylamide-beta methacryloyloxethyltrimethylammonium methyl sulfate copolymer, AM/ADMMS - Acrylamide/acryloxymethyl dimethylsulfonium methylsulfate copolymer

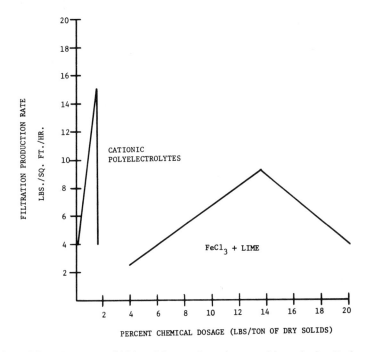

Fig. 15.   Vacuum filtration of primary digested sludges.

acrylamide - β-methacryloyloxyethyltrimethylammonium methylsulfate
copolymer, poly(vinylbenzyltrimethylammonium chloride) plus poly(sodium styrenesulfonate), and polyvinylimidazoline vs a combined ferric
chloride and lime treatment to dewater primary and secondary digested
sludge was reported by Schaper (74), Cook et al (73), Burd (81),
Priesing et al (63), and Buhl (82). For comparative purposes, the
performance of vacuum filters with cationic polymers vs a combined
ferric chloride and lime treatment is shown in Table VIII. As
indicated, the average performance parameters with cationic polymers
were: 7.1 lb/dry ton and 8.5 $lb/ft^2/hr$ vs lime and ferric chloride
at 304 lb/dry ton and 6.4 $lb/ft^2/hr$.

   Figure 16 illustrates the vacuum filtration of primary + secondary digested (elutriated) sludges. The ferric chloride and lime
treatment dosage required to dewater primary + secondary digested
(elutriated) sludges is between 8% and 20% (160-400 lb/ton), which
resulted in filtration production rates from a low of 3.4 $lb/ft^2/hr$
to a high of 5.0 $lb/ft^2/hr$. The average ferric chloride and lime
treatment dosage required is 15%, which has a filtration production
rate of 4.5 $lb/ft^2/hr$ (75,76,77). The cationic polyelectrolyte
dosage required to dewater primary + secondary digested sludge is
0.5% to 2.0% (10-40 lb/ton), which results in filtration production
rates between 4 and 8 $lb/ft^2/hr$ (78).

TABLE VIII

DEWATERING PRIMARY AND SECONDARY DIGESTED SLUDGE BY VACUUM FILTRATION

CATIONIC POLYELECTROLYTES VS. FERRIC CHLORIDE AND LIME

	PERFORMANCE OF CATIONIC POLYELECTROLYTES				PERFORMANCE OF $FeCl_3$ AND LIME				
TYPE	DOSAGE LBS/TON	PRODUCTION LBS/FT.^{2}/ HR.	CAKE SOLIDS PERCENT	COST $/TON	DOSAGE LBS/TON	PRODUCTION LBS/FT.^{2}/ HR.	CAKE SOLIDS PERCENT	COST $/TON	REFERENCE
AM/DMPAC	8.3	----	---	-----	180	---	---	---	SCHAPER
P.A.	----	4.6	---	11.97	---	---	---	---	CROOK
P.V.I.	----	4.5	---	9.98	---	---	---	---	CROOK
P.A.	13.0	11.0	19	-----	480	8.5	19	---	BURD
P.E.I.	8.0	7.0	---	-----	310	4.0	---	---	PRIESING
PVBTAC + SPSS	4.0	18.6	32	-----	335	9.2	32	---	PRIESING
AM/MTMMS	2.2	5.0	---	-----	216	3.7	---	---	BUHL
MEAN	7.1	8.5			304	6.4			

P.A. - Polyamine, P.E.I. - Polyethylenimine, P.V.I. - Polyvinylimidazoline, PVBTAC - Polyvinylbenzyl trimethyl ammonium chloride, SPSS - Sodium polystyrene sulfonate, AM/MTMMS - Acrylamide-beta methacryloyloxyethyltrimethyl ammonium methyl sulfate copolymer, AM/DMPAC - Acrylamide-diallylmethyl B-propionamido ammonium chloride copolymer

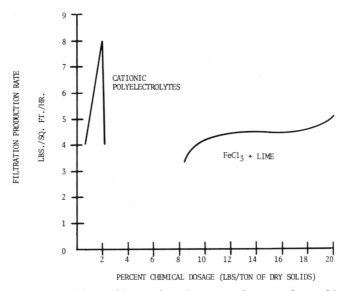

Fig. 16.   Vacuum filtration of primary and secondary digested
            sludges.

From the above reported operating experiences and those of
others, it can be concluded that cationic polyelectrolytes are the
most effective additives to increase the efficiency of sludge dewatering by vacuum filtration. When compared to inorganic chemicals,
the polyelectrolytes have higher filtration production rates with
similar filter cake solids at lower costs.

Sludge Dewatering by Centrifugation

The utilization of the centrifuge for waste sludge dewatering
was a failure for five decades, but the simplicity of operation and
small area requirements were intriguing enough to encourage new
approaches. Today, the deficiencies of the earlier units have been
overcome by modern materials, improved designs, and new process
technology.

Centrifuges are used to concentrate and dewater digested and
nondigested primary sludge, primary + activated sludge and straight
waste activated sludge. The centrifuges being employed to dewater
municipal sludges are the basket, disc and solid bowl conveyer
types.

To improve the efficiency of sludge solids recovery and increase the production of sludge cake solids during centrifugation,
a dilute polymer feed solution containing about 0.05-0.1% active

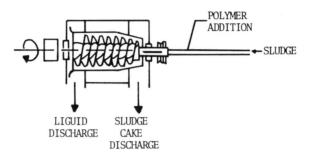

Fig. 17.   Polyelectrolyte addition to a solid bowl centrifuge.

polymer is applied (Figure 17) to the sludge feed line. This polymer application point ensures that the sludge and polymer are homogeneously mixed before entering the centrifuge.

The types of anionic, nonionic, and cationic polyelectrolytes
that have been reported in the literature as flocculants to centrifuge sludge solids follow.

Anionic Polyelectrolytes. Anionics have no known utility as
a flocculant in sludge centrifuging. However, Thomas et al (85)
claimed that hydrolyzed polyacrylamide would increase the efficiency
of a $FeCl_3$ and lime treatment to dewater sludges in laboratory
centrifuge tests. In plant application, Schaffer (86) found that
the addition of poly(sodium styrenesulfonate) would increase the
efficiency of a polyamine treatment to centrifuge sludges. Also,
Kumke et al (87) found that the addition of an anionic polymer would
improve the efficiency of cationic polymers to centrifuge waste
activated sludge.

Nonionic Polymers. Nonionics have only limited application as
flocculants to centrifuge sludge.

Schaffer (86) has reported that with the use of polyacrylamide
at 2-8 lb/dry ton of solids, when centrifuging primary and primary
+ secondary sludge blends, the solids recovery could be increased
by 11% (40% to 51%) to 55% (40% to 95%). In one plant, at a dosage
of 8 lb/dry ton of polyacrylamide, the sludge production was doubled
from 25,000 lb/day for the control to 50,000/day with the polymer
treatment.

Cationic Polyelectrolytes. Cationics have demonstrated the
greatest utilization in most centrifuge applications.

Vesilind (88) reported that the cationic polymers most commonly
used for centrifuging sewage sludges are polyamines, polyvinylimid-

azoline, acrylamide - β-methacryloyloxyethyltrimethylammonium methylsulfate copolymer, dimethyldiallylammonium chloride - acrylamide copolymers, and polyethylenimine.   In one case, Figure 18, he cited the thickening of waste activated sludge. When these cationic polymers were applied to waste activated sludge at 10 lb/dry ton (solids), the cake solids increased 6% over no chemical treatment (2% vs 8%) with the solids recovery for both being 95%. Also, these cationic polymers at 25 lb/ton increased the cake solids by 8% over that found without any chemical (2% vs 10%) with the solids recovery for both being 95%.

Montanaro (89) has reported that the use of a polyamine at 5-6 lb/dry ton of solids in centrifuging a primary digested sludge increased solids recovery from 70% to 90% at similar percent cake solids. He has found that the increased solids recovery with cationic polymers can be obtained at a cost from $2.00-10.00/dry ton of solids.

Crowe (9) reported the use of centrifuges to dewater primary and secondary digested sludge with and without a polyamine treatment. Without polymer treatment, solids recovery was between 42 and 47% with the percent cake solids being between 15 and 22%. The addition of a polyamine at 10-11.5 lb/dry ton of solids increased solids recovery to 98% while cake solids remained between 15 and 22%.

Albertson and Guidi (90) reported dewatering waste sludges with polyvinylimidazoline and polyamines. The dewatering of primary

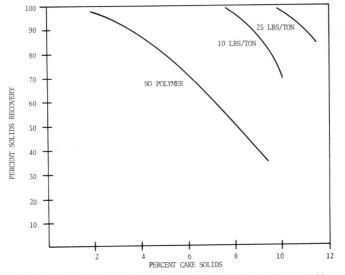

Fig. 18.   Centrifugation of waste activated sludge with cationic polyelectrolytes.

sludges with these polymers showed that the cake solids and the
solids recovery could be increased by an average of 5 and 9%,
respectively, over the control at polymer costs between $2.00-7.00/
dry ton of solids. Figure 19 shows their performance results in
centrifuging primary sludges with and without polymer treatment.
Without polymer treatment, solids recovery was between 78 and 88%
with cake solids between 25 and 32%. The addition of the above
polymers at $2.00-7.00/dry ton of solids increased solids recovery
to 93 and 96% while the cake solids were increased to 28 and 37%.

The dewatering of primary digested sludge with these polymers
showed that the cake solids and the solids recovery were increased
by an average of 8% and 10%, respectively, over the control without
treatment at a polymer cost of $3.00-8.00/dry ton of solids.
Figure 20 compares the performance of centrifuging primary digested
sludge with and without polymer treatment. Without polymer treatment, solids recovery was between 22 and 24%. The addition of the
above polymers at $3.00-8.00/dry ton of solids, increased solids
recovery to 90 and 95%, while the cake solids were increased to
30 and 32%.

The dewatering of primary + secondary sludge with these polymers showed that the cake solids and solids capture were increased
by an average of 5% and 10%, respectively, over the control without
treatment at a polymer cost of $5.00-15.00/dry ton of solids.
Figure 21 compares the performance of centrifuging primary and
secondary sludge with and without polymer treatment. Without polymer
treatment, solids recovery was between 82 and 85% with the cake

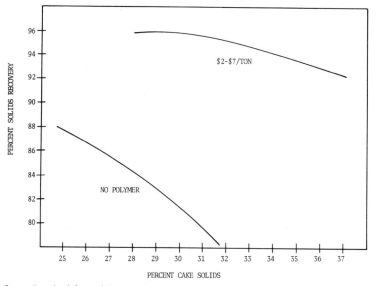

Fig. 19. Centrifugation of primary sludge with cationic
          polyelectrolytes.

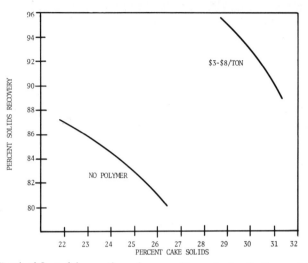

Fig. 20. Centrifugation of primary digested sludge with cationic polyelectrolytes.

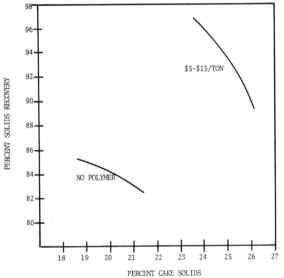

Fig. 21. Centrifugation of primary and secondary sludge with cationic polyelectrolytes.

solids between 19 and 21%. The addition of the polymers at $4.00-15.00/dry ton of solids increased solids recovery to 90 and 96% while the cake solids were increased to 24 and 26%.

Albertson and Sherwood (91) reported the use of centrifuges to dewater primary and secondary digested sludge with and without polyvinylimidazoline and polyamines. Without polymers, solids recovery was between 18 and 24%. The addition of either of the above polymers at $6.00-20.00/dry ton of solids increased solids recovery to above 95%, while cake solids remained between 18 and 24%.

From the above reported studies it can be concluded that cationic polyelectrolytes have the greatest flexibility in dewatering all types of sewage sludges by centrifugation. The use of cationic polyelectrolytes has shown that solids recovery in a centrifuge can be increased by 15 to 40% while sludge solids can be increased up to 8% at a polymer dosage generally between 5 and 25 lb/dry ton of solids.

SUMMARY

The application, performance, and costs of polyelectrolytes have been reviewed in the major liquid - solids separation application areas of municipal waste treatment. Also, performance comparisons between the classes and types of polyelectrolytes and/or between polyelectrolytes and inorganic chemicals were reviewed.

From the data published, it is apparent that anionic and cationic polyelectrolytes have found the greatest utility as flocculants to improve or upgrade existing municipal waste-treatment liquid - solids separation processes. Although nonionic polyelectrolytes have utility as flocculants in municipal waste treatment, results in most cases have been marginal compared to anionic or cationic polyelectrolytes.

The liquid - solids separation processes where anionic and cationic polyelectrolytes have found the greatest utility in municipal waste treatment plants are summarized below.

1. Conventional Primary Treatment. Anionic polyelectrolytes that consist of hydrolyzed polyacrylamide, sodium acrylate - acrylamide copolymers, or poly(sodium styrenesulfonate) provide increased removal efficiencies for suspended solids, BOD, and phosphate in the primary sedimentation basin.

2. Conventional Secondary Treatment. Cationic polyelectrolytes that consist of polyamines or polyquaternaries provide increased removal efficiencies for suspended solids and BOD in the secondary sedimentation basin.

Anionic polyelectrolytes that consist of hydrolyzed polyacrylamide provide increased phosphate removal efficiencies in the secondary sedimentation basin.

3. Sludge Concentration and Dewatering. Cationic polyelectrolytes that consist of polyamines or polyquaternaries provide increased solids capture, higher sludge solids, reduce filtration costs, and increase overall plant efficiencies when they are applied in these waste treatment liquid - solids separation processes.

BIBLIOGRAPHY

1. S. Gutcho, Waste Treatment with Polyelectrolytes, Noyes Data Corporation, Park Ridge, N. J., 1972.

2. M. F. Hoover, "Cationic Quaternary Polyelectrolytes - A Literature Review," Presented at 158th National ACS Meeting, New York, Sept. 9, 1969; J. Macromol. Sci.-Chem., $\underline{A4}$, 1327 (1970).

3. W. M. Chamont and J. T. Burke (to Nalco Chemical Co.), U. S. Pat. 3,479,282 (Nov. 18, 1969).

4. F. J. Glavis, D. H. Clemens, and R. P. McDonnell (to Rohm & Haas Co.), U. S. Pat. 3,442,799 (May 6, 1969).

5. J. R. Harrison and D. J. Monagle (to Hercules Inc.), U. S. Pat. 3,479,283 (Nov. 18, 1972).

6. C. P. Priesing, S. J. Mogelnicki, G. J. Schwark, and S. L. Daniels (to Dow Chemical Co.), U. S. Pat. 3,539,510 (Nov. 10, 1970).

7. J. C. Eck and W. E. Zegel (to Allied Chemical Corp.), U. S. Pat. 3,537,990 (Nov. 3, 1970).

8. J. C. Eck (to Allied Chemical Corp.), U. S. Pat. 3,493,501 (Feb. 3, 1970).

9. R. E. Crowe, "Recent Developments in the Application of Organic Polyelectrolytes in Sewage Treatment Practices," American Chemical Society National Meeting, Chicago, Sept. 2, 1964.

10. P. V. Freese and E. Hicks, "Full-Scale Raw Wastewater Flocculation with Polymers," Environmental Protection Agency, Program #17050 E.J.B., Grant #WPRD 53-01-67, Nov. 1970.

11. S. Mogelnicki, "Experiences in Polymer Applications to Several Solids-Liquids Separation Process," Proceedings - Tenth Sanitary Engineering Conference - Waste Disposal from Water and Wastewater Treatment Processes, University of Illinois, Feb. 6-7, 1968.

12. J. F. Rizzo and R. E. Schade, "Secondary Treatment with Gran-
    ular Activated Carbon," Water and Sewage Works, 116, 307-12
    (Aug. 1969).

13. J. J. Wirts, "The use of Organic Polyelectrolyte for Opera-
    tional Improvement of Waste Treatment Processes," Federal
    Water Pollution Control Administration, Grant No. W.P.R.D.
    102-01-68, May 1969.

14. Effects of Raw Sewage Flocculation in Secondary Waste Treatment
    Plants, The Dow Chemical Co., Midland, Michigan.

15. G. Thorpe, "Polyelectrolytes Help Water-Pollution Control,"
    Hercules Chemist, 55, 13-20 (Sept. 1967).

16. H. Akyel and M. Neven, "High Molecular Weight Synthetic Floc-
    culants in Modern Wastewater Treatment," Chem. Ingr. Tech.
    (Germany), 39, 172 (1967); Chem. Abstracts, 67, 25248 (1967).

17. K. D. Larson, R. E. Crowe, D. A. Maulwurf, and J. L. Witherow,
    "Use of Polyelectrolytes in Treatment of Combined Meat-Packing
    and Domestic Wastes," Water Pollution Control Federation
    Journal, 43, No. 11, 2218-2228 (Nov. 1971).

18. Environmental Protection Agency-Water Quality Office, Chemical
    Treatment of Combined Sewer Overflows, Water Pollution Control
    Research Series, 11023FDB, Sept. 1970.

19. E. Asendorf, "A New Way of Treating Combined Domestic and
    Industrial Sewages," Industrieabwaesser, 30-2 (May 1962).

20. D. E. Nagy (to American Cyanamid Co.), U. S. Pat. 3,567,659
    (Mar. 2, 1971).

21. C. C. Thurman, J. S. Scruggs, and F. N. Teumac (to Dow Chemical
    Co.), U. S. Pat. 3,341,476 (Sept. 12, 1967).

22. T. J. Suen and A. M. Schiller (to American Cyanamid Co.),
    U. S. Pat. 3,171,805 (Mar. 2, 1965).

23. C. S. Scanley and H. P. Panzer (to American Cyanamid Co.),
    U. S. Pat. 3,503,946 (Mar. 31, 1970).

24. D. I. Hoke (to Lubrizol Corp.), U. S. Pat. 3,666,810
    (May 30, 1972).

25. W. A. Rogers and J. E. Woehst (to Dow Chemical Co.), U. S.
    Pat. 3,320,317 (May 16, 1967).

26. R. D. Lees (to Hercules Inc.), U. S. Pat. 3,480,541 (Nov. 25, 1969).

27. R. H. Jones, "Liquid-Solids Separation in Domestic Waste with a Cationic Polyelectrolyte," Thesis, University of Florida, 1966.

28. J. J. Odom, T. P. Schumaker, and P. R. Bloomquist (to Reichhold Chemicals, Inc.), U. S. Pat. 3,484,837 (Dec. 16, 1969).

29. R. F. Wukash (to Dow Chemical Co.), U. S. Pat. 3,506,570 (Apr. 14, 1970).

30. R. F. Wukash and R. D. Goodenough (to Dow Chemical Co.), U. S. Pat. 3,488,717 (Jan. 6, 1970).

31. S. L. Daniels and D. G. Parker (to Dow Chemical Co.), U. S. Pat. 3,617,569 (July 31, 1970).

32. R. F. Wukash, "Phosphate Removal, Grayling, Michigan, and Lake Odesso, Michigan," Presented at Michigan Water Pollution Control Association Annual Conference, June 16-18, 1968.

33. R. C. Brenner, "Phosphate Removal by Chemical Addition During Primary Treatment," Proc. of Eighth Pittsburgh Sanitary Engineering Conference - Nutrient Removal and Advanced Waste Treatment, University of Pittsburgh, Feb. 17, 1970.

34. H. G. Flock and F. E. Bernardin, "Phosphate Removal by Chemical Precipitation and Activated Carbon Filtration - Denitrification Through Activated Carbon Filtration," 41st Annual Conference of the Water Pollution Control Association of Pennsylvania, August 6-8, 1969.

35. H. G. Flock and E. G. Rausch (to Calgon Corp.), U. S. Pat. 3,655,552 (Apr. 11, 1972).

36. J. C. Eck and W. C. Zegel (to Allied Chemical Corp.), U. S. Pat. 3,453,207 (July 1, 1969).

37. P. C. Singer, W. O. Pipes, and E. R. Hermann, "Flocculation of Bulked Activated Sludge with Polyelectrolytes," Water Pollution Control Federation Journal, 40, No. 2, 21-129 (1968).

38. M. L. Woldman, J. H. Dougherty, and A. F. Gaudy, "Effect of Selected Organic Polyelectrolytes on Activated Sludge Systems," Division of Water, Air and Waste Chemistry, American Chemical Society, Minneapolis, Apr. 1969.

39.   P. L. Busch and W. Stumm, "Chemical Interactions in the
      Aggregation of Bacteria-Bioflocculation in Waste Treatment,"
      Environmental Science and Technology, 2, No. 1, 49-53 (1968).

40.   K. M. Ries, "Application of Polymers in Existing Municipal
      Sewage Treatment Plants," Water and Sewage Works, R 195-R 197
      (Nov. 29, 1968).

41.   J. F. Walker and J. H. Dougherty, "Use of Polyelectrolytes
      Coagulants to Enhance Settling Characteristics of Activated
      Sludge," Proceedings of the 20th Annual Purdue Industrial
      Waste Conference, 52, No. 3, 715-723 (July 1968).

42.   M. W. Tenney and W. Stumm, "Chemical Flocculation of Micro-
      organisms in Biological Waste Treatment," Water Pollution
      Control Federation Journal, 37, No. 10, 1370-1388 (1965).

43.   M. J. Hurwitz and H. Aschkenasy (to Rohm & Haas Co.), U. S.
      Pat. 3,288,707 (Nov. 29, 1966).

44.   V. J. Jordan and C. H. Scherer, "Gravity Thickening Techniques
      at a Water Reclamation Plant," Water Pollution Control
      Federation Journal, 42, No. 2, Part 1 (Feb. 1970).

45.   B. L. Goodman, "Chemical Conditioning of Sludges," Water and
      Wastes Engineering, 3, No. 2, 62-64 (Feb. 1966).

46.   B. L. Goodman and K. A. Mikkelson, Advanced Wastewater Treat-
      ment, Chemical Engineering/Deskbook Issue, 77, No. 9,
      (Apr. 27, 1970).

47.   Water and Sewage Works, 111, 64-66 (Jan. 1964).

48.   R. A. Boehler and M. R. Purvis (to Nalco Chemical Co.),
      U. S. Pat. 3,617,542 (Nov. 2, 1971).

49.   B. Davis and S. Love, "Biological Chemical Combination
      Dramatically Cuts Treatment Costs," Water and Pollution
      Control, 35 (May 1969).

50.   J. H. Duff, R. Dvorin, and E. Salem, "Phosphate Removal by
      Chemical Precipitation," Industrial Water Engineering, 19-22,
      (Aug. 1968).

51.   J. W. Ducan, "The Use of Polyelectrolytes as an Aid to Sewage
      Sludge Thickening," Nigeria Journal of Science, 3, No. 1,
      65-68 (Mar. 1969).

52. M. A. Hilson, "Sludge Conditioning by Polyelectrolytes,"
     Journal of the Institution of Water Engineers, 25, 402-417
     (1971).

53. R. D. Crowe and E. E. Johnson, "Organic Polyelectrolytes-A
     New Approach to Raw Sewage Flocculation," Paper Presented at
     the 38th W.P.C.F. Conference (Oct. 1965).

54. J. R. Harrison, D. J. Monagle, and R. W. Smith (to Hercules
     Inc.), Canadian Pat. 893,945 (Feb. 22, 1972).

55. G. A. Ettelt and T. J. Kennedy, "Research and Operation
     Experience in Sludge Dewatering," Journal Water Pollution
     Control Federation, 38, No. 2, 250 (Feb. 1966).

56. R. L. Braithwaite, "Polymers as Aids to the Pressure Flotation
     of Waste Activated Sludge," Water and Sewage Works, 11, No. 12,
     545-547 (1964).

57. G. A. Ettelt, "Activated Sludge Thickening by Dissolved Air
     Flotation," Proc. of the 9th Purdue Industrial Waste Con-
     ference, 210-244 (1964).

58. W. J. Gatz and A. Geinopolis, "Sludge Thickening by Dissolved
     Air Flotation," Journal Water Pollution Control Federation,
     946-957 (June 1967).

59. C. P. Priesing and S. Mogelnicki (to Dow Chemical Co.), U. S.
     Pat. 3,247,102 (Apr. 19, 1966).

60. C. P. Priesing and S. Mogelnicki (to Dow Chemical Co.), U. S.
     Pat. 3,259,570 (July 5, 1966).

61. F. C. Buhl and R. D. Lees (to Hercules Inc.), U. S. Pat.
     3,414,513 (Dec. 3, 1968).

62. B. W. Dahl, J. W. Zelinski, and O. W. Taylor, "Polymer Aids
     in Dewatering and Elutriation," Journal Water Pollution
     Control Federation, 44, No. 2, 201-211 (1972).

63. C. P. Priesing and S. Mogelnicki (to Dow Chemical Co.), U. S.
     Pat. 3,300,407 (Jan. 24, 1967).

64. J. M. Sherbeck, "Synthetic Organic Flocculants Used for Sludge
     Conditioning," Journal Water Pollution Control Federation,
     37, No. 8, 1180-1183 (1965).

65. P. B. Reilly, "A Discussion on Guidelines for Applying Organic
     Polymers in Sewage Sludge Conditioning," Presented at Ohio
     Water Pollution Control Conference Northeast Section, Akron,
     Ohio, April 28, 1966.

66. G. J. Hopkins and R. L. Jackson, "Polymers in the Filtration of Raw Sludge," Journal Water Pollution Control Federation, 43, No. 4, 689-698 (Apr. 1971).

67. J. E. Boothe, H. G. Flock, and M. F. Hoover, "Some Homo- and Copolymerization Studies of Dimethyl Diallyl Ammonium Chloride," Presented at 158th ACS National Meeting, New York, Sept. 9, 1969.

68. R. A. LoSasso and E. G. Rausch (to Calgon Corp.), U. S. Pat. 3,642,619 (Feb. 15, 1972).

69. R. D. Bargman, W. F. Garber, and J. Nagano, "Sludge Filtration and the Use of Synthetic Organic Coagulants at Hyperion," Sewage and Industrial Wastes, 30, No. 9, 1079-1099 (Sept. 1958).

70. F. E. Bailey and E. M. LaCombe (to Union Carbide Corp.), U. S. Pat. 3,214,370 (Oct. 26, 1965).

71. M. F. Hoover, R. J. Schaper, and J. E. Boothe (to Calgon Corp.), U. S. Pat. 3,412,019 (Nov. 19, 1968).

72. W. J. Levy, J. P. Newport, and R. J. W. Reynolds (to Imperial Chemical Industries Ltd., England), U. S. Pat. 3,432,430 (Mar. 11, 1969).

73. E. H. Crook and F. X. Pollio, "Removal of Soluble Organic and Insoluble Organic and Inorganic Materials by Flocculation," Presented at the 26th Annual Meeting of the International Water Conference, Pittsburgh, Pa. (Oct. 20-26, 1965).

74. R. J. Schaper (to Calgon Corp.), U. S. Pat. 3,514,398 (May 26, 1970).

75. B. A. Schepman, "Designing Vacuum Filter Systems to Fit the Type of Sludge," Wastes Engineering, 162-165 (Apr. 1956).

76. J. M. Brown, "Vacuum Filtration of Digested Sludge," Water and Sewage Works, 107, No. 5, 193-195 (1960).

77. E. H. Trubnick, "Vacuum Filtration of Raw Sludge," Water and Sewage Works, 107, R-287-R-292 (1960).

78. R. S. Burd, The Dow Chemical Co., Personal Communication, 1964.

79. L. Sharman, Water and Wastes Engineering, 4, No. 8, 50-52 (Aug. 1967).

80. R. A. Conway, "Cationic Organic Coagulants in Wastewater Treatment," Water and Sewage Works, 109, No. 9, 342-344 (Sept. 1962).

81. R. S. Burd, Proc. 2nd Annual San. Eng. Conf., Vanderbuilt University, Nashville, Tenn., 76-85, 1963.

82. R. C. Buhl (to Hercules Inc.), U. S. Pat. 3,414,514 (Dec. 3, 1968).

83. R. V. Larvey, "Use of Polymers in Sludge Dewatering," W.P.C.F. Highlights, 8, No. 7, pD-6, (July 1971).

84. R. H. Morris, "Polymer Conditioned Sludge Filtration," Water and Wastes Engineering (Mar. 1965).

85. G. Thomas, W. Wisfeld, and G. Derenk, Brit. Pat. 1,225,635 (Mar. 17, 1971).

86. R. B. Schaffer, "Polyelectrolytes in Industrial Waste Treatment," Proc. of the 18th Purdue Industrial Wastes Conference, 1963, pp. 447-459.

87. G. W. Kumke, J. F. Hall, and R. W. Oeben, "Conversion to Activated Sludge at Union Carbide's Institute Plant, Water Pollution Control Federation Journal, 40, 1408 (Aug. 1968).

88. P. A. Vesilind, "Polymer Usage Gaining for Sludge Dewatering," Water and Waste Engineering, 8, No. 4 (Apr. 1969).

89. W. J. Montanaro, "Sludge Dewatering by Centrifuges Recent Operating Experiences," Presented at the 37th Annual Conf. Water Pollution Control Association of Pennsylvania, Penn. State University (Aug. 5, 1965).

90. O. E. Albertson and E. E. Guidi, "Centrifugation of Waste Sludges," Water Pollution Control Federation Jounral, 607-628 (Apr. 1969).

91. O. E. Albertson and R. J. Sherwood, "Centrifuge for Dewatering Sludges," Water and Wastes Engineering, 5, No. 4, 56-58 (Apr. 1968).

FLOCCULATION OF CHROME-PLATING WASTES WITH POLYELECTROLYTES[a]

Donald L. Sussman and Edward Chun-Chin Wang

University of Rhode Island

Kingston, Rhode Island

One of the largest industries in southeast New England is the metal-finishing industry. Normally a metal is pretreated before the actual finishing step. In these operations there is usually a dip in organic solvents and/or detergents to remove grease, oils, dirt, etc. This is followed by a few rinse baths. The metal is then pickled in a strong acid bath, rinsed again several times, followed by the plating or finishing operation and more rinses. A major problem associated with these industries is the treatment or lack of treatment of the prodigious volumes of waste water from the many rinses. These waste waters contain either acids or alkalies, metal ions (such as Cr, Cu, Zn, Ag, Pb, Fe, and Mn), cyanides, organics, chemical oxygen demand, and a high solids content of dissolved, suspended, and settleable material. Proper treatment can reduce or eliminate the impurities from the waste waters. However, there are usually problems of cost, time, ease of operation, and ultimate disposal of sludges.

In this study we tried to develop a new method for treating chrome-plating wastes. Our approach was to flocculate directly a turbid chrome-plating waste by using polyelectrolytes with and without calcium and cupric ions bound to the polymer. We intended to compare this method with the most commonly used reduction-precipitation method using ferrous sulfate and lime.

- - - - - - -

[a]This chapter is based on a paper presented to the Division of Water, Air, and Waste Chemistry, American Chemical Society, New York, Aug. 28, 1972.

THEORETICAL BACKGROUND

The destabilization of colloids involves a two-step process. First, the particles to be aggregated must be able to adhere to each other when brought into contact. Second the destabilized particles must be transported into contact to effect the formation of larger particles. The extent to which the aggregation of particles can be accomplished is usually determined by chemical factors, while the rate at which aggregation occurs is determined by the rate of interparticle contact. The addition of a chemical coagulant to such a system makes coagulation possible and the rate of coagulation can be enhanced by means of agitation. Destabilization can be accomplished by the following mechanisms: 1. by an electrolyte through compression of the electrical double layer surrounding the colloidal particles; 2. by the specific adsorption of a coagulant of opposite charge, and through the formation of a bridge between two colloidal particles by a polymeric molecule.

Colloidal particles possess electrical properties that strongly influence their behavior. Charges located on particle surfaces establish an electrostatic field that is a major factor in determining the stability of the colloid system. In the double layer surrounding a colloid, ions of opposite charge to the particle are electrostatically attracted to the particle surface, so that their concentration near the particle surface is higher than in the bulk solution.

If an indifferent electrolyte (for example, a salt whose ions do not specifically adsorb or otherwise chemically interact with colloid particles) is added to a colloidal dispersion, the counterions are attracted toward the particle surface and can enter into the double layer. Increasing the concentration of ions in the diffuse layer reduces the distance over which the primary charge on the colloidal particle can exert coulombic effects, and compresses the double layer. The repulsive interaction between similar colloidal particles is thus reduced. If sufficient compression occurs, the van der Waals attraction forces can bring about interparticle attachment when an opportunity of contact is provided.

The flocculation of colloidally dispersed solids by polyelectrolytes has been postulated by a process of adsorption of polymers on the colloid surface, and of bridging of polymer chains between solid particles (1,8). In order to be effective in destabilization a polymer molecular must contain the polarizable or ionizable groups which can interact with sites on the surface of the colloidal particles. An explanation of the bridging of polymers is as follows:

When a polymer molecule comes into contact with a colloid particle, some of these groups are adsorbed at the particle surface,

leaving the remainder of the molecule extending out into the solution. If a second particle with some vacant adsorption sites comes
in contact with these extended segments, floc formation occurs. If
a second particle is not available, the extended segments may eventually adsorb on other sites on the original particle. This restabi-
lizes the colloidal particle. Excess polymer dosages produce a restabilized colloid, since no sites are available for the formation
of polymer bridges. Extended agitation can break the polymer - surface bonds and fold back the extended segments on the surface of the
particles. This ruptures the flocs and permits secondary adsorption.
A direct relationship exists between the available surface area in
the colloidal system and the quantity of polymer required to produce
optimum destabilization.

The performance of polyelectrolyte flocculants is in general
agreement with the predictions of the bridging model; there is a
direct relationship between optimum polymer dosage and colloid concentration, and restabilization due to overdosing can occur (2).
The presence of divalent metal ions greatly enhances the ability of
polymers to adsorb on colloids.

The purpose of the study was to define the chemical and physical
factors involved in the destabilization of colloids in an industrial
waste by different synthetic organic polyelectrolytes. The polyelectrolytes were used alone or with the divalent counterions calcium and copper.

It has been shown by Sommerauer, Sussman, and Stumm (3) that
the anionic polymers form 1:1 coordination complexes with cations,
like calcium or copper. At polymer concentrations larger than those
required for flocculation, restabilization or colloid protection
occurs.

EXPERIMENTAL

Three separate determinations were made. The first was the
measurement of the change in turbidity with time. The second was
the refiltration rate of a filtrate through a noncompressed sludge.
The third was the measurement of the specific resistance of the
sludge.

The change in absolute turbidity with time was made by means
of a light-scattering photometer. The solutions were rapidly mixed
for 1 minute, then slowly mixed for 60 minutes. Readings were taken
initially at every other minute for 10 min. and then every 5 min.
until 60 min. Two parameters were used to study the effectiveness
of our flocculation system. These were the initial change in turbidity with time dT/dt_{10}, and the final turbidity reported as T_{60}.

The first determination gave us an indication of the rate of destabilization and adsorption. The second gave us an indication of
the floc size.

The second determination that we used was the application of
refiltration rate measurements to study the flocculation process.
This was developed by LaMer and co-workers (4), by the application
of the Kozeny-Carman equation for flow through a porous media.
This application of filtration hydraulics allows us to get an idea
of the floc size. The flocs are precipitated on a membrane filter
at a constant pressure of 80 mmHg, and the filtrate is collected
and passed through the flocs, and the time is measured for a given
volume to pass through the filter.

The Kozeny-Carman equation (Equation 1) relates the rate of
filtration through a filter cake to the specific surface area of
particles in the cake,

$$Q = \frac{\Delta pgA\epsilon^3}{K\eta LS^2} \tag{1}$$

where Q = volume rate of flow of filtrate in cm^3/sec; p = pressure
drop across the bed (filter cake) in g/cm^2; g = acceleration of
gravity in 980 cm/sec^2; A = cross-sectional area of the bed in
cm^2; ϵ = porosity of the bed or $1-W/AL\rho$, where W is the weight
of solids in the bed in g, L is the depth or thickness of bed in
cm and ρ is the bulk density of solid in g/cm^3; η = viscosity of
filtrate (centipoise); K = constant $\cong 5$; S = surface area of
particles in unit volume of bed in cm^2/cm^3.

The Kozeny-Carman equation predicts that the filtration rate
Q will be inversely proportional to the square of the surface area
of the particles in the cake. If the flocs are considered as
essentially spherical in shape and made up of closely packed primary particles, it may be shown that the other factors in the
Kozeny-Carman equation will be essentially constant.

The relationship between Q and S can be expressed as follows:

$$Q = KS^{-2} \tag{2}$$

The efficiency of a flocculation system could be gauged by
the size of the flocs produced. The more thorough the flocculation,
the larger would be the floc size and the greater would be the rate
of refiltration.

Specific resistance is defined as the resistance of a unit
weight of cake per unit area at a given pressure and is expressed
in centimeters per gram. In order to compare the dewaterability

of the resulting sludge which was produced in the flocculation process, specific resistance was selected as the primary dependent variable. A modified Buchner-funnel technique was developed by Coackley and Jones (5) which made it possible to compare sludges having small or large differences in dewaterability. The same investigators concluded that specific resistance is a valid indicator of sludge filterability and is a better parameter for evaluating the effect of chemical conditioning than is the time required for the cracking of a filter cake.

In the case of the modified Buchner-funnel technique the tests were carried out as follows. A moistened filter paper was placed in the funnel and the vacuum applied so that the paper was in close contact with the funnel. The sample of sludge under test was poured into the funnel and the vacuum was slowly applied until the required pressure was attained. The vacuum (80 mmHg) was kept constant by a Cartesian manostat throughout the full experiment. The volume of filtrate coming off the sample was then measured, after a suitable time had been allowed for a cake to form, every 25 ml for 250 ml. It was considered that, in general, the time needed for the first 50 ml of filtrate to come off the sample was a suitable length of time to allow for the formation of a cake with a resistance equal to that of the filter medium.

The specific resistance can be calculated by using Equation 3:

$$r = \frac{2PA^2b}{\eta C} \qquad (3)$$

where r = specific resistance (sec^2/g); A = filter area (cm^2); P = pressure (g/cm^2); η = filtrate viscosity ($gm/cm\text{-}sec$); C = solids content of feed sludge ($g\ solid/ml\ liquid$); b = filtering rate or the slope of plot of t/V vs. V (sec/cm^6) where V is ml of filtrate collected in t seconds and t is time in seconds. The graph of t/V versus V is plotted from the experiment results.

The polyelectrolytes used in this experiment were: poly(sodium styrenesulfonate) (NaPSS), Purifloc A21, Dow Chemical Co.; Hydrolyzed copolymer of methyl vinyl ether and maleic anhydride (MVE-MA), Gantrez AN 169, GAF Corp.; polyvinylpyrrolidone (PVP), NP-K90, GAF Corp.; polyacrylamide (PAM), Cyanamer P-250, American Cyanamid Co.; and two polymers of dimethyldiallyl ammonium chloride and acrylamide, WT-2640 and WT-2580, Calgon Corp.

RESULTS AND DISCUSSION

The results of our experiments are calculated in terms of flocculation kinetics - as measured by the rate of change of absolute turbidity and the degree of turbidity removal.

The absolute turbidity, $(-dT/dt)$, was generally taken after ten minutes to offset any possible human error that was susceptible in earlier readings. The turbidity of the flocculation system changed drastically during the first ten minutes and $(-dT/dt)_{10\ min}$ is chosen as a comparing factor in the effectiveness of the flocculation system (Table I).

TABLE I.   Turbidity after 10 Minutes, $(-dT/dt)_{10\ min}$, for the Flocculation of Solutions Containing 25 ppm Cr(VI) and 50 ppm Turbidity

Dosage, %	$[-dT/dt_{10\ min}]; \times 10^7/min$					
	PNaSS	MVE-MA	PVP	PAM	WT-2580	WT-2640
10^{-1}	2.3	7.0	8.0	1.5		
10^{-2}	2.4	6.0	7.2	7.0	5.0	4.0
10^{-3}	4.0	7.2	7.5	11.2	3.5	6.0
10^{-4}	7.0	7.5	8.1	14.0	5.0	2.5

In flocculating chrome wastes with polysodium(styrenesulfonate) (PNaSS), the largest flocs and the best clarity were observed in system with $10^{-4}\%$ (w/v%) of polymer. Beyond the optimum dosage the effectiveness of the polyelectrolytes begins to wane as redispersion of the colloidal particles is established. The optimum dosage of methyl vinyl ether - maleic anhydride copolymer (MVE-MA) for the flocculation of chrome waste is $10^{-3}\%$.

The rate of turbidity removal increases with decreasing polymer dosage. Slate and Kitchner (6) have postulated that at higher concentrations polyelectrolytes act as protective colloids, i.e., to keep the flocculation system dispersed. The viscosity of these two anionic polymers increases with polymer concentration. The viscosity at high polymer concentration impaired the mobility of the particle dispersed in the flocculation system.

With the two nonionic polymers investigated in this study, polyvinylpyrrolidone (PVP) and polyacrylamide (PAM), the largest flocs and the best clarity were obtained at the optimum dosage of $10^{-4}\%$. As seen before, as the polymer concentration decreased the rate of change of absolute turbidity, $(-dT/dt)_{10\ min}$ increased for PAM. For PVP the fastest rate occurred at $10^{-4}\%$. At high polymer concentration, the polymer is adsorbed in a thin molecular layer

on the surface of the individual colloidal particles to prevent
contact and overcome the tendency to form the flocs.

Of the two cationic polymers investigated in this study,
WT-2640 and WT-2580, WT-2640 is the more cationic. The average
molecular weight of WT-2640 is ca. 500,000, and that of WT-2580
ca. 2,000,000. In these two flocculation kinetics systems, a polymer
dosage of $10^{-3}\%$ for WT-2580 and $10^{-2}\%$ for WT-2640 produced the
greatest clarity and biggest flocs.

Black and co-workers (7) have postulated that cationic polyelectrolytes serve a dual function in the destabilization reaction
of colloid particles. First, they serve as very effective coagulants either by compressing the double layers that surround the
negatively charged colloids or by reducing the negative surface
potentials of the colloidal particles. Second, if the polymer molecules are of adequate length, they may also serve an interparticle
bridging function by physically binding the discrete colloidal par-
icles together into multiparticle aggregates. Anionic and nonionic
polymers would not be expected to exhibit the dual functionality in
the destabilization of negatively charged colloids, and Black and
co-workers have postulated that anionic polymers function exclusively as bridging molecules.

Effect of Divalent Metal Ions

The effect of the addition of divalent metal ions upon the
flocculation of chrome waste by polyelectrolytes was also investigated. Complex formation between the divalent metal ions in the
double layer and the functional groups of the anionic polymers
lowers the electrochemical free energy of attachment and the extent
of adsorption is increased. There were increases in flocculation
efficiency attributable to the polyelectrolytes in cases involving
Ca^{++} and Cu^{++}. When an industrial chrome waste was flocculated in
the presence of metal ions and polyelectrolytes, flocculation was
almost completed in the first five minutes and turbidity removal
was substantial. This waste contained phenolics, paint immiscible
solvents, as well as $Cr(VI)$, $Cu(II)$ and turbidity.

The parameter T_{60} was taken as a residual turbidity and a
measure of the effectiveness of the flocculation kinetics. The
residual turbidity for different polymers with the industrial
waste-water sample is listed in Table II. The data clearly indicate that the use of polymer results in good turbidity removal.
One would expect to observe the best improvement in turbidity removal as the optimum polymer dosage is approached. When the optimum dosage is exceeded, the surfaces of the colliding particles become crowded with adsorbed polymer segments and the interparticle
bridging function is hindered, because bridging sites are less

TABLE II.   Residual Turbidity for Different Polymers with an
            Industrial Waste-Water Sample

| Dosage, % | Residual Turbidity[a], T_{60}; x 10^2 | | | | | |
	NaPSS	WT-2640	WT-2580	PAM	PVP	MVE-MA
10^{-1}	36			25	51	38
10^{-2}	34	22	31	43	37	28
10^{-3}	27	19	22	30	39	23
10^{-4}	32	18	15	9	22	22

- - - - - - -

[a]Initial turbidity = 67 x 10^{-2}; pH = 9.0 $\pm$ 0.2

abundant on the particle surface.   Consequently, the colloidal
particles become restabilized at polymer dosage in excess of the
optimum dosage.

When a 10^{-3} M Ca^{++} or Cu^{++} solution was added, there was a
decrease from 10% to 50% in the absolute turbidity.   This was due
to the fact that with the anionic polyelectrolytes a new calcium-
bound polymer is formed with cationic character.   With the other
polyelectrolytes, the Ca^{++} and Cu^{++} act as a secondary flocculant.

The effect of different polymer types and dosages, and the
effect of Ca^{++} and Cu^{++} on the rate of refiltration and specific
resistance of the sludge produced in the flocculation process is
given in Tables III-VI.

On the basis of the Kozeny-Carman equation (Equation 1) for
flow through porous media, the flow rate during the refiltration
step is inversely proportional to the square of the effective particle surface area in a unit volume of filter cake.   The refiltration time is therefore directly proportional to S^2 if our experimental conditions are kept constant.   The more thorough the flocculation, the larger would be the floc size and therefore the
greater would be the rate of refiltration.   The data listed in
Table III comply with the Kozeny-Carman equation.   The improvement
in filterability resulting from the polymer conditioning is very
clearly evident from these data.   The lower the filtering rate and
the lower the specific resistance, the larger the particle size.
The specific resistance data are shown in Tables IV-VI and Figures
1-6.

TABLE III. Effect of Metal Ions on Rate of Refiltration

| | | Rate of Refiltration, Q, ml/min | | |
| | | Polymer and no salt | Polymer and 10^{-3} $\underline{M}$ CaCl$_2$ | Polymer and 10^{-3} $\underline{M}$ CuCl$_2$ |
Polymer	Dosage, %			
NaPSS	10^{-2}	2.60	15.0	
	10^{-3}	12.3	28.3	12.2
	0.5×10^{-3}	28.8	75.0	20.0
	10^{-4}	18.8	60.0	13.3
WT-2640	10^{-1}	20.1	7.79	
	10^{-2}	23.5	11.5	50.0
	10^{-3}	37.3	75.0	40.0
	0.5×10^{-3}	45.8	8.57	17.6
	10^{-4}	30.1	11.3	20.0
WT-2580	10^{-1}	23.8	25.0	45.0
	10^{-2}	30.0	83.2	60.0
	10^{-3}	78.9	100.	33.3
	0.5×10^{-3}	28.8	88.2	24.3
	10^{-4}	45.5	100.	27.3
PAM	10^{-1}	2.06	4.86	6.81
	10^{-2}	2.60	33.3	46.9
	10^{-3}	60.0	75.0	60.0
	0.5×10^{-3}	50.0	30.0	25.3
	10^{-4}	7.89	68.2	38.5
PVP	10^{-1}	1.29	6.25	2.14
	10^{-2}	5.00	25.0	2.50
	10^{-3}	5.76	27.0	2.72
	0.5×10^{-3}	6.50	31.4	3.75
	10^{-4}	15.0	37.5	4.28
MVE-MA	10^{-1}		7.79	
	10^{-2}	4.44	12.0	17.6
	10^{-3}	57.7	75.0	20.0
	0.5×10^{-3}	50.0	7.79	50.0
	10^{-4}	10.5	11.3	50.0

TABLE IV. Effect of Different Polymers on Specific Resistance

Polymer	Dosage, %	Filtrating Rate[a] sec/cm^6; x 10^3	Specific Resistance[b] sec^2/min; x 10^{-6}
NaPSS	10^{-3}	23.4	26.0
	$0.5x10^{-3}$	16.7	18.5
	10^{-4}	9.2	10.2
WT-2640	10^{-1}	5.05	5.60
	10^{-2}	1.67	1.84
	10^{-3}	4.55	5.05
	10^{-4}	9.70	10.6
WT-2580	10^{-1}	6.12	6.75
	10^{-2}	11.7	12.8
	10^{-3}	1.45	1.60
	$0.5x10^{-3}$	7.33	8.05
	10^{-4}	4.35	4.79
PAM	10^{-3}	41.0	45.6
	$0.5x10^{-3}$	1.19	1.31
	10^{-4}	4.40	4.85
PVP	10^{-1}	31.3	34.2
	10^{-2}	3.66	4.06
	10^{-3}	1.66	1.83
	$0.5x10^{-3}$	3.66	4.06
	10^{-4}	5.10	5.66
MVE-MA	10^{-2}	114	126
	10^{-3}	32.0	35.5
	$0.5x10^{-3}$	4.07	4.55
	10^{-4}	4.70	5.20

- - - - - - -

[a]Slope t/V vs V

[b]Without metal ion

TABLE V. Effect of Calcium Ion and Polymer on Specific Resistance

Polymer	Dosage, %	Filtrating Rate[a] sec/cm^6; x 10^3	Specific Resistance[b] sec^2/min; x 10^{-6}
NaPSS	10^{-2}	22.2	24.6
	10^{-3}	6.0	6.66
	$0.5x10^{-3}$	1.9	2.14
	10^{-4}	3.5	3.88
WT-2640	10^{-2}	3.00	3.33
	10^{-3}	0.80	0.88
	$0.5x10^{-3}$	1.53	1.09
	10^{-4}	1.47	1.67
WT-2580	10^{-1}	5.00	5.55
	10^{-2}	2.32	2.57
	10^{-3}	0.50	0.55
	$0.5x10^{-3}$	1.14	1.26
	10^{-4}	1.14	1.26
PAM	10^{-2}	4.08	4.52
	10^{-3}	1.02	1.13
	$0.5x10^{-3}$	4.68	5.19
	10^{-4}	1.81	2.00
PVP	10^{-1}	4.02	4.46
	10^{-2}	1.33	1.47
	10^{-3}	1.33	1.47
	10^{-3}	1.08	1.19
	10^{-4}	0.50	0.55
MVE-MA	10^{-2}	73.0	81.0
	10^{-3}	1.00	1.11
	$0.5x10^{-3}$	51.6	57.3
	10^{-4}	32.0	35.5

- - - - - - -

[a]Slope t/V vs V

[b]With 10^{-3} $\underline{M}$ $CaCl_2$

TABLE VI. Effect of Copper Ion and Polymer on Specific Resistance

Polymer	Dosage, %	Filtrating Rate[a] sec/cm^6; x 10^3	Specific Resistance[b] sec^2/min; x 10^{-6}
NaPSS	10^{-2}	22.2	24.6
	10^{-3}	21.3	23.6
	0.5×10^{-3}	9.52	10.6
	10^{-4}	19.4	21.5
WT-2640	10^{-1}	2.22	2.46
	10^{-2}	1.17	1.29
	10^{-3}	3.00	3.33
	0.5×10^{-3}	14.0	15.5
	10^{-4}	12.5	13.9
WT-2580	10^{-1}	5.50	6.10
	10^{-2}	3.03	3.19
	10^{-3}	11.3	12.5
	0.5×10^{-3}	16.1	17.9
	10^{-4}	21.5	23.9
PAM	10^{-2}	5.00	5.55
	10^{-3}	2.3	2.55
	0.5×10^{-3}	6.60	7.39
	10^{-4}	8.88	9.85
PVP	10^{-1}	17.6	19.5
	10^{-3}	12.5	13.9
	0.5×10^{-3}	8.00	8.88
	10^{-4}	10.5	11.7
MVE-MA	10^{-2}	10.8	12.0
	10^{-3}	18.7	20.8
	0.5×10^{-3}	5.35	5.93
	10^{-4}	5.76	6.39

- - - - -

[a]Slope t/V vs V

[b]With 10^{-3} $\underline{M}$ $CuCl_2$

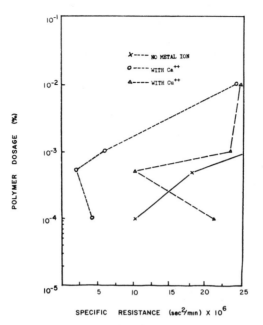

Fig. 1.   Effect of metal ions on specific resistance of anionic
polymer, poly(sodium styrenesulfonate).

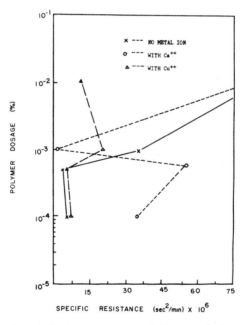

Fig. 2.   Effect of metal ions on specific resistance of anionic
hydrolyzed copolymer of methyl vinyl ether and maleic
anhydride.

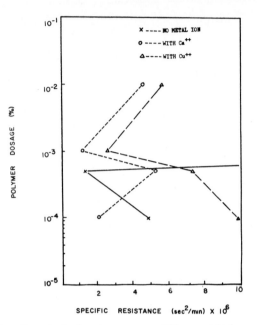

Fig. 3.   Effect of metal ions on specific resistance of nonionic
polymer, polyacrylamide.

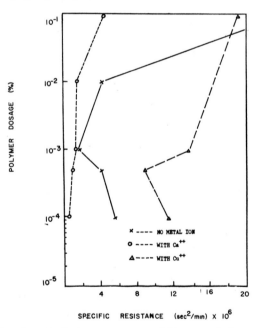

Fig. 4.   Effect of metal ions on specific resistance of nonionic
polymer, polyvinylpyrrolidone.

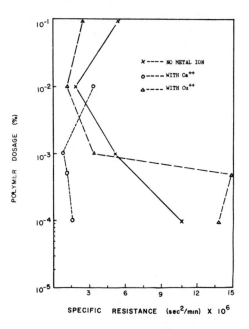

Fig. 5. Effect of metal ions on specific resistance of cationic polymer WT-2640.

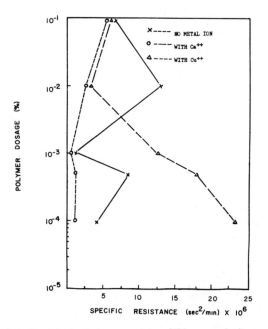

Fig. 6. Effect of metal ions on specific resistance of cationic polymer WT-2580.

The polymers were effective over a resonably wide pH range, i.e. pH 3-11. The pH of all experiments was standardized to 9.0 $\pm$ 0.2 by the addition of sodium hydroxide. The effectivness of the divalent metal ion (Ca^{++}, Cu^{++}) is attributed to the formation of a very stable metal ion - polymer complex. Ca^{++} forms a 1:1 complex with polymer and was found to be much more effective than Cu^{++} in the flocculation of chrome waste.

All the polymers investigated show the ability of removing metal ions which were contained in the metal plating wastewater. And also the polymers improved the drainage rates or filterability of the sludge.

Cr(VI) Removal by Fe(II) Reduction and Precipitation

Reduction and precipitation is a process whereby the soluble metallic ions are reduced through an oxidation-reduction reaction and then precipitated by conversion to an insoluble metallic compound.

In order to compare the efficiency of flocculation between the reduction-precipitation and polymeric flocculation method, ferrous sulfate ($FeSO_4 \cdot 7H_2O$) was used as the reducing agent to treat the same chrome waste. Although the ferrous sulfate reduced the hexavalent chromium to a trivalent state and almost had the same sludge dewaterability as that of the sludge produced by using polymer, the volume of sludge produced by the reduction method was three times as great as that produced by using a polymeric flocculant.

CONCLUSION

We have looked at solutions containing chromium ions and colloidal turbidity for the purpose of removing both by means of flocculation with polyelectrolytes. The three measurement parameters we chose, namely rate of turbidity removal, refiltration rate, and the specific resistance of the sludge gave us an indication of the process. Usually a 10^{-3} to 10^{-4} percent polymer concentration with a divalent cation (such as Ca^{++}) gave the best turbidity removal and dichromate removal. This is due to the pronounced effect of these ions on the dielectric properties of the polymer adsorbed on a colloid surface (3). The polymer-calcium complex $(R\text{-}SO_3\text{-}Ca)^+$ will seek preferable positions close to the colloid surface. This orientation ability appears to be more susceptible to bridging between polymer and colloid. This was confirmed by the fact that this system produced the fastest rate of turbidity removal, the best refiltration rate (largest flocs), and the least specific resistance (best floc solid development).

In comparison to the oxidation and precipitation method ("iron and lime" method), there was no difference in removal rate, re-filtration rate, and specific resistance; however, the sludge volume produced by the iron and lime method was approximately three times as great as that produced by polyelectrolyte flocculation.

The authors wish to thank the Rhode Island Water Resources Center for their support by research contract 14-31-0001-3240.

BIBLIOGRAPHY

1. V. K. LaMer and T. W. Healy, J. Phys. Chem., 67, 2417 (1963).

2. R. H. Smellie and V. K. LaMer, J. Colloid Sci., 23, 589 (1958).

3. A. Sommerauer, D. L. Sussman, and W. Stumm, Kolloid Z., 225 (2), 147 (1968).

4. V. K. LaMer, R. H. Smellie, and P. K. Lee, J. Colloid Sci., 21, 566 (1957).

5. P. Coackley and B. R. S. Jones, Sewage and Industrial Wastes, 28, 963 (1956).

6. R. W. Slater and J. A. Kitchner, Disc. Farad. Soc., 42, 224 (1966).

7. A. P. Black, F. B. Birkner, and J. J. Morgan, J. Am. Water-works Assoc., 57, 1547 (1965).

8. J. K. Dixon, "Flocculation" in N. M. Bikales, Ed., Encyclopedia of Polymer Science and Technology, Interscience Publishers, New York, Vol. 7, 1967, pp. 64-78.

PARTICLE SIZE VERSUS MOLECULAR WEIGHT RELATIONSHIPS IN CATIONIC FLOCCULATION

Monica A. Yorke

Calgon Corporation

Pittsburgh, Pennsylvania

Most studies concerning the flocculation of kaolin suspensions with polyelectrolytes have been performed on suspensions having very broad particle-size distributions (1,2,3). A few investigators have referred to particle size as an influencing parameter in determining the optimum polymer dosage required to flocculate a suspension (4,5). Black and Vilaret (6) report that the optimum polymer dosage is proportional to the surface area of the colloidal system. Optimum polymer dosage is defined in this paper as that amount of polymer required to produce the least residual turbidity.

LaMer and Healy (7) postulate that there exists an optimum polymer molecular weight dependent upon the surface area of the colloidal system being investigated. If the molecular weight is too high, the bridging may be hindered by steric stabilization (which is the extension of segments into solution preventing close approach of colloidal particles) or coverage of a large surface area by a single polymer molecular on a single particle. If the molecular weight is too low, the bridging will be ineffective because the extended segments will be too short (8).

In this study, the relationship between polymer molecular weight and colloid particle size has been investigated. A polydisperse cationic polyelectrolyte, poly(diallyldimethylammonium chloride), was fractionated by thin-channel ultrafiltration. The flocculation efficiency of the resultant fractions was tested by a jar test procedure on suspensions of specially-sized kaolin clays. Since the effects of pH, ionic strength, and mixing conditions have been well documented in the literature (9-13), these variables were held constant throughout this investigation.

EXPERIMENTAL

Kaolin clays with particle size ranges of 0.1-0.4 μm, 0.15-1.5 μm, and 2.0-20.0 μm, were obtained from Particle Information Service - Palo Alto, California. The 2.0-20.0 μm clay samples were analyzed for particle size distribution by Coulter Counter Methods (14). The remaining clays were analyzed by the scanning electron microscopic method developed by White and co-workers (15). Table I contains a listing of clay samples used and their average particle size as ascertained by the above analyses.

Clay suspensions were prepared by hydrating the clays at 100 mg/l concentration in deionized water containing 50 mg/l $NaHCO_3$ (ionic strength 6×10^{-4}). The pH remained constant throughout the experiments at 7.5 ± 0.2 units. The suspensions were hydrated for four hours prior to testing. This time period was chosen because at this point there is no further increase in the electrophoretic mobility of the particles.

TABLE I.   Mean Particle Diameter of Kaolin Clay Samples

Sample	A	B	C	D	E	F	G	H
Diameter, μm	0.09	0.23	0.22	0.70	0.70	9.8	9.5	9.5

Poly(diallyldimethylammonium chloride), poly(DADMAC), is a polyquaternary ammonium salt produced by the free-radical polymerization of diallyldimethylammonium chloride (16). The structural unit of the polymer is proposed by Butler (17) to be as follows:

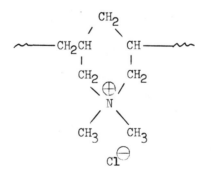

The polymer was fractionated through a series of membranes of decreasing pore size using an Amicon TC-1D ultrafiltration unit. Four liters of a 0.5% solution of poly(DADMAC), 20.0 g, in deionized water was fractionated at an effective membrane pressure of 15 psi. Two methods of diafiltration were employed (18,19). The discontinuous method involves repetitive concentration and dilution of the fractionating solution. The o.5% polymer solution was concentrated until 500 ml of filtrate was collected. At this point, 500 ml of deionized water was added to the fractionation cell. This procedure was followed until 6 l of filtrate had been obtained. The fractionation solution was then concentrated to a volume of 300 ml and then drained from the apparatus. The filtrate collected was then passed through the next successive membrane. This process was repeated until the desired number of fractions had been obtained.

The second fractionation method, the continuous method, uses an additional reservoir and valve assembly which allows replacement of solvent, deionized water, to the fractionation cell at the same rate at which filtrate is removed from the system. Three volumes of solution were passed through the cell and the resultant fraction was then refractionated until more than 90% of the material was retained by the membrane. This procedure was then repeated with the combined filtrates of the fractionation-refractionation step to obtain subsequent fractions on successively smaller size membranes. The continuous method was used in an attempt to obtain fractions of narrower molecular-weight distributions.

Concentrations were determined on a Beckman Total Organic Carbon Analyzer. A portion of each fraction solution was concentrated, precipitated in acetone, filtered and dried at $80^{\circ}C$ for 16 hours. Polymer fractions were characterized by membrane osmometry in 1 $\underline{N}$ NaNO$_3$. Table II shows the $\bar{M}_n$ values obtained for the polymer fractions.

The one-liter clay suspensions (100 mg/l) were mixed by a Phipps & Bird Multiple Stirrer and polymer added at premeasured dosages. The test suspensions were stirred at 100 rpm for 20 min, 15 rpm for 20 min, and then allowed to settle for 30 min undisturbed. An untreated one-liter sample of the clay suspension was used as a blank for the settling period. Residual turbidities were measured by a Hach turbidimeter on 25-ml aliquots of the supernatent liquids. Aliquots were removed 1-1/2 in. below the test solution surface. Percent residual turbidity was determined by dividing the residual turbidity of the test soltuion by the residual turbidity of the initial clay suspension, or blank, measured after the settling period. Also, percent residual turbidity was used to facilitate comparisons between the differently sized suspensions.

TABLE II.   Molecular Weights of Fractionated PDADMAC

Polymer Sample	$\bar{M}_n; \times 10^{-5}$
Discontinuous Method	
Fractionation No. 1	
Fraction 1	1.13
Fraction 2	0.35
Fraction 3	0.11
Fraction 4	0.07
Fractionation No. 2	
Fraction 1	1.10
Fraction 2	0.26
Continuous Method	
Fraction 1	1.00
Fraction 2	0.70
Fraction 3	0.13

All tests were duplicated to ensure reproducibility. Both liquid and precipitated forms of the polymer fractions were tested to compare differences in flocculation efficiency. Deionized water and the highest purity analytical reagents commercially available were used throughout the study.

RESULTS AND DISCUSSION

Four polymer fractions were obtained for the first fractionation by the discontinuous method. Their flocculation efficiencies were evaluated via the jar-test procedure, clay suspensions of 0.09 μm, 0.7 μm, and 9.8 μm mean particle diameters. In the 9.8 μm clay suspension, optimum polymer dosage decreased with increasing molecular weight (see Figure 1) which corresponds to findings reported by Rembaum (3) and Kasper (10). For the 0.09 μm clay sol, however, Fraction 2 ($\bar{M}_n$ = 35,000) required the lowest optimum dosage (50 μg/l) while Fraction 1 ($\bar{M}_n$ = 113,000) only performed as well as Fractions 3 and 4 (with $\bar{M}_n$ = 11,000 and 7,000 respectively)

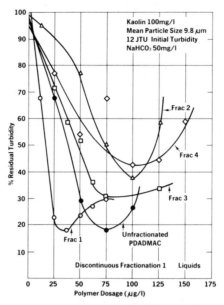

Fig. 1.   Flocculation performance of discontinuous fractionation 1 (liquids) in kaolin suspensions of 9.8 μm mean particle diameter (100 mg/l); ionic strength 6 x 10^{-4}; pH = 7.5; initial turbidity 12 JTU (Jackson turbidity units[a]).

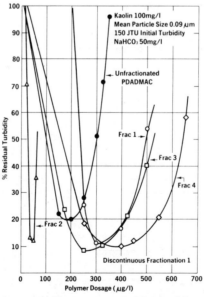

Fig. 2.   Flocculation performance of discontinuous fractionation 1 (liquids) in kaolin suspension of 0.09 μm mean particle diameter; ionic strength 6 x 10^{-4}; pH = 7.5: initial turbidity 150 JTU (Jackson turbidity units).

[a]ASTM D 1889

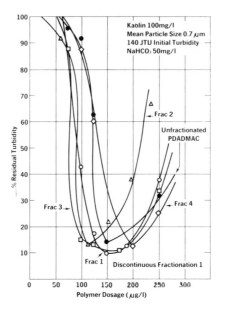

Fig. 3.   Flocculation performance of discontinuous fractionation
             1 (liquids) in kaolin suspension of 0.7 μm mean particle
             diameter; ionic strength 6 x 10^{-4}; pH = 7.5; initial
             turbidity 140 JTU (Jackson turbidity units).

(see Figure 2).   Also, for 0.7 μm particle diameter clay suspension,
Fractions 1 and 2 performed comparably to unfractionated PDADMAC
(see Figure 3).

     To test the reproducibility of the fractionation process, a
second 20.0 g sample of poly(DADMAC) was fractionated by the discontinuous method. Since Fractions 1 and 2 of the first fractionation were the only samples to show incongruous behavior in the flocculation tests, only the flocculation performance of Fraction 1 and
2 of the second fractionation were evaluated. The same behavior
exhibited by the first and second fractions of fractionation No. 1
were again observed for Fractions 1 and 2 of fractionation No. 2.

     Black and co-workers (2) postulate that the range of polymer
concentration over which efficient turbidity removal occurs is dependent upon the number of adsorption sites available in a given
colloidal system; i.e., the flocculation zone increases as the
number of adsorption sites, or surface area, increases. Gregory
(9) indicates that the width of the flocculation zone increases as
the polymer molecular weight increases. Neither appears to be the
case as evidenced in Figures 1 to 3.

    Since a high degree of polydispersity was suspected for those
polymer fractions produced by the discontinuous method, another
20.0 g sample of poly(DADMAC) was fractionated, this time by the
continuous method to reduce the polydispersity of the fractions.
The first three fractions of the continuous fractionation were
tested on clay suspensions of 0.22 μm, 0.70 μm and 9.5 μm. Figure
4 shows different flocculation performance on the 0.7 μm clay than
was observed for the discontinuous fractionations. The flocculation performance of Fraction 1 is as efficient as the unfractionated
sample, while Fraction 2 is less than Fraction 1 and Fraction 3 is
less than Fraction 2. The flocculation performance exhibited by
these fractions on the 9.5 μm clay suspensions differs only in the
amount of residual turbidity remaining in the solution and the
shapes of the curves, but the optimum polymer dosage is the same
for all fractions (25 μg/l) (see Figure 5). In the flocculation
of a 0.22 μm clay almost identical dosages (ca. 150 μg/l) were again
observed for the fractions obtained by the continuous fractionation
method. The breadth of the flocculation curves obtained makes graph-
ical representation difficult. Therefore, the data obtained for the
flocculation of the 0.22 μm clay are listed in Table III.

    The differences observed between the continuous and discon-
tinuous fractions are unexplainable at this time. Further information concerning polymer molecular weight and structural properties is necessary before any conclusions can be drawn.

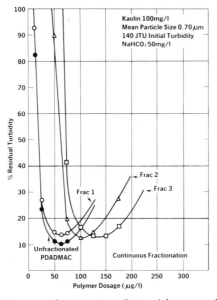

Fig. 4. Flocculation performance of continuous fractionation 1 in
         kaolin suspension of 0.7 μm mean particle diameter (liquids);
         ionic strength 6 x 10^{-4}; pH = 75; initial turbidity 140
         JTU (Jackson turbidity units).

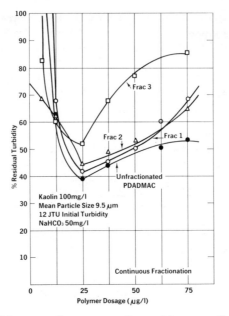

Fig. 5.   Flocculation performance of continuous fractionation in
          kaolin suspension of 9.5 µm mean particle diamer (liquids);
          ionic strength 6 x 10^{-4}; pH = 7.5; initial turbidity 12
          JTU (Jackson turbidity units).

TABLE III.   Flocculation Performance Data of Continuous Fractiona-
             tion (Liquids) in Kaolin Suspension[a]

Polymer Sample	Optimum Polymer Dosage (µg/l)	% Residual Turbidity[b]
Fraction 1	175	10
Fraction 2	150	9
Fraction 3	150	10
Unfractionated PDADMAC	200	9

- - - - - -

[a]kaolin 100 mg/l; mean particle diameter 0.22 µm; NaHCO$_3$ 50 mg/l

[b]initial turbidity: 150 Jackson turbidity units

A plot of the optimum dosage and molecular weight (number average) for the performance data obtained with the discontinuous fractionation series on the 0.09 μm clay was made to determine if any correlation might exist between those two parameters (see Figure 6). Although the $\bar{M}_n$ values were determined on precipitated samples and are, therefore, in error, the same trend would be indicated. However, since the optimum polymer dosage, as determined by the jar test, is not an exact value, an optimum dosage range was plotted, arbitrarily choosing a breadth of ± 5% of the initial turbidity of the clay suspension for this range. According to Healy and LaMer (7) an optimum molecular weight exists for a colloidal system of a given surface area. From Figure 6 this appears to be the case. A similar plot may be drawn for each of the other clays tested by this fractionation series (Figure 7 upper and lower diagrams).

Apparently, in Figure 7 the polymer sample employed for this investigation does not contain fractions of sufficiently high molecular weight to determine whether a minimum exists for the larger-sized clay suspensions. That there is a molecular weight below which flocculation efficiency decreases is clearly indicated.

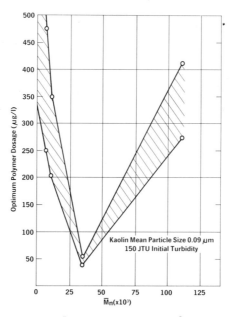

Fig. 6. Optimum polymer dosage versus number-average molecular weight for discontinuous fractionation 1 in kaolin suspension of 0.09 μm mean particle diameter.

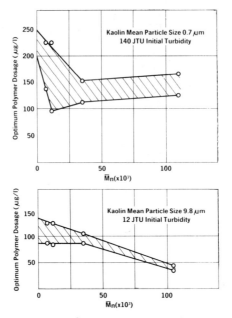

Fig. 7.   Optimum polymer dosage versus number-average molecular
weight for discontinuous fractional 1 in kaolin suspensions of 0.7 μm (upper diagram) and 9.5 μm (lower diagram)
mean particle diameter.

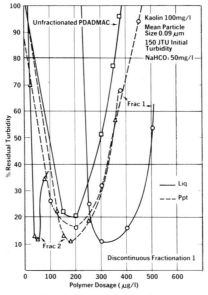

Fig. 8.   Comparison of flocculation performance of liquid and
precipitated samples of discontinuous fractional 1 in
kaolin suspension of 0.09 μm mean particle diameter;
ionic strength 6 x 10⁻⁴; pH = 7.5; initial turbidity
150 JTU (Jackson turbidity units).

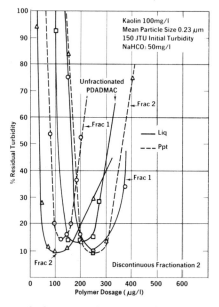

Fig. 9. Comparison of flocculation performance of liquid and
precipitated samples of discontinuous fractionation 2
in kaolin suspension of 0.23 μm mean particle diameter;
ionic strength 6 x 10⁻⁴; pH = 7.5; initial turbidity
150 JTU (Jackson turbidity units).

        Since poly(DADMAC) is generally employed in the liquid form
and since precipitation was desirable for the characterization work,
the flocculation performance of the precipitates and liquid forms
of all fractions was compared. Figures 8 and 9 show the erratic
behavior observed with the precipitated fractions. The relative
positions of the flocculation curves for the liquid samples remain
the same, but those of the precipitates do not. This was found to
occur for all fractions regardless of fractionation method. These
results are unexplainable at this time.

I wish to thank Francis Mangravite for his helpful advise in the
preparation of this paper. Appreciation is also acknowledged to
all of the technical personnel whose efforts helped make this
publication possible.

BIBLIOGRAPHY

1. A. P. Black, F. B. Birkner, and J. J. Morgan, J. Amer. Water
    Works Ass., 57, 1547 (1965).

2. A. P. Black, F. B. Birkner, and J. J. Morgan, J. Colloid
    Interface Sci., 21, 626 (1966).

3. A. Rembaum and D. Casson, Polymer Letters, 8, 733 (1970).

4. W. F. Langelier, H. F. Ludwig, A. M. Asch, and R. G. Ludwig,
    Amer. Soc. Civil Eng. Proc., 78, 147 (1952).

5. J. C. Kane, V. K. LaMer, and H. B. Linford, J. Phys. Chem.,
    68, 3539 (1964).

6. A. P. Black and M. R. Vilaret, J. Amer. Water Works Ass., 61,
    209 (1969).

7. T. W. Healy and V. K. LaMer, Rev. Pure Appl. Chem. 13, 112
    (1963).

8. A. S. Teot, Anal. N. Y. Acad. Sci., 155, 593 (1969).

9. J. Gregoy, Trans. Faraday Soc., 65, 2260 (1969).

10. D. R. Kasper, Ph.D. Thesis, California Institute of Technology,
    Pasadena, California (1971).

11. H. S. Posselt, A. H. Reidies, and W. J. Weber, Jr., J. Amer.
    Water Works Ass., 60, 48 (1968).

12. M. Pressman, J. Amer. Water Works Ass., 59, 169 (1967).

13. A. S. Teot and S. L. Daniels, Environ. Sci. Tech., 3, 825 (1969).

14. W. H. Coulter (to Coulter Electronics), U. S. Pat. 2,656,508
    (Oct. 20, 1953).

15. N. Thaulow and E. W. White, Powder Techn., 5, 377 (1971).

16. J. E. Boothe, H. G. Flock, and M. F. Hoover, J. Macromol Sci.
    Chem., A4, 1419 (1970).

17. G. B. Butler (to Peninsular ChemResearch), U. S. Pat. ,
    3,288,770 (Nov. 29, 1966).

18. W. F. Blatt, S. M. Robinson, and H. J. Bixler, Analyt. Biochem.,
    26, 151 (1968).

19. W. F. Blatt, Agri. Food Chem., 19, 589 (1971).

WATER-SOLUBLE POLYMERS IN PETROLEUM RECOVERY

D. C. MacWilliams, J. H. Rogers, and T. J. West

Dow Chemical U.S.A.

Walnut Creek, California

Water-soluble polymers have found a broad range of application in the production of petroleum. Table I shows the polymer types commonly employed in each application (1,2,3). The presence of an entry indicates that the polymer in question has received some commercially significant utilization in the indicated application. The lack of an entry does not imply that the polymer shows no activity in that application, but rather that to the authors' knowledge no significant commercialization has been realized. The polymers serve one or more of six functions:

    1. Water-loss control
    2. Viscosity control
    3. Flocculation
    4. Suspension (dispersion)
    5. Turbulent friction reduction
    6. Mobility control.

The polymers come from only four families: polyamines, vinyl polymers, modified celluloses, and naturally occurring polysaccharides. Notably absent are members of the condensation polymer family.

Drilling fluids ("Drilling" in Table I) range from clear water through conventional clay suspensions to barite-loaded heavy muds. The wide range of properties required for these fluids has led to a proliferation of additives. For example, the polyacrylates are favored for high-temperature operation but are not effective in the presence of calcium ion. Likewise hydroxyalkylcellulose derivatives are used to replace the familiar carboxymethylcellulose if calcium ion is present. Xanthan gum (used in this paper to describe the

TABLE I.   Utility of Polymers in Petroleum Production

Polymer	Drilling	Cementing	Acidizing	Fracturing	Recovery
Polyamines		1,4			
Polyacrylate salts	1,2,3				
Polyacrylamides	3		5	1,2,5	6
Poly(vinyl acetate-co-maleic anhydride)	2				
Carboxymethyl-cellulose	1,4				
Hydroxyalkyl-cellulose	1,4	1,4	1	2	
Guar gum	1,3,4		5	1,2,5	
Xanthan gum	2,4				6

extracellular heteropolysaccharide produced by the bacterium Xan-thomonus campestris) and guar gum are tolerant of salts but are readily degraded by biological attack, a problem not encountered with the other polymers.

Cementing of casing into oil wells requires good disperant ac-tion and control of the water loss from the slurry to the surround-ing formation. In the acidizing of wells, water-soluble polymers have been used to reduce pump horsepower requirements by decreasing line pressure losses by friction reduction. In some wells, water loss must also be controlled. Hydraulic fracturing requires that a large volume of fluid be pumped very rapidly into the well to se-parate the rock layers mechanically; hence, friction reduction by addition of water-soluble polymers is practiced routinely. The same polymers usually aid in suspending the "propping agent," if one is used. Secondary and tertiary recovery techniques which involve the injection of water or a brine into a series of wells to push the oil toward other production wells frequently require adjustment of the mobility of the water or brine to prevent bypassing of the oil by the injected fluids. See also the chapter by J. W. Hoyt and Robert H. Wade for a general discussion of friction reduction.

The Dow Chemical Company has carried out an extensive development program on the application of polymers, and in particular, polyacrylamides to secondary and tertiary oil recovery. The remainder of this discussion will be confined to this application since it represents a new and potentially large market for water-soluble polymers. Our understanding of the process is still incomplete, but it is possible to set some of the criteria for the polymers which are used as mobility control agents. A polymer to be useful in secondary recovery must possess a workable combination of the following properties. It must reduce the mobility of the injected water; propagate through the rock without plugging the rock and with a minimum of loss by adsorption on the rock; have adequate mechanical, chemical, thermal, and biological stability; and confer these properties at costs that are economically feasible.

SECONDARY OIL RECOVERY

An oil reservoir consists of a porous rock or sand matrix containing a mixture of oil, water (or brine), and usually some gas. Oil does not occur in pools independent of the rock matrix. Gas, natural water, or compression provide the energy to drive the oil to the well bore. If the pressure is great enough, the well will flow naturally but usually a pump is required to raise the oil to the surface. Once the natural drive dissipates, usually with only one-fifth to one-third of the oil-in-place recovered, the period of primary production is at an end.

At this point, and preferably earlier, some method of repressuring or pressure maintenance must be used if more oil is to be recovered. The most widely used methods are waterflooding and gas injection. Steam injection enjoys a limited use primarily in some California fields which have very viscous oils. In situ combustion processes wherein air or oxygen is pumped into the formation to burn a portion of the oil have been attempted in a number of fields. Waterflooding is the dominant method in the East, Plains states, and West Coast. Gas injection and waterflooding are practiced on the Gulf Coast and in Western Canada dependent upon the economics.

At the beginning of a waterflood some of the wells in the field may be converted or additional wells drilled for water injection. As water is forced into the formation through the injectors, the gas distributed throughout the oil zone and the gas cap, if any, are first compressed. Eventually sufficient pressure builds up and increased flow is observed into the producing wells. Water entering the rock follows the path of least resistance and it is not long into the life of a flood that "fingers" of water start to form and propagate more rapidly than the main body of water. Eventually, the injected water breaks through into the producers in large quantity and the waterflood becomes economically unattractive. The proportion of water produced relative to oil is called the water-oil ratio. Typically, if the water-oil ratio is greater than 20:1 the operation of the well becomes uneconomic. At this point, usually about one-fifth to one-third of the oil which remained after primary production has been produced. By increasing the resistance to flow of the water relative to that of the oil, fingering is reduced. In many cases, the amount of oil recovered is increased by 50 to 100% over that which would have been realized with water alone.

This effect is illustrated in Figure 1. The oil produced by straight waterflooding is compared to the oil which was recovered using a 30% pore volume slug of 0.05% solution of HPAM-500 (a partially hydrolyzed polyacrylamide of intermediate molecular weight) followed by water (4). The water in this case contained 1.1% total dissolved solids. A comparison of the oil recovered at 20:1 water-oil ratio by the two methods shows the polymer flood produced 80% more oil. See also the chapter by Robert H. Friedman.

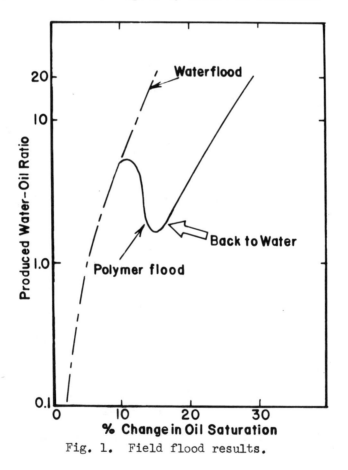

Fig. 1.   Field flood results.

Miscible flooding is an important modification of waterflooding. Such processes involve the injection of a small amount of material miscible with the oil followed by a less expensive displacing fluid. The efficiency of these processes is dependent upon effective mobility control to prevent the displacing fluid from fingering through the miscible fluid. Of particular current interest are the "microemulsion," surfactant, and surfactant-solvent systems. For these processes, control of the mobility of the following aqueous displacement fluid is essential.

MOBILITY CONTROL

The average velocity, $\bar{v}$, of a fluid flowing in a porous medium such as an oil-containing rock or sand is usually described in terms of the Darcy permeability, k, as

$$\bar{v} = \frac{1}{\epsilon} \frac{k}{\eta} \frac{\Delta p}{L} \tag{1}$$

where $\Delta p/L$ is the pressure gradient, ϵ is the fractional porosity of the medium, and η is the fluid viscosity. The magnitude of k depends on the size, shape, distribution, and tortuosity of the pores making up the flow paths. The value of k for a porous medium containing a single fluid is not a function of the fluid unless the fluid interacts with the prous medium. The commonly used unit of permeability is the darcy (9.87×10^{-9} cm^2). Oil-bearing rocks are typically in the 10^{-3} to 10 darcy range which is equivalent to average pore sizes ranging from 1μ to about 30μ.

When two or more fluid phases are present in the porous medium, the value of k for each fluid is different and varies with the amounts (saturations) of other fluids present. Hence, under any conditions, the velocity of any phase is defined by its mobility, k/η. The ratio of the mobility of the injected water to the mobility of the oil is referred to as the mobility ratio, M (Equation 2).

$$M = (k_w/\eta_w)/(k_o/\eta_o) \tag{2}$$

For our consideration here, the mobility ratio is not a point value but rather the ratio of water mobility at the irreducible oil saturation to the oil mobility at the irreducible water saturation.

The lower the mobility ratio the more nearly the velocity of the water approaches that of the oil and, consequently, the displacement is more efficient as has been shown in Figure 1.

The major benefits are derived when the mobility ratio can be reduced from an unfavorably high value (10-30) conducive to finger-

ing to a value of 1 or 2. Reduction of the mobility to values below unity is seldom required. The irreducible oil saturation is
determined by the wettability and interfacial tension forces and is
not affected by changes in the mobility ratio. In miscible flooding the miscible fluid does affect these forces and the oil saturation may be reduced to virtually zero.

Soluble polymers are added to injection water with the inten-
tion of reducing the mobility of the aqueous phase without significantly affecting the oil mobility. A convenient parameter (4) is
the resistance factor, R, the ratio of water mobility to polymer
solution mobility in the porous medium at the same oil saturation.

$$R = (k_w/\eta_w)/(k_p/\eta_p) \tag{3}$$

where the subscripts w and p refer to water and polymer solution,
respectively. While the individual mobilities of the aqueous fluids
are functions of the oil saturation, the ratio R is not greatly
different in the same porous medium whether it has an irreducible
oil saturation or is in an oil-free condition. This simplifies the
practical evaluation of the resistance factor so that the mobility
ratio for polymer solution - oil can then be estimated from the
known mobility ratio for water - oil. Polymers appropriate to this
application provide resistance factors of five or greater at concentrations of 200-500 ppm polymer in the injected fluid.

Mechanisms of Mobility Control

Decreased mobility ratio (R>1) is obtained with polymer solutions by virtue of increased viscosity or decreased permeability,
or both. The nature of the polymer and the electrolyte content of
the water dictate which effect predominates.

In cases where the polymer affects the permeability, a relatively persistent resistance to the subsequent flow of water but
not to oil is observed. A residual resistance factor (3) has been
defined as

$$R_r = \frac{(k_w/\eta_w) \text{ before polymer}}{(k_w/\eta_w) \text{ after polymer}} \tag{4}$$

This expression has the same form as Equation 3 but differs in that
it describes the resistance to the flow of water injected behind a
polymer solution.

Most moederately high molecular weight polymers do not exhibit
sufficient solution viscosity to warrant consideration on that basis
alone, for example, guar gum and modified cellulosics. Two very
high molecular weight polymers which are of interest are partially

hydrolyzed polyacrylamide (HPAM-700) and xanthan gum. The typical pseudoplastic viscosity curves exhibited by 500 ppm solutions of these polymers are shown in Figure 2.

In the shear rate range of greatest interest in waterflooding, 0.1-10 sec^{-1}, the xanthan gum has a high enough viscosity to be of interest, and it has been shown (3) by estimating a shear rate based on a simple model for the porous medium that this viscosity is essentially equal to the measured resistance factor. A relatively small salt effect causes a decrease in the viscosity in 3% brine. Generally, xanthan gums do not significantly reduce the permeability of the rock to water and little or no residual resistance factor is observed.

A large salt effect is seen with hydrolyzed polyacrylamide. A highly viscous, highly pseudoplastic solution in deionized water becomes a relatively low viscosity, almost Newtonian solution in brine. Despite this low viscosity, hydrolyzed polyacrylamide solutions in brine do show large resistance factors in a wide range of porous media.

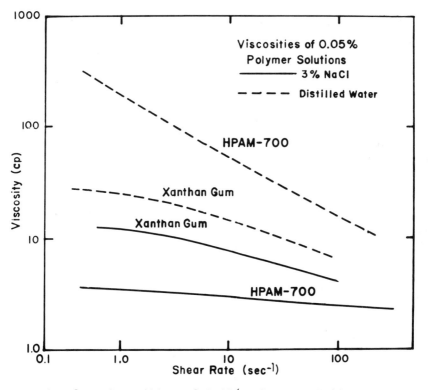

Fig. 2. Viscosities of 0.05% polymer solutions.

Typical resistance factor behavior over a range of flow rates for a brine solution of hydrolyzed polyacrylamide is shown in Figure 3. There are two distinctly different flow regions found for hydrolyzed polyacrylamide solutions in porous media in the permeability ranges normally found in oil reservoirs. Of the most practical significance is the low-flow region, i.e., a fluid advance rate under 10 ft/day. In this region hydrolyzed polyacrylamide interacts with the porous media to alter the permeability to water to such an extent that the resistance factor is significantly higher than the viscosity of the solution (3,4). This effect is further illustrated by the existence of a significant and relatively persistent residual resistance factor, shown by the curve "brine after HPAM." Fresh-water solutions of hydrolyzed polyacrylamide show a combination of permeability alteration and shear dependent viscosity effects which are difficult to separate (5). The permeability alteration effect is, like the solution viscosity, highly dependent upon the molecular weight of the polymer.

In the second flow region, generally occurring above about 10 ft/day (0.2 cm/min), the resistance factor becomes a strong function of flow rate. This type of flow, which is characteristic of certain viscoelastic fluids flowing in media with some degree of tortuosity of the flow paths, is discussed elsewhere (3,6). The phenomenon occurs outside of the normal range of flow rates encountered in

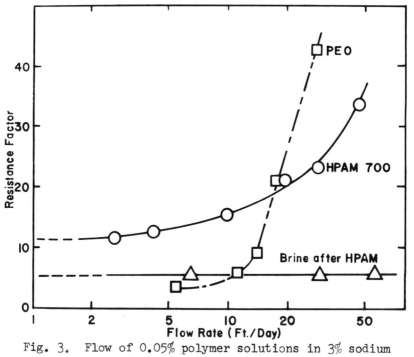

Fig. 3.   Flow of 0.05% polymer solutions in 3% sodium chloride in sandstone.

waterflooding and consequently is of little direct practical significance. For hydrolyzed polyacrylamide solutions, however, the magnitude of the viscoelastic flow resistance appears to be related to
macromolecular size in much the same way as is the low flow rate
resistance. This provides a basis for a correlation which will be
discussed later.

Viscoelasticity is not a sufficient criterion for mobility reduction by permeability alteration as shown by the behavior of a
poly(ethylene oxide) polymer (Figure 3) which gives little resistance to flow in a porous medium at low flow rates while showing a
rather large viscoelastic effect (7). No viscoelastic flow region
has been observed for xanthan gums.

There has not been complete agreement on the exact mechanism
of mobility reduction with hydrolyzed polyacrylamide solutions.
Increased apparent viscosities due to capillary constriction by
adsorbed polymer films have been claimed (8). However, these effects are generally smaller than those observed in porous media.
Moreover, no correlation has been found between the resistance factor and the amount of polymer adsorbed. In certain cases it has
been possible to eliminate more than 95% of the polymer adsorption
corresponding to less than 2 μg/g (the limit of the analytical
method) by pretreatment of the sand with other adsorbing materials.
In these cases there was no loss in resistance factor or oil-dis-
placing activity of the polymer (3). Thus the desired effects are
obtained by the retention of very small amounts of polymer even
though there may be a larger but unproductive retention of polymer
by adsorption on the surfaces of the rock. Since the retained polymer affects the permeability to water but does not usually affect
the permeability to oil, it may be presumed that the polymer is associated with the rock surface at critical points in the flow paths
for the water which are not necessarily the same as the oil flow
paths (9). It has been proposed (10) that the polymer may be lodged
in the apexes of what are generally triangular-shaped capillary
cross sections.

The resistance effect does not appear to be due to the blocking of a large amount of pore space (3) and is not analagous to pro-
gressive filtration-type plugging since constant flow resistances
are obtained. Polymer solutions which contain significant amounts
of gels or other particulate matter do produce filtration-type plugging characterized by steadily increasing pressures as the solution
is injected.

The major variables affecting mobility control by hydrolyzed
polyacrylamide are the molecular weight and composition of the polymer, the water composition and polymer concentration, and the permeability and porosity of the sand (average pore size and pore size

distribution). Generally, for a given sand, the higher the polymer molecular weight, the higher the resistance factor. The hydrolyzed polyacrylamide molecular weights are typically 1-5 million. The maximum dimensions of hydrolyzed polyacrylamide molecules in this molecular weight range have been estimated in 3% sodium chloride solution (11). The values range from 0.2 to 0.8 μ. Hydrolyzed poly-acrylamide polymers having 20-40% hydrolysis are usually adsorbed to a lower degree than low-hydrolysis polyacrylamide (Figure 4) (12) and do not precipitate in the presence of calcium ion as does poly(sodium acrylate).

Resistance factors decrease moderately with increasing salinity of the injection water. Resistance factors decrease slowly as the polymer concentration is reduced and then decline rapidly at concentrations of less than 100 ppm (Figure 5). The viscosities decrease in a much more uniform manner. For the same type of sand, the effectiveness of a polymer is found to decrease as the mean pore size increases (3). As a practical matter it has been found necessary to experimentally establish the resistance factor using actual core samples, synthetic water samples matching the composition of the water to be injected, and an oil from the field under study. Once the mobilities of the various fluids are established, an extensive computation is necessary to establish the economic feasibility of the project. The programs used for these computations are generally proprietary and there are many variants. The results of a large number of mobility control projects using hydrolyzed polyacrylamide have been reported (13).

THE SCREEN VISCOMETER

Laboratory measurements of resistance factors in rock or sand samples are a time-consuming operation. It was therefore desirable to have some method for evaluating the properties of hydrolyzed polyacrylamide solutions independent of measurements in rocks or sands. The screen viscometer was developed specifically to provide a rapid method of evaluation. The screen viscometer responds primarily to the viscoelastic behavior of the polymer. In this instrument, wire screens are used as the porous media and the time for a given volume of polymer solution to flow through the screens is measured.

Fluid flow through screens, grids, or fiber mats has been studied by a number of investigators (14). It has been found that when water or brine flows through a wire screen, the pressure drop is due to both viscous and inertial effects. Under turbulent flow conditions the total pressure drop, Δp, for a screen is given by

$$\Delta p = a\eta\bar{v} + b\rho\bar{v}^2 \qquad (5)$$

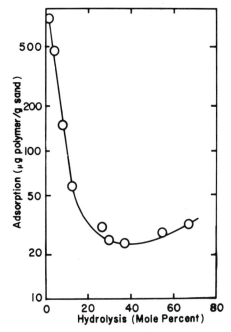

Fig. 4. Adsorption of hydrolyzed polyacrylamide in 2.2% sodium chloride solution on Miocene sand (12).

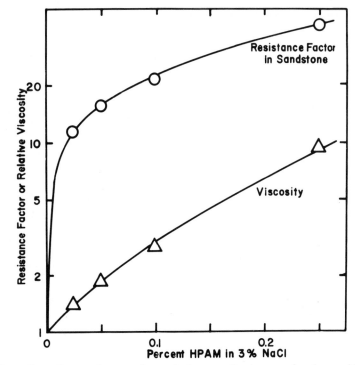

Fig. 5. Comparison of resistance factor and viscosity.

where η is viscosity, $\bar{v}$ is the flow velocity, ρ is density, and a and b are constants characteristic of the screen. We have found the same law applies to stacks of screens where a and b are evaluated for the stack.

The flow resistance, $\Delta p/\bar{v}$, is linear in $\bar{v}$, i.e.

$$\frac{\Delta p}{\bar{v}} = a\eta + b\rho\bar{v} \tag{6}$$

as shown in Figure 6 for the flow of brine in stacks of 5 screens. The viscous resistance given by the intercept a is of the same order of magnitude as the inertial resistance, $b\rho\bar{v}$. The viscous resistance increases with the decreasing size of the screen opening, as might be expected, while the inertial resistance shows little change with the screen opening.

The flow behavior of 500 ppm hydrolyzed polyacrylamide in 3% sodium chloride solution was quite different (see Figure 7). Two results for brine taken from Figure 6 are shown for comparison. The flow resistances of the polymer solution are an order of magnitude higher than those for brine. The flow resistance of the polymer solution varies with flow velocity in the same manner as the brine resistance but differs in that the polymer solution resistance is a strong function of the size of the screen openings. A simplified expression for the resistance to flow of a viscoelastic fluid, $\Delta p/\bar{v}$, through a series of constrictions has been given (3) as the sum of viscous, kinetic, and viscoelastic terms

$$\frac{\Delta p}{\bar{v}} = a\eta L + b\rho\bar{v} + c\eta\theta_f \frac{\bar{v}}{d^2} \tag{7}$$

where $\bar{v}$ is the average velocity in the orifice, d is the diameter, and θ is the relaxation time of the fluid. From this equation both simple and viscolastic fluids would show linear plots for $\Delta p/\bar{v}$ vs. $\bar{v}$ if the relaxation time of the fluid is constant with velocity.

It was found that the flow resistance, $\Delta p/\bar{v}$, for stacks of screens from 3 to 60 could be described by Equation 8,

$$\frac{\Delta p}{\bar{v}} = \frac{c'N\bar{v}}{A} \tag{8}$$

where A is the area of a screen opening, N is the number of screens in the stack, and c' is a constant. This term has the same form as the viscoelastic term of Equation 7. Thus, it appears that the flow behavior of a dilute polyacrylamide solution in a porous medium

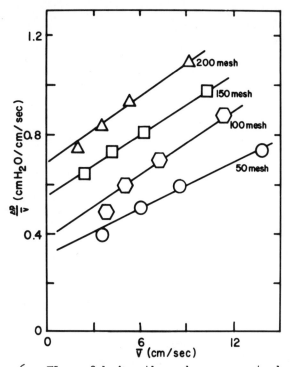

Fig. 6. Flow of brine through screen stacks.

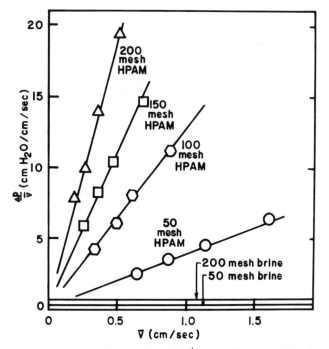

Fig. 7. Flow of brine and 0.05% solutions of hydro-
         lyzed polyacrylamide through screen stacks.

composed of screens is consistent with Equation 7, and a stack of screens can be used as a porous medium to qualitatively measure viscoelastic effects (3).

A simple instrument was designed to hold a given volume of fluid so that the flow time of a polymer solution through a five-screen stack could be determined, see Figure 8. This instrument, because of its similarity to a capillary viscometer, has been called a "screen viscometer." The screen viscometer is a pipette, with a volume of about 30 ml, with a specially prepared nylon coupling containing a pack of five 0.25-inch diameter, 100-mesh stainless-steel screens. Measurements are made by timing the fall of the fluid level between two reference marks. The fluids are first filtered through a 200-mesh screen to preclude plugging of the instrument. The results of the measurement are reported as the "screen factor" which is the ratio of the flow time of the polymer solution to that of the solvent.

The screen factor has been found useful in the control and evaluation of polyacrylamides for mobility control. An empirical correlation can be developed between screen factor and resistance factor measured in cores as is shown in Figure 9. A further indication of the utility of screen factor is shown in Figure 10 where the same resistance factors are plotted against the solution viscosities for two HPAM polymers of different molecular weight. It can be seen that the two polymers give a different correlation with viscosity for each rock permeability while for the correlation with screen factor both molecular weights give single curves varying only with the rock permeability. A screen factor-resistance factor correlation applies only to the polymer family and rock permeability for which it was obtained. Screen factors cannot be used to compare the resistance factor activity of chemically different polymers. For example, poly(ethylene oxide) solutions have large screen factors but have small resistance factors in medium permeability sandstone. The screen factor does correlate with resistance factor within a given polymer type, and it is sensitive to changes in molecular weight and structure, both of importance in mobility control.

The screen viscometer has proven to be a valuable tool in both the laboratory and the field. It is sensitive to changes in polymer activity and is useful in determining the effects of variables in the handling of polymer solutions. The screen factor is not a quantitative measure of resistance factor activity since resistance factor is in part an interaction of the polymer and the rock matrix, but it does correlate resistance factor activity when properly used.

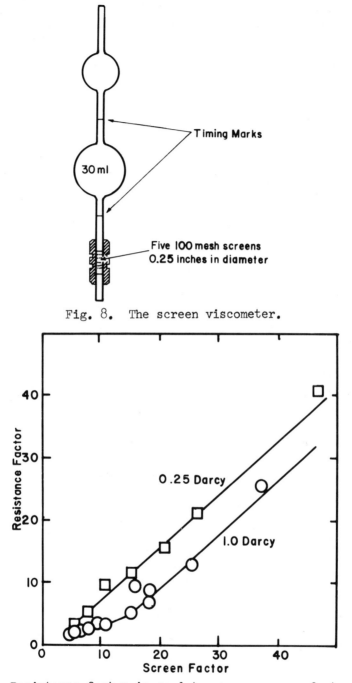

Fig. 8.   The screen viscometer.

Fig. 9.   Resistance factor in sandstones vs. screen factor.

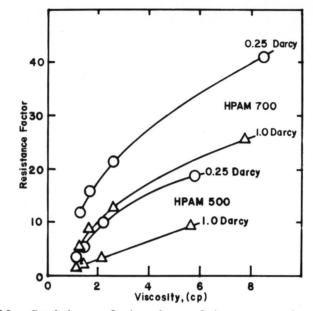

Fig. 10.   Resistance factor in sandstones vs. viscosity.

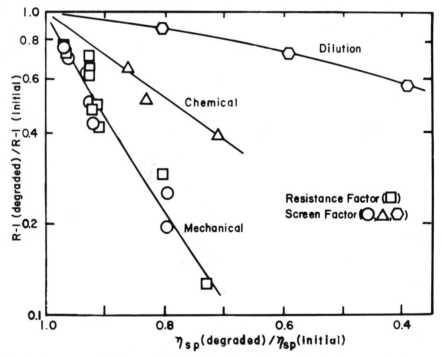

Fig. 11.   Degradation of hydrolyzed polyacrylamide: resistance
factor and screen factor vs. viscosity.

POLYMER STABILITY

The stability of dilute polymer solutions used in secondary oil recovery is of major importance since the polymers must function for long periods of time, frequently at elevated temperatures. Mechanical (shear), chemical, and thermal stability have been considered.

The difference between mechanical and chemical stability of HPAM-500 is shown in Figure 11. The rate of screen factor loss and viscosity loss are compared for simple dilution and for degradation produced by mechanical shear and chemical oxidation (sodium hypochlorite) methods. The mechanical shear produced experimentally by a Waring Blender is believed to be selective toward higher molecular weight molecules (15) and the screen factor falls much more rapidly than the viscosity. The chemical degradation is not molecular-size selective so the screen factor does not decline as rapidly relative to the viscosity.

The mechanical stability of a partially hydrolyzed polyacrylamide and poly(ethylene oxide) have been compared (16). The poly(ethylene oxide) was shown to be more shear sensitive then the hydrolyzed polyacrylamides in this study. Xanthan gums are claimed to be shear resistant (17). The shear problem can be minimized by proper engineering design to eliminate high pressure gradients.

The chemical and thermal stability of all water-soluble polymers is limited. In the introduction, the relatively high thermal stability of acrylic polymers was noted. This statement is true only so long as there are no sources of free radicals in the system. The primary source of such radicals is dissolved oxygen combined with a reducing agent, usually a trace of iron in a lower valence state. In addition, the presence of ferric iron must usually be avoided since this compound forms insoluble ferric hydroxide which will plug the well face. The solution to both of these problems is to add sodium hydrosulfite to the injection fluid prior to the polymer addition (18).

The beneficial effect of air (oxygen) removal is shown in Figure 12. The polymer is 500 ppm HPAM-700 in 3% sodium chloride. All experiments were run at $70^{\circ}C$ in flame-sealed glass ampoules with protection against the hazards of overpressure. Analyses were done at $25^{\circ}C$ under a nitrogen blanket. As the exclusion of oxygen is improved, the stability systematically improves as indicated by the screen factor data. Under optimum conditions 84% of the activity is retained. Sodium sulfite may be used to control oxygen but this reagent will not reduce any ferric hydroxide that may be present to the more soluble ferrous hydroxide. If hydrosulfite is added after the polymer, and oxygen is present, a redox couple is created and the polymer will be degraded. See also the chapter by N. M. Bikales.

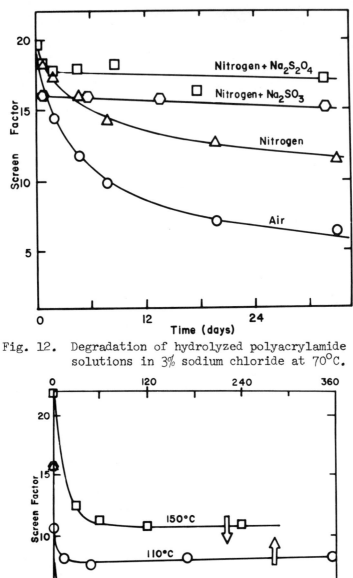

Fig. 12. Degradation of hydrolyzed polyacrylamide
         solutions in 3% sodium chloride at 70°C.

Fig. 13. Degradation of sodium hydrosulfite stabilized
         solutions of hydrolyzed polyacrylamide in 3%
         sodium chloride at high temperatures.

Long-term stability studies were run at a higher temperature, 110°C instead of 70°C, with the result shown in Figure 13. The polymer in these studies had a lower initial screen factor and the solution was treated with hydrosulfite. Eighty percent of the initial activity was retained. Studies were conducted with the higher molecular weight product at 150°C. The degradation was more severe and 50% of the activity was retained. At temperatures of 204°C and 253°C, 90% of the activity was lost within 24 hours. These data suggest a long-term exposure temperature ceiling of about 150°C (300°F) for the conditions employed. This result is consistent with that reported by Mungan (19). Under specific conditions a higher temperature ceiling may be observed but each example must be reviewed. The maximum temperature for stability for acrylic polymers used in drilling fluids where time requirements are not so severe has been given as 350°F (177°C) (1). Similar studies of partially hydrolyzed polyacrylamide in fresh water have been reported by Knight (20). Ninety-four percent of the screen factor was retained after 32 days at 50°C.

We have examined the thermal stability of a number of polymers of potential interest in secondary oil recovery. The results are summarized in Table II. For HPAM-700 the retention of screen factor at room temperature, 50°C, and 70°C, is compared with sodium hydrosulfite deoxygenation, deaeration with nitrogen by sparging, and air saturation. The excellent retention of screen factor, especially in distilled water and at lower temperatures in the presence of hydrosulfite, is shown by the results in the last column. Two other polyacrylamides examined at 110°C in sodium hydrosulfite medium show good stability.

The thermal stability of two xanthan gums was examined by viscosity rather than by screen factor. These polymers give poor stability performance at elevated temperatures in dilute brines and distilled water but for some reason which is not understood the performance in 3% salt solution is good. Xanthan gum B is very air sensitive compared to xanthan gum A. Jeanes and co-workers have reported that the stability of the xanthan gum produced by X. campestris NRRL B-1459 is much improved by the presence of potassium chloride at concentrations greater than 0.2% (21). The temperature ceiling for xanthan gums is obviously dependent on salt concentration so it is difficult to state a single value as can be done for partially hydrolyzed polyacrylamides. In drilling mud applications the maximum operating temperature has been stated to be 250°F (121°C) (1) as compared to 350°F for acrylics. See also the chapter by A. Jeanes.

Biological attack on polyacrylamide polymers is negligible. However, there may be biological activity in the solutions if nutrients are available. The xanthan gums are nutrients and the addition

TABLE II.	Degradation of 0.05% Solutions of Water-Soluble Polymers at Elevated Temperatures

Polymer	Medium	Temp, $^{\circ}C$	Time, Days	Screen Factor Init	Screen Factor Final	% Activity Retained
HPAM-700	Dist.H_2O[a] ($Na_2S_2O_4$)	50	32	24.0	22.5	94
	2% NaCl[a] ($Na_2S_2O_4$)	50	26	14.5	11.4	79
	3% NaCl ($Na_2S_2O_4$)	70	32	19.3	16.2	84
	3% NaCl (N_2)	70	32	19.3	11.5	60
	3% NcCl (air)	70	32	19.3	6.3	33
HPAM-500	3% NaCl ($Na_2S_2O_4$)	110	352	10.2	8.1	79
Low-Hydrolysis Polyacrylamide	3% NaCl ($Na_2S_2O_4$)	110	352	8.6	6.5	76

				Viscosity (cP at 6.6 sec^{-1}) Init	Final	
Xanthan Gum A	0.03% NaCl ($Na_2S_2O_4$)	65	20	15.0	6.25	40
	3% NaCl ($Na_2S_2O_4$)	65	20	15.3	14.0	92
	Dist. H_2O ($Na_2S_2O_4$)	85	7	16.0	2.5	16
	3% NaCl ($Na_2S_2O_4$)	85	7	14.8	14.3	97
	Dist. H_2O (air)	85	7	40.0	4.0	10
	3% NaCl (air)	85	7	15.0	12.8	85
Xanthan Gum B	3% NaCl (air)	23	20	13.3	12.2	92
	0.3% NaCl (air)	23	20	13.7	12.0	88
	3% NaCl ($Na_2S_2O_4$)	85	20	14.5	12.3	85
	0.3% NaCl (N_2)	85	20	13.3	5.6	42
	3% NaCl (N_2)	85	20	13.7	11.6	85
	0.3% NaCl (air)	85	20	13.3	1.0	8
	3% NaCl (air)	85	20	13.7	1.2	9

[a]Reference 20.

of a biocide is essential unless the solutions are sterile. Sodium trichlorophenate (365 ppm) has been recommended (17).

CONCLUSIONS

The use of water-soluble polymers of the polyacrylamide and xanthan gum families as mobility control agents is a significant area of application for these materials. These are the only two families of polymers to date that have survived the rigorous screening as mobility control agents. The two polymers operate by different mechanisms. In fresh water the polyacrylamides produce both viscosity improvement and permeability alteration effects. In brine the polyacrylamides operate primarily by permeability alteration and special methods must be used for evaluation. The xanthan gums effect mobility control in fresh or salt water by viscosity alone. The elimination of particulate insolubles in solutions of both types of polymer is essential to prevent filtration-type plugging. This quality requirement presents a major manufacturing problem. Having obtained a product of adequate quality, much effort must then be expended to evaluate the economics of each oil-recovery prospect and finally to engineer the flooding operation to get maximum mobility control activity from the polymer. The production and application of water-soluble polymers to secondary recovery is, clearly, intensely technological.

The use of polymers in waterflooding is expected to increase in the coming years since the major source of oil remaining in the Continental United States is in fields that have already reached or will soon reach the economic limit on primary production.

BIBLIOGRAPHY

1. G. R. Gray, Symposium "Oil Field Chemicals Natural Gas Liquids in the Chemical Industry," Div. of Marketing Economics and Petroleum Chemistry, 159th Meeting of the American Chemical Society, 1970, p. 1.

2. R. E. Hurst, ibid, p. 14.

3. R. R. Jennings, J. H. Rogers, and T. J. West, J. Petrol. Technol., 23, 391 (1971).

4. D. J. Pye, J. Petrol. Technol., 16, 911 (1964).

5. W. B. Gogarty, J. Petrol. Technol., 19, 161 (1967).

6. J. G. Savins, Ind. Eng. Chem., 61, 18 (1969).

7.   D. L. Dauben and D. E. Menzie, J. Petrol. Technol., 19, 1065 (1967).

8.   F. Rowland, R. Bulas, E. Rothstein, and F. R. Eirich, Ind. Eng. Chem., 57, 46 (1965).

9.   M. R. J. Wyllie in T. C. Frick and R. W. Taylor, Eds., Petroleum Production Handbook, McGraw-Hill Book Co., New York, Vol. 2, 1962, p. 25-2.

10. R. E. Harrington and B. H. Zimm, J. Polym. Sci., Part A-2, 6, 294 (1968).

11. E. J. Lynch and D. C. MacWilliams, J. Petrol. Technol., 21, 1247 (1969).

12. K. R. McKennon (to Dow Chemical Company), U. S. Patent 3,039,529 (1962).

13. R. L. Jewett and G. F. Schurz, J. Petrol. Technol., 22, 675 (1970).

14. W. L. Ingmanson, S. T. Han, H. D. Wilder, and W. T. Myers, Jr., Tappi, 44, 47 (Jan. 1961).

15. H. Fujiwara and K. Goto, Kogyo Kagaku Zasshi, 71, 1430 (1968); Chem. Abstr., 70, 38340 (1969).

16. J. H. Elliott and F. S. Stow, Jr., J. Polym. Sci., 15, 2743 (1971).

17. F. H. Deily, G. P. Lindblom, J. T. Patton, and W. E. Holman, Oil and Gas J., 62 (June 26, 1967).

18. D. J. Pye (to Dow Chemical Company), U. S. Patent 3,343,601 (1967).

19. N. Mungan, J. Can. Petroleun Technol., 8, 45 (1969).

20. B. L. Knight, Symposium "Polymers in Oil Recovery," AIChE, Dallas, Texas, Feb. 1972, Preprint 45d.

21. A. Jeanes, J. E. Pittsley, and F. R. Senti, J. Appl. Poly. Sci., 5, 519 (1961).

REVERSIBLE CROSSLINKING OF POLYMERS IN OIL RECOVERY

Robert H. Friedman

Getty Oil Company

Houston, Texas

This chapter describes a technique for producing an inexpensive fluid that will exhibit different viscosities in the varying environments in a petroleum reservoir and will thus assist in increasing the recovery.

PETROLEUM RECOVERY

During the early exploitation of a petroleum reservoir, oil is produced using the natural energy present in the reservoir. When this energy is exhausted, some fluid, usually water, can be injected into some of the wells in order that the remaining producing wells can continue to be profitably operated. At the economic limit, when the producing wells produce so little oil and so much water that operations cannot profitably continue, 20-80% of the original oil may remain in the reservoir.

Much research in recent years has been done on the development of injection fluids which will act to recover more oil. So far, none of these more sophisticated fluids have come into common use. There are two principal reasons. The first of these is economic: the selling price of crude oil has been low because of a plentiful supply produced by cheap, "primary," methods. This low price has put a severe constraint on the cost of any method used to recover oil by postprimary means. The second reason, though obviously not of lesser importance, is the fact that no fluid yet devised has been able to perform all the complex tasks involved in causing the oil to move through the formation to the producing wells. Ideally, the injected fluid should traverse all of the oil-bearing strata

and displace all of the contained oil. Several methods efficiently
displace oil from the portions of the reservoir which they enter,
but tend to go into regions which have already been reduced to low
oil saturation. These methods depend upon the fact that a hydrocarbon gas liquified by pressure (1,2), an oil-in-water emulsion
(3) or water containing a surfactant (4) are sufficiently miscible
with the oil so that the latter is completely displaced from the
rock matrix. They are thus more efficient than water, which usually
leaves 20-25% of the oil in place in the volume traversed.

At the economic limit of a water flood, more than 50% of the
original oil usually remains in the reservoir. Two-thirds or more
of the unrecovered oil is in portions of the reservoir untouched by
the water flood. The more complex fluids eventually develop, like
water itself, a preferred path from injection well to producing well
and bypass much of the remaining oil (5). In addition, these fluids
are quite expensive while oil is a remarkably cheap product, selling for about one cent per pound.

It is easy to see that if water could be made to permeate the
entire reservoir, it would recover more oil than a fluid which per-
meated only a minor fraction of the volume. In recent years, work
has been done with high-molecular-weight polymers in water, which
have been shown to aid oil recovery by causing some of the injected
water to enter previously untouched regions (6,7,8). While the
method is being tested extensively, it has not been generally accepted. It should be noted that the diversion of the water direction caused by the polymer is to a large degree nonselective.

The solution to be discussed here is the creation of an inexpensive fluid which will self-conform to the reservoir, i.e., go
into the unswept portions of the reservoir and assist in the recovery of oil. It is selective in the sense that it exhibits oleo-
tropism, or a turning toward the oil.

Four major factors influence the way water traverses a porous
medium. They are the pressure gradient; the permeability, or the
ability of a porous medium to conduct fluid; the viscosity of both
the displacing fluid and the resident oil; and the area open to
flow. Of these, only the viscosity of the displacing fluid and the
injection pressure are under the control of the operator once the
oil field is producing. Pressure can only be controlled at the surface and cannot be varied to compensate for variable permeability
in the reservoir. A high-viscosity fluid will tend to lessen "fingering" due to viscosity differences between the driving fluid and
the oil, but will not of itself compensate for permeability variation.

A fluid that is highly viscous in water zones and markedly
less viscous in oil zones should help to compensate for permeability

variation. The components of a fluid to adjust viscosity for varying oil saturations caused by variable permeability can be:

1. an agent to increase the viscosity of water, such as a hydrating
    polymer;
2. a crosslinking agent for the polymer to increase the viscosity
    still further;
3. an agent to control the crosslinking and the viscosity, to be
    acted upon by something in the geological formation which
    effects the change in the desired direction. The presence of
    oil is a logical choice to control the agent.

USE OF CROSSLINKED POLYMERS

Our first attempt to make a self-conforming fluid utilized
guar gum. Guar is inexpensive and can be crosslinked with borate
ion. The desired viscosity change was obtained by varying the pH
of the fluid. The change from $BO_3^=$ to $HBO_3^=$ occurs over about
1.5 pH units. It was not possible to produce a pH change over
about 0.8 unit by extraction of an acid or base from water with oil.
Guar also has certain inherent deficiencies; the quality of the product varies markedly and it is extremely susceptible both to bacterial degradation and to interference by trace ions. This makes guar
unsatisfactory as the polymer for the proposed applications.

Carboxymethylcellulose

The polymer which proves to be the most satisfactory is sodium
carboxymethylcellulose (CMC). Moderate (3-4 cP) viscosities can be
developed with very low concentrations (0.06-0.08%). The material
is relatively inexpensive and has good quality control, and large
supplies are available. CMC can be crosslinked with several divalent and trivalent ions to give a solution with a viscosity many
times that of the unlinked polymer.

The rate and degree of crosslinking of CMC are greater at high
than at low pH and when a trivalent rather than a divalent ion is
used for crosslinking. At a sufficiently low pH, about 6.7, crosslinking does not occur or is destroyed if it has previously occurred.
The viscosity buildup can be made to take place slowly by using a
complex ion for the crosslinking. The ability to control viscosity
buildup gives greater versatility in that the low-viscosity fluid
can be pumped at lower pressures and also provides a method to use
extremely-high-viscosity fluids for plugging. Chromium ion, because of its tendency to form complex ions, gives the best results
as the crosslinking agent.

Variation in viscosity can be obtained by reversing the crosslinking. The rate of crosslinking is base dependent. In a moderately basic environment, crosslinking and the concomitant viscosity buildup will occur relatively slowly. Acidifying the viscous solution will cause the crosslinking to reverse and the viscosity to diminish. A suitable triggering agent thus appeared to be a weak base with a high oil-water partition. The latter feature was desirable so that oil would extract the base, thus effecting the viscosity reduction.

Our first work was with aqueous fluids having high concentrations of CMC. For such fluids, all bases tested proved to react too rapidly at any concentration that would effect crosslinking. However, a slow viscosity buildup is necessary because rapid crosslinking creates very large particles which surface plug the well bore. Also, the resulting high viscosity in the high-velocity region near the well bore requires high injection pressures. Several weak acids were also tested because, in the highly acidic environment created by chromium, a weak acid could serve as the "base." Ortho-cresol proved to be very satisfactory; it can serve the additional function of bactericide for CMC. In dilute systems, some amines (tert-butylamine and dicyclohexylamine, for example) and even aldehydes perform the trigger function.

The cost of the fluid depends upon the viscosity level. As Figure 1 shows, considerable economic advantage is obtained by using crosslinking to obtain the desired high viscosity. Calculations made at the beginning of the project resulted in setting a target of 25¢ per 42-gallon barrel for the chemical cost for the fluid. The actual cost in field tests is less than the target by a small margin.

Crosslinking reversal was tested by measuring the viscosities of fluid samples that had been in contact with oil for specified times. Tests were conducted by placing a small quantity of oil on top of a sample of fluid in a beaker and withdrawing the fluid for viscosity measurement at the end of the time interval. The small surface area available in a beaker makes this type of test quite severe. Results of one such set of tests are shown in Figure 2. These particular tests were with a fluid with undeveloped viscosity. Though the viscosity of the oil-contacted material increased temporarily, the viscosity buildup was slower and the maximum achieved was much lower. Similar tests were conducted with fluids that had already developed viscosity. Those in contact with oil lost viscosity after a few days.

Fluids which are to be injected into oil reservoirs should be stable for extended periods. Figure 3 shows a viscosity stability curve for one of the earlier fluids, indicating viscosity integrity for over sixteen months.

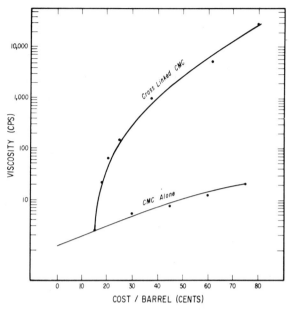

Fig. 1. Cost of thickened solutions as a function of viscosity.

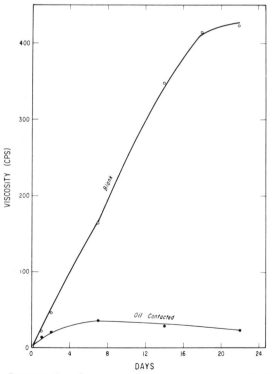

Fig. 2. Reversal of crosslinking as measured by viscosity of fluid samples.

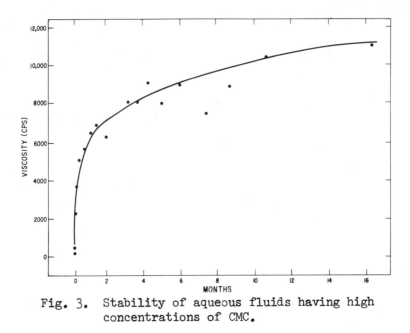

Fig. 3. Stability of aqueous fluids having high
concentrations of CMC.

        Development of a fluid having optimum characteristics was con-
ducted principally using many small bottle tests. To simulate
large-scale mixing, 500-gallon batches were prepared and tested.
Behavior of the fluid in an oil reservoir was simulated by injection
of the fluid into sand-packed tubes. The experimental arrangement
simulated two noncommunicating zones in a formation penetrated by
an injection well. One zone was at residual oil saturation (i.e.,
flooded to the point where no more oil could be produced) while the
other was at irreducible water saturation (i.e., charged to the
maximum oil saturation but still retaining residual water). The experiment, described in detail below, showed that use of the fluid
resulted in greater oil recovery. This occurred because of the increased tendency of the fluid to enter regions of high oil saturation.

        Testing. Displacement tests were made in sand-packed poly-
(vinyl chloride) tubes which were about 50 in. long and 3-1/4 in.
in diameter. Two tubes were packed with 80-100 mesh sand and two
others with 100-150 mesh sand, to obtain a permeability contrast.
The tubes with 80-100 mesh sand had a water permeability of about
20 darcies, the others of about 7 darcies. The low-permeability
tubes were water saturated and then oil flooded to irreducible
water saturation (~5-6%) of pore volume). The two high permeability tubes were then water flooded to residual oil saturation
(~23% of pore volume). The oil and water charging were accomplished with the tubes in a vertical position using 60 psig pressure because this procedure gave more efficient displacement of
the light crude oil and tap water used. The flow tubes were then

placed in a horizontal position and paired - one high permeability tube at residual oil saturation (referred to as a water tube) and one low permeability tube at irreducible water saturation (referred to as an oil tube). The inlets of each pair were connected for common injection.

To provide a base for comparison, water was initially injected into one pair of tubes and the CMC-based recovery fluid was injected into the other pair. Using two small pumps for injection, continuous injection provided sufficient residence time in the sand pack for viscosity buildup to occur. The results of injection into each pair of tubes are discussed below. Tap water was injected into one pair of tubes at a rate of about 0.04 pore volume per 24 hours, with a total of 0.6 pore volume injected. This was the same rate used for the CMC-based fluid. The tubes were rotated 180° every 24 hours to alleviate gravity effects. Little oil was recovered from this pair of tubes, as all the injected water flowed in the water tube. Of the original oil in place, 2.5% was produced from the oil tube, because there was a hold-up volume at the inlet of the tube. No oil was recovered from the water tube. To determine whether viscosity alone would be responsible for increasing recovery, a thickened water (34% glycerin) with a viscosity of 2.8 cP was then injected into the tubes. A total of 0.57 pore volume of thickened water was injected and very little additional oil was produced. The total amount of oil recovered by water and thickened water was 3.1% of the original oil in place.

The CMC-based fluid was injected into the other pair of tubes at a rate of about 0.04 pore volume every 24 hours. The injection was slow, to provide residence time for viscosity buildup to occur. As before, the tubes were rotated 180° every 24 hours to minimize the gravity effects. The fluid was taken from the 500-gallon tank in the pilot mixing facility. Every 24 hours, the pH of the mixed fluid was adjusted to 7.25 and the fluid pumped through a positive displacement pump prior to injection into the tubes. Bottle tests indicated the viscosity of this fluid increased from 3 cP to about 15 cP in 10 days. Initially, the fluid flowed into the water tube, i.e., where the resistance to flow is the lowest. However, after injection of about 0.17 pore volume (0.28 pore volume total injection) and time for viscosity buildup to occur, the resistance to flow in the water tube had increased to the point where some of the fluid entered the oil tube. As the fluid contacted the oil, the viscosity reduced as one of the ingredients necessary for viscosity buildup was removed. Thus, more fluid tended to flow into the oil tube (where the resistance to flow was lowest). The oil production showed a drastic increase at this time, shown by the current water-oil ratio (Figure 4). The data from the pair of tubes into which water was injected are shown for comparison. The oil production continued at a relatively low water-oil ratio until injected fluid

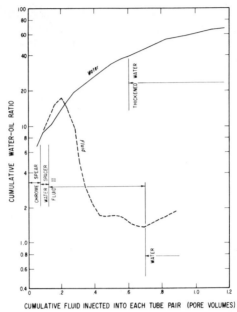

Fig. 4.   Cumulative water-oil ratios obtained from
          experiments with pairs of tubes.

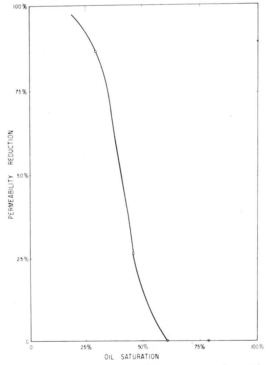

Fig. 5.   Permeability reduction as a function of oil
          saturation.

breakthrough occurred in the oil tube. At this point, injection of the fluid was ceased and water injected into the pair of tubes. Modest water production from the water tube continued throughout the entire experiment, i.e., the water tube did not plug. The viscosity buildup in the water tube only slowed water production and caused the oil to be produced at a very efficient water-oil ratio. All the oil production was from the oil tube; no oil was produced from the water tube. Fifty-three percent of the original oil in place was recovered by the injection of 0.58 pore volume of the fluid. The selective character of the fluid is demonstrated in Figure 5. The fluid used to obtain these data was more concentrated and designed for selective plugging discussed in the next paragraph. The same phenomenon was operative in the experiment just described. When the fluid was injected into a sand containing both oil and water, permeability reduction resulted, depending upon the water saturation. Reduction was zero if oil saturation was greater than about 65% and was greater as oil saturation diminished.

Extreme Permeability Variations. The previous discussion considered fluids to be used in the recovery of oil to compensate for moderate permeability variations. Another application would be to correct for extreme permeability variations. When a very high permeability layer (a "thief zone") extends from an injection well, it is desirable that this zone be plugged off because such a stratum results in much of the oil being bypassed and in extremely high water-oil ratios at the producing wells. Using the same chemicals, but in greater concentrations, a fluid was devised that could be pumped into the well at relatively low pressures and would selectively build up only in the water zones but not in the oil zones. Final viscosities in the range of 20,000-60,000 centipoises have been obtained at a cost of about $1.00 per barrel of fluid. This fluid has been tested in four water-flood injection wells in three widely separated locations. The treatment successfully altered the flow of water within the formation, as evidenced by changes in water production at the nearby producing wells.

Limitations to the present processes for both oil recovery and selective plugging include an upper temperature limitation of about 120°F and a requirement for fresh, soft water. The temperature ceiling limits the process to wells shallower than about 5000 feet.

The general method chosen, i.e., reversible crosslinking, appears to be the right course, though much work needs to be done to assure practical operation of the designed systems. What has been accomplished so far is the creation of a fluid system that changes viscosity in a desirable fashion to control fluid flow in varying permeability. It utilizes a chemical cybernetic system involving viscosity control by the reversible crosslinking of carboxymethylcellulose.

BIBLIOGRAPHY

1. H. L. Stone and H. L. Crump, AIME Trans., <u>207</u>, 105 (1956).

2. J. L. Mahaffey, W. M. Rutherford, and C. S. Matthews, Soc.
    Pet. Eng. Trans., <u>237</u>, 73 (1966).

3. W. B. Gogarty and W. C. Tosch, Soc. Pet. Eng. Trans., <u>243</u>,
    1407 (1968).

4. C. G. Inks and R. I. Lahring, Soc. Pet. Eng. Trans., <u>243</u>,
    1320 (1968).

5. F. F. Craig, Jr., <u>The Reservoir Engineering Aspects of
    Waterflooding</u>, Monograph No. 3, Henry F. Doherty Series,
    Society of Petroleum Engineers, New York, 1971.

6. D. J. Pye, Soc. Pet. Eng. Trans., <u>231</u>, 911 (1964).

7. N. Mungan and F. W. Smith, Soc. Pet. Eng. Trans., <u>237</u>,
    1143 (1966).

8. D. L. Dauben and D. E. Menzie, Soc. Pet. Eng. Trans., <u>240</u>,
    1065 (1967).

9. R. L. Whistler and J. N. BeMiller, <u>Industrial Gums</u>, Academic
    Press, New York, 1959.

10. Hercules Incorporated, <u>Cellulose Gum</u>, Wilmington, Del., 1971.

TURBULENT FRICTION REDUCTION BY POLYMER SOLUTIONS

J. W. Hoyt and Robert H. Wade

Naval Undersea Research and Development Center

Pasadena, California

It has been found experimentally that very small concentrations of dissolved high-polymeric substances can reduce the frictional resistance in turbulent flow to as low as one-fourth that of the pure solvent. The viscosities of these solutions are always somewhat higher than the pure solvent and so the fact that the turbulent friction is reduced has been a surprising technological development. Important summaries of the status of the drag-reduction effect have been given by Lumley (1), Patterson, Zakin, and Rodriguez (2), Hoyt (3), and Gadd (4). The emphasis in this paper will be on chemical engineering and applications aspects of the phenomenon. See also the chapters by D. C. MacWilliams, J. H. Rogers, and T. J. West; R. H. Friedman; and O. K. Kim and R. Y. Ting.

HISTORY

The earliest published data on the friction-reducing effect is found in the paper by B. A. Toms (5). Toms obtained friction reductions of up to 50% compared with the pure solvent with a 0.25% solution of poly(methyl methacrylate) in chlorobenzene.

During the Second World War, K. J. Mysels of Edgewood Arsenal measured the pressure drop in small pipelines containing either pure gasoline or gasoline thickened to a jelly-like substance with aluminum soaps. In turbulent flow, the pressure loss per unit length of pipe when the thickened gasoline was used was much lower than the pure solvent (6). The phenomena of a jelly having a lower frictional resistance in turbulent pipe flow than the low viscosity liquid from which it was made illustrates the interesting fluid mechanics involved with the flow of polymer solutions.

At about the same time, workers in the oil industry noticed
that when the plant derivative, guar gum, was used to suspend sand
in the high pressure sand-water mixtures employed in oil-well techniques, the friction was greatly decreased. Solutions of this common industrial product are extremely stable and resistant to breakdown, and are among the best friction-reducing natural products known
today. The experience of the oil-well industry with guar gum (7),
and in particular the contributions of H. R. Crawford of the Westco
Research Company, led to exploration of the friction-reducing effect
for other applications.

In 1962, A. G. Fabula identified the spectacular friction-reducing ability of poly(ethylene oxide), still the most effective
friction-reducing material known. Only a few parts per million of
this chemical are required to lower friction coefficients apprecia-
bly. Poly(ethylene oxide) has become a standard material for studies
of reduced turbulence.

Maximum Drag Reduction

Experimentally it has been found that the drag reduction for
any given pipe-flow Reynolds number reaches a maximum for some conditions of polymer concentration or pipe size, for each polymer
type. Expressed another way, any given drag-reducing polymer can
produce a maximum value of friction reduction for a given pipe size
and velocity, if the concentration is adjusted. This maximum was
first shown by Hoyt and Fabula (8) in a plot of friction reduction
as a function of solvent-based Reynolds number (Figure 1).

The maximum friction reduction achievable is about 80% of the
friction reduction that would be attained if the flow were made
completely laminar. This remarkable result seems to be independent
of polymer, pipe size, or Reynolds number, and may be a valuable clue
in providing an understanding of the mechanism of polymer friction
reduction.

The "Onset" of Drag Reduction

As sketched in Figure 2 there appears to be a "threshold"
shear stress which must be exceeded in order for the friction-reduction effect to manifest itself. In flows of the same fluid
through pipes of increasing diameter, a larger Reynolds number is
required to achieve the same shear stress. Hence the effect shown
in Figure 2 implies that the onset of friction reduction corresponds to some threshold shear stress.

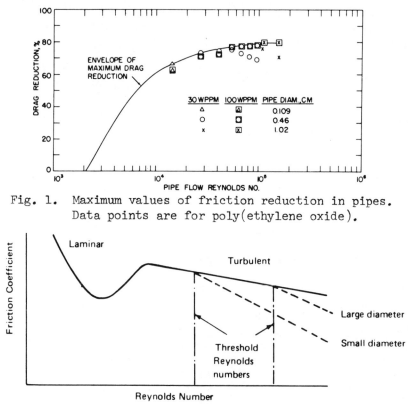

Fig. 1.   Maximum values of friction reduction in pipes.
Data points are for poly(ethylene oxide).

Fig. 2.   Sketch of the threshold effect of drag reduction
in pipes of different diameters.

Many polymer-solvent pairs can produce pipe-flow friction coefficients markedly less than those of the pure solvents alone.
The study of the friction-changing ability of these polymers has
lead to the conclusion that any macromolecular substance of sufficiently high molecular weight (50,000 or preferably more) with a
generally linear structure will lower the friction characteristics
of any fluid which is a solvent.

THEORIES

One goal of the study of friction-reducing polymer solutions
is the understanding of the interaction of macromolecules on the
flow process. Since many aspects of the turbulent flow of Newtonian
fluids are still poorly understood, the insights gained through
study of a flow that has been profoundly altered by polymer molecules
may eventually aid in the grasp of the entire realm of flow funda-

mentals. From the viewpoint of the applied fluid dynamicist, understanding the fundamental principles underlying friction reduction
may aid in more effective utilization of the principle in industry.

A number of theoretical explanations have been offered for the
friction-reducing effect. Gadd (4) was among the first to hypothe-
size that the mechanism involved in drag reduction was not turbulence dissipation, but rather a reduced production of turbulence.
This concept was a major step in the progress toward understanding
friction reduction. Johnson and Barchi were the first to show experimentally the decreased production of small eddies in a developing boundary layer containing polymer.

Walsh (10) has provided the most comprehensive theory of friction reduction to date. Walsh considers the small disturbances in
the viscous layer which grow by extracting energy from the local
velocity profile through the action of the Reynolds stresses. The
disturbances tend to lose energy due to viscous dissipation. In
polymer solutions, Walsh believes these small disturbances will tend
to store energy in the polymer molecules.

By decreasing the number of disturbances which grow per unit
area and time and move out from the edge of the viscous sublayer,
the addition of the polymer molecules ultimately changes the structure of the turbulence in the outer part of the boundary layer.
This change results in lower Reynolds stresses and hence there is a
reduction of friction.

Out of all work to date, one major theme seems to emerge: the
polymer molecules must interfere with the primary generation of turbulent eddies in a way that is substantially less dissipative than
the eddy production itself. Rather than absorbing energy, or storing
it, or radiating it away from the boundary layer, there remains the
possibility that the polymer molecules are purely passive, remaining
in the flow while mechanically interfering with disturbances which
would create turbulent eddies.

USEFUL POLYMERS

Guar. Guar is a complex plant polysaccharide used commercially
as a food additive and thickener. In petroleum technology, the use
of this natural gum as a suspending agent for sand in drilling muds
and fracturing fluids led to the discovery that guar solutions had
a lower friction than water. Guar has become a widely used material
for both industrial and experimental friction reduction because of
the robust characteristics of its molecule which tend to prevent
degradation due to high shear froces.

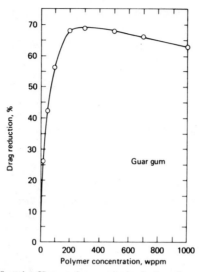

Fig. 3. Turbulent-flow rheometer data for guar; Reynolds number, 14,000.

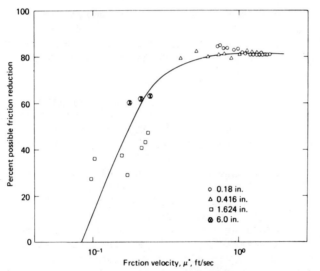

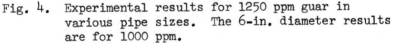

Fig. 4. Experimental results for 1250 ppm guar in various pipe sizes. The 6-in. diameter results are for 1000 ppm.

Figure 3 is a drag-reduction chart for guar taken in a turbulent-flow rheometer (essentially a miniature pipe-flow apparatus (11)). Drag reduction in pipe flow is defined as follows:

$$\% \text{ Drag Reduction} = \frac{\Delta Ps - \Delta Pp}{\Delta Ps} \times 100$$

where ΔPs = pressure loss due to friction in unit length of pipe,
           solvent only

  ΔPp = pressure loss due to friction in unit length of pipe,
           polymer solution.

Tests of guar solutions in pipe diameters up to 6 inches in diameter are reported by Whitsitt, Harrington, and Crawford (12). Their results with friction reduction shown as a function of friction velocity ($\mu*$), are given in Figure 4.

Poly(ethylene oxide). Fabula (13) was the first to observe that high-molecular-weight poly(ethylene oxide) is an extremely effective water-soluble friction-reducing polymer. A friction reduction of 20% was noted in pipe flow with a polymer concentration of 1 part per million (ppm). The viscosity and density properties of such a solution are indistinguishable from water. At higher concentrations, 25 ppm for example, Fabula found a 75% friction reduction in a 1.02-cm inside diameter pipe at a Reynolds number of 10^5.

The effect of poly(ethylene oxide) on drag reduction in the turbulent-flow rheometer is given in Figure 5. Solutions of this substance (Polyox 301) give remarkable friction reduction when used in the ppm ranges. Even solutions as dilute as 0.5 ppm give a noticeable friction reduction.

Poly(ethylene oxide), which has a completely linear structure, is an effective turbulent-flow friction-reducing agent in any solvent in which it is soluble. Its ability in this regard has made it a widely used material for all types of experimental work.

Polyacrylamide. Dilute solutions of polyacrylamide are effective friction reducers compared with pure water. Metzner and Park (14) have shown how the product "J-100" departs spectacularly from the Metzner-Reed correlation for purely viscous non-Newtonian fluids. Figure 6 indicates the drag-reduction effect which was observed for solutions having a n' of about 0.55. Metzner and Park attribute the departure from the purely viscous correlation to the effects of viscoelasticity. Data for a different polyacrylamide "AP-30" over a range of pipe sizes from 0.18 inch to 6.0 inches are given in Figure 7 (from (12)) and it can be seen that friction reduction of 80% or more is readily obtained.

In terms of effectiveness (concentration required for a given drag-reduction) polyacrylamide appears to lie between guar and poly-(ethylene oxide). Since it dissolves more readily than poly(ethylene oxide), polyacrylamide has found favor with experimenters requiring rapid solubility and a large friction-reduction effect.

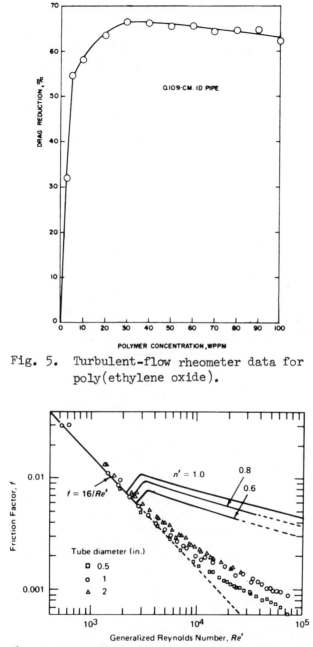

Fig. 5.   Turbulent-flow rheometer data for poly(ethylene oxide).

Fig. 6.   Friction factor as a function of generalized Reynolds number for purely viscous fluids (solid lines) and for 0.3% polyacrylamide J-100 in water (data points).

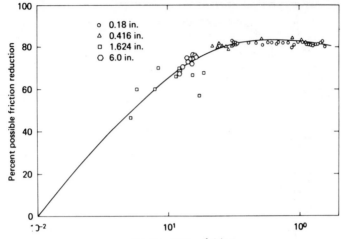

Fig. 7. Experimental data for 250 ppm polyacrylamide
         AP-30 solutions.

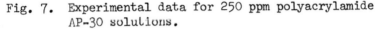
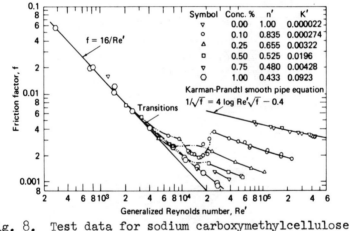

Fig. 8. Test data for sodium carboxymethylcellulose
         CMC-7HSCP solutions flowing in a 0.9-in.
         diameter pipe.

Sodium Carboxymethylcellulose (CMC). Both Dodge and Metzner
(15) and Shaver and Merrill (16) found the friction coefficients
to be "anomalously" low for CMC. Although CMC is less effective
on a weight basis than even, say, guar, several other experimenters
have also utilized it.

Ripkin and Pilch (17) were among the first to make a comprehensive study. Figure 8 shows some of their data for various concentrations of CMC-7HSCP (Hercules Incorporated) in a 0.9-inch diameter

pipe. Although the required concentrations are high compared with other materials covered in this section, large reductions in turbulent friction can be obtained with CMC. In this case the flow index n' is also very low, as noted on the figure.

Other Polymers and Solvents. Many other substances have been found to give friction-reducing properties to their solvents. Some of these are given in Table I. More extensive tables can be found in Reference 18.

TABLE I. Some Friction-Reducing Fluids

Solute	Solvent
Guar	Water
Poly(ethylene oxide)	Water
Polyacrylamide	Water
Sodium carboxymethylcellulose	Water
Flax meal	Sea water
Hydroxyethylcellulose	Water
Cetyltrimethylammonium bromide - 1-naphthol	Water
Polyisobutylene	Cyclohexane
Poly(methyl methacrylate)	Toluene
Polyisobutylene	Crude oil
Polyisobutylene	Kerosene
Proprietary	Hydraulic fluid
Poly(sodium styrenesulfonate)	Water

Some work has been done with poly(ethylene oxide) in solvents other than water. Exploratory tests with blood transfusion fluids, (dextran, saline, and plasma) indicate the turbulent-flow friction to be greatly reduced. Sea water as the solvent shows results similar to pure water. Poly(ethylene oxide) in organic solvents such as benzene, dioxane, carbon tetrachloride, nitromethane, chloroform, methylene chloride, and anisole give drag-reduction results comparable to water.

APPLICATIONS

Since the study of friction-reducing polymers is in its in-
fancy, it is of course too early to do more than sketch the possible
courses of their development into useful applications. In this
section we will outline areas in which some progress has occurred.

Experimental investigations with a fire-department pumper and
hose show that the pressure drop through 700 ft of hose could be
reduced 40%, thus allowing a 50% further throw of the water stream,
and a 10% increase in flow volume, when 200 ppm of poly(ethylene
oxide) was added - the fire pump operating at constant speed. The
scheme apparently will be adopted in New York City. In fact, calculations indicate that polymer addition to an entire city water
supply in times of high demand, such as a major fire, may be an
attractive method of providing an emergency increase in capacity of
the system. In addition to providing a greater flow, or less pressure drop in a fire-fighting hose line, it has been shown that the
polymer jet from a nozzle tends to be more cohesive, more resistant
to break-up by the wind, and to concentrate the stream in a smaller
area, all of which factors should be important in fire fighting.
Irrigation systems have also been tested and, likewise, show improvements when polymer is added to the flow.

Pumping Water and Oil

The earliest commercial development in the use of friction-reducing additives has been the utilization of high-polymeric materials, principally guar, in the hydraulic "fracturing" of oil wells.
In this technique, after the well is drilled, a suspension of sand
in water is pumped under high pressure into the oil-bearing rock
formation; the formation is physically cracked and broken open by
the high pressure, and the sand prevents the cracks from closing.
Gums and thickening agents are sometimes added to aid in suspending
the same; the friction-reducing effectiveness of guar was discovered
in this manner.

The use of guar in oil-well fracturing now appears to be well
established. Using the optimum concentration, the horsepower required is reduced by a factor of over 5 for the fracturing operation
compared to the use of water alone.

Success in utilization of polymer additives in pipe-line situations such as those found in oil-well fracturing depends to a
large extent upon rapid mixing of the polymer solutions, and consequently a number of studies have been made to determined the
effect of injector geometry. Both centerline and pipe-wall slot
injectors have been studied. When a 0.1% solution of guar is in-

injected into flowing water at the center-line of the pipe, about 20 diameters are required before the fluid approaches uniform mixing, as judged by the friction reduction, but wall friction was reduced below the uniform making level as early as 10 diameters downstream, indicating impingement of the more concentrated guar solution on the wall. When the slot injector was used, immediate friction reduction was found, with the values at first much higher than the uniform mixing value, but after 5 diameters or so approaching the uniform mixing value.

Although most experimental work to date has been carried out using water-based solutions, Ram, Finkelstein, and Elata (19) showed that both crude oil and kerosene could be modified by the addition of high-molecular-weight polyisobutylene so as to greatly reduce the turbulent friction. Substantial friction reductions could be obtained in crude oil. The principal barrier to widespread use of the technique in crude-oil pipelines seems to be degradation of the polymer. A new, as yet undisclosed material, has shown little degradation in diesel-oil pipelines, in tests of a 6.4-inch diameter, 6.5-mile-long Army pipeline, although rather severe degradation was caused by multistage centrifugal pumps (20). Resistance reduction of up to 37.5%, reflecting pump power savings of around 30%, were noted in the tests of a 12-inch, 32 mile long crude-oil line (21). As robust polymers appear, the technique may find widespread utilization.

Scientific Studies

The use of friction reduction to assist in the study of molecular properties is another area which promises to have an ever widening impact. The effect has shown to be useful in the determination of molecular weight and in molecular configuration investigations. Kenis (22) has suggested the friction-reduction effect as a method of following polymerization - depolymerization reactions in biological systems. Other applications of the effect in scientific studies are certain to appear.

Experiments have shown that the drag of small shapes can be reduced by means of a soluble coating attached to the nose which can wash off during the water travel. Thurston and Jones (23) found that the drag of a 1.5-inch diameter, one-foot-long cylindrical body could be reduced by 25% by polymer-containing soluble disks attached to the nose. The development of rapid-sinking expendable oceanographic instruments could follow, utilizing this technique.

In summary, the scientific and technological advances which have already been brought, and which will undoubtedly come forth in the future through the study of friction-reducing polymers, make

this area of research an exciting blend of increased insight into
the working of nature and applications for the benefit of people
everywhere.

BIBLIOGRAPHY

1. J. L. Lumley, "Drag Reduction by Additives," in W. R. Sears,
    Ed. Annual Review of Fluid Mechanics, Annual Reviews, Inc.,
    Palo Alto, Vol. 1, 1969, p. 367.

2. G. K. Patterson, J. L. Zakin, and J. M. Rodriguez, "Drag
    Reduction: Polymer Solutions, Soap Solutions and Solid
    Particle Suspensions in Pipe Flow," Ind. Eng. Chem. 61, 22,
    (1969).

3. J. W. Hoyt, "The Effect of Additives on Fluid Friction," Trans.
    ASME, J. Basic Engineering, 94D, 258 (1972).

4. G. E. Gadd, "Reduction of Turbulent Fraction by Dissolved
    Additives," Nature, 212, 874 (1966); "Friction Reduction" in
    N. M. Bikales, Ed., Encyclopedia of Polymer Science and
    Technology, Interscience Publishers, New York, Vol. 15, 1971,
    pp. 224-253.

5. B. A. Toms, "Some Observations on the Flow of Linear Polymer
    Solutions Through Straight Tubes at Large Reynolds Numbers,"
    Proceedings International Congress on Rheology, 1948, Vol. ii,
    North Holland Publishing Co., Amsterdam, p. 135, 1949.

6. K. J. Mysels, Flow of Thickened Fluids, U. S. Pat. 2,492,173
    (Dec. 27. 1949).

7. J. G. Savins, "Drag Reduction Characteristics of Solutions of
    Macromolecules in Turbulent Pipe Flow," Society of Petroleum
    Engineering Journal, 4, 203 (1964).

8. J. W. Hoyt and A. G. Fabula, "The Effect of Additives on Fluid
    Friction," Proceedings Fifth Symposium on Naval Hydrodynamics,
    Bergen, Norway, Office of Naval Research ACR-112, 1964, p. 947.

9. B. Johnson and R. H. Barchi, "Effect of Drag Reducing Additives
    on Boundary-Layer Turbulence," J. Hydronautics, 2, 108 (1968).

10. M. Walsh, "Theory of Drag Reduction in Dilute High-Polymer
    Flows," International Shipbuilding Progress, 14, 134 (1967).

11. J. W. Hoyt, "A Turbulent-Flow Rheometer," in A. W. Marris and
    J. T. S.Wang, Eds., Symposium on Rheology, ASME, New York, 1965,
    p. 71.

12. N. F. Whitsitt, L. J. Harrington, and H. R. Crawford, "Effect of Wall Shear Stress on Drag Reduction of Viscoelastic Fluids," Western Co. Report No. DTMB-3 (1968). See also same authors and title in C. S. Wells, Ed., Viscous Drag Reduction, Plenum Press, New York, 1969, p. 265.

13. A. G. Fabula, "The Toms Phenomenon in the Turbulent Flow of Very Dilute Polymer Solutions," in E. H. Lee, Ed., Proceedings Fourth International Congress on Rheology, 1963, Part 3, Interscience Publishers, New York, 1965, p. 455.

14. A. B. Metzner and M. G. Park, "Turbulent Flow Characteristics of Viscoelastic Fluids," J. Fluid Mechanics, 20, 291 (1964).

15. D. W. Dodge and A. B. Metzner, "Turbulent Flow of Non-Newtonian Systems," AIChE Journal, 5, 189 (1959).

16. R. G. Shaver and E. W. Merrill, "Turbulent Flow of Pseudoplastic Polymer Solutions in Straight Cylindrical Tubes," AIChE Journal, 5, 181 (1959).

17. J. F. Ripkin and M. Pilch, "Studies of the Reduction of Pipe Friction with the Non-Newtonian Additive CMC," St. Anthony Falls Hydraulic Laboratory, Technical Paper No. 42, Series B, 1963.

18. J. W. Hoyt, "Drag Reduction Effectiveness of Polymer Solutions; A Catalog," Polymer Letters, 9, 851 (1971).

19. A. Ram, E. Finkelstein, and C. Elata, "Reduction of Friction in Oil Pipelines by Polymer Additives," Ind. Eng. Chem. Process Design and Development, 6, 309 (1967).

20. K. L. Treiber and L. M. Sieracki, "The Effect of Non-Newtonian Friction Reducing Additives in a Diesel Fuel Pipeline," Columbia Research Corp. Report No. 101-2, 1970.

21. J. A. Lescarboura, J. D. Culter, and H. A. Wahl, "Drag Reduction with a Polymeric Additive in Crude Oil Pipelines," Soc. Pet. Engrs, Preprint SPE 3087, 1970.

22. P. R. Kenis, "Drag Reduction by Bacterial Metabolites," Nature, 217, 940 (1968).

23. S. Thurston and R. D. Jones, "Experimental Model Studies of Non-Newtonian Soluble Coatings for Drag Reduction," AIAA J. of Aircraft (March/April 1965).

DRAG-REDUCTION PROPERTIES OF ULTRA-HIGH-MOLECULAR-WEIGHT

POLYACRYLAMIDE AND RELATED POLYMERS

Robert Y. Ting and Oh-Kil Kim

Naval Research Laboratory

Washington, D. C.

The drag-reduction phenomenon has been the subject of intense interest and activity among scientists and engineers for the past decade. This phenomenon is observed when solutions of very small amounts of high-molecular-weight linear polymers are subjected to turbulent pipe flow. The resultant effect is that the pressure gradient required to move the fluid is substantially reduced at a given flow rate. Toms (1) gave the first clear description of this phenomenon in his study of the turbulent flow of poly(methyl methacrylate) in monochlorobenzene. Many investigators since then have confirmed such effects in aqueous solutions of guar gum, carboxymethylcellulose, poly(acrylic acid), polyacrylamide, and poly(ethylene oxide). The bulk of the materials which have been tested to date are commercial samples, among which the poly(ethylene oxides) and polyacrylamides are the most widely used. These polymers are inexpensive, easy to handle and are extremely effective agents; for example, a 44 percent drag reduction is possible by the addition of 10 parts per million by weight (ppmw) of poly(ethylene oxide) of molecular weight 900,000 (2).

In spite of the extensive research activity in drag reduction during the past decade, there is still no agreed interpretation of the mechanism of drag reduction. It is generally accepted, however, that the factors contributing to the effectiveness of drag-reducing polymers are: molecular flexibility and linearity, high molecular weight, and good solubility (3). Because of the sensitivity of drag-reduction behavior to molecular properties, a systematic investigation of the relationship between drag reduction and these molecular parameters is not only desirable but necessary. An experimental plan suited to this purpose was designed by systemati-

cally synthesizing drag-reducing agents of varying molecular structure. Since aqueous drag reduction systems are of prime interest
in marine applications, the present research effort emphasizes the
synthesis of water-soluble polymers. This report is concerned with
the drag reduction properties of high-molecular-weight polyacrylamides and selected derivatives.

EXPERIMENTAL

Preparation of Polymers

Polyacrylamide. Acrylamide (Eastman) was polymerized without
further purification in aqueous solution with ammonium persulfate
as the initiator under various conditions to yield a series of
high-molecular-weight polyacrylamides. In the case of the ultra-high-molecular-weight polymerization, a proprietary catalyst
system (4) was employed. The polymer solutions were made directly
by dissolving the gel-like polymerization products (12 - 15% by
weight) without fractionation. Viscosity measurements of the polyacrylamide solutions were made in water at 25°C. The results are
summarized in Table I.

Poly(acrylic Acid). The polymerization of acrylic acid (freshly distilled) was carried out in an aqueous solution of pH ~ 4.2
with an ammonium persulfate - sodium bisulfite redox catalyst. The
resulting polymer solution was diluted with additional water, and
the pH was raised to 10 for 100% neutralization. The polymer was
precipitated by the addition of acetone to the aqueous solution to
remove the lower range of molecular weights. It was reprecipitated
twice in essentially the same manner.

Hydrolysis of Polyacrylamide. Three 650 ml solutions contain-
ing 0.15% by weight of polyacrylamide ($\bar{M}_V = 4.7 \times 10^6$) and 2 g sodium hydroxide were allowed to react at 75°C for 9 hrs, 5 hrs, and
2 hrs; these conditions resulted in 63%, 57%, and 47% hydrolysis
of amide groups to carboxylic acid groups, respectively. The polymers were separated as their sodium salts by the addition of acetone.

Glyoxal Modification of Polyacrylamide. The reaction of glyoxal with polyacrylamide ($\bar{M}_V = 4.7 \times 10^6$) was conducted with a
variety of molar ratios and reaction times. After a 50 ml solution
of 0.15% polyacrylamide was treated with sufficient 30% aqueous glyoxal at a selected molar ratio, the resulting solution was buffered
to pH 9.0 with carbonate. It was then allowed to react at 31°C for
20 min or 60 min. At the end of the reaction time, the resultant
gel-like product was immediately diluted with 100 ml water, and the
pH was adjusted to neutral. Reduced viscosities of the products
(20 min reaction) at Gly/PAM = 1/2 and 1/3 (molar basis) were 11.48
and 13.17 (c = 0.0296 g/dl in water, 25°C) respectively.

Drag-Reduction Measurements

The drag-reduction properties of the polymer solutions were characterized through use of a simple turbulent pipe flow system shown in Figure 1 (5). The device is basically a metal syringe controlled by a DC motor which drives the test liquid through a 0.62 cm diameter pipe. The flow rate was monitored by a small DC generator coupled to the motor drive. Two pressure taps were placed at approximately 135 and 175 diameters from the upstream end of the flow. The pressure difference between taps was measured by a differential pressure transducer. The outputs from the DC generator and the transducer were recorded continuously. The flow rate and the wall shear stresses were then calculated by using the calibrated constants of the apparatus. The percent drag reduction was computed by using the following relationship:

$$\text{percent drag reduction} = \left(1 - \frac{f_{polymer}}{f_{water}} \right) \times 100$$

where $f_{(\,)}$ is the friction coefficient, defined as the ratio of wall shear stress to the mean dynamic pressure head of turbulent flow, $1/2\,\rho(\bar{u})^2$, where ρ is the density of the fluid and $\bar{u}$ the mean velocity of the solution.

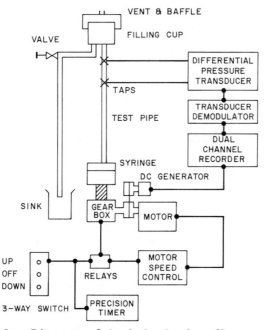

Fig. 1. Diagram of turbulent pipe flow system.

RESULTS AND DISCUSSION

As shown in Table I, a series of acrylamide homopolymers cover-
ing a wide range of molecular weights have been successfully synthesized. Preliminary measurements show that drag reduction increases rapidly when the molecular weight exceeds values of 2×10^6.
These results are in agreement with the conclusions of Merrill et al
(6), who analyzed solutions of polyisobutylene and poly(ethylene
oxide). These workers found that the drag reduction effect scales
with increased polymer chain length.

The drag reduction results of a polyacrylamide sample ($\bar{M}_v = 4.7 \times 10^6$) and three hydrolyzed derivatives are presented in
Figure 2. All these materials exhibit an increasing degree of drag
reduction with increasing flow rates approaching Virk's maximum
drag reduction asymptote at high flow rates (7). The hydrolyzed
polyacrylamide samples show a significant improvement in drag reduction ability over the basic material, polyacrylamide itself.
Figure 3 specially illustrates the effect of hydrolysis and indicates that the drag reduction is maximized when approximately 55%
of the polyacrylamide is hydrolyzed. A poly(acrylic acid) sample
($\bar{M}_v = 3.7 \times 10^6$) was also tested but it was noticeably less effective. Until further study is completed, it is not yet certain
whether the reduced effectiveness is due to the lower molecular
weight or the excessive presence of charged groups on the polymer
backbone.

The enhancement of drag reduction in the case of polyelectrolytes may be interpreted as a result of further extension of the
polymer main chain caused by the repulsion of similarly charged
groups. Clarke's work (8) in relating drag-reduction characteristics
to polymer conformation showed that polyacrylamide molecules were
in more collapsed state than the hydrolyzed polyacrylamide molecules
in 0.02 $\underline{M}$ NaCl. Drag reduction was subsequently found to be less
in the polyacrylamide solutions - apparently a result of the more
compact configuration. This is exactly in the same direction indicated by the present data. The existence of an optimum content
of ionic groups in the polyacrylamide polymer to the drag reduction
observed seems to be related to the literature results reported for
poly(acrylic acid) (9) and poly(methacrylic acid) (10). These
results showed that the reduced viscosity of both poly(acrylic acid)
and poly(methacrylic acid) increased sharply with an increase of
neutralization up to 40% and then decreased with further neutralization. To further clarify this point, however, it would be desirable to test the drag-reduction properties of high-molecular-weight
poly(acrylic acid) and of polyacrylamide samples with higher degrees
of hydrolysis.

Figure 4 is a plot of the drag-reduction data of glyoxal-modified polyacrylamide samples over a range of flow rates. Samples

TABLE I. Molecular Weight of Polymers

Sample	Type of Polymer	$[\eta]$; (dl/g)	$\bar{M}_v \times 10^{-6}$
02	polyacrylamide	5.54[a]	1.5
17-2	polyacrylamide	8.05[a]	2.4
18	polyacrylamide	13.20[a]	4.4
05	polyacrylamide	13.75[a]	4.7
32-a	polyacrylamide	16.20[a]	5.8
32-b	polyacrylamide	18.30[a]	6.7
47	poly(acrylic acid)	2.98[b]	3.3
44	poly(acrylic acid)	3.59[b]	3.7
22-1	hydrolyzed polyacrylamide	-	∼4.7
23-2	hydrolyzed polyacrylamide	-	∼4.7
23-3	hydrolyzed polyacrylamide	-	∼4.7
43-1	Gly/PAM(1/2)[c]	11.48[d]	>4.7
43-2	Gly/PAM(1/3)[c]	13.17[d]	>4.7

- - - - - -

a. $[\eta] = 6.31 \times 10^{-5} M^{0.80}$ (dl/g); in water, 25°C (18).

b. $[\eta] = 5.39 \times 10^{-3} M^{0.43}$ (dl/g); in 2 $\underline{N}$ NaOH, 30°C (19).

c. Gly = glyoxal; PAM = polyacrylamide.

d. Reduced viscosity (c = 0.0296 g/dl , in water, 25°C).

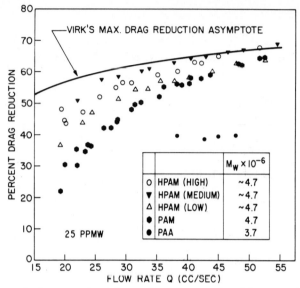

Fig. 2. Percent drag reduction of polyacrylamide, partially hydro-
         lyzed polyacrylamide, and poly(acrylic acid) at different
         flow rates.

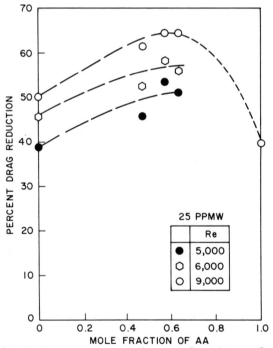

Fig. 3. Relation between percent drag reduction and degree of
         hydrolysis of polyacrylamide at different flow rates.

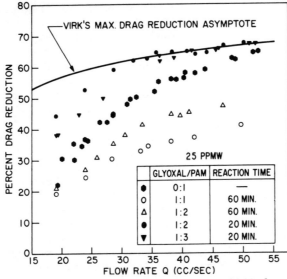

Fig. 4. Percent drag reduction of glyoxal-modified polyacrylamides
for different compositions and reaction times.

prepared with a long reaction time (60 minutes) show a definite
decrease in drag reduction when compared with the basic material,
polyacrylamide itself. On the other hand, samples reacted for only
20 minutes showed a distinct improvement in drag reduction - especially at lower flow rates. In the former case, it is entirely possible
that the overreacted glyoxal-polyacrylamide samples represent the
adverse effect of crosslinking on drag reduction (11). Merrill et al
(6) found that for the same number of monomer units in the main chain,
the concentration of polyisobutylene required for a given effect
was substantially greater than that of poly(ethylene oxide). This
indicates that the effectiveness of polymers on drag reduction is
dependent upon backbone flexibility, and may explain in part why
the crosslinked polyacrylamide samples are less effective. By comparison, the drag-reduction results obtained with samples having a
short reaction time might be due to the existence of the bulky
hydrated glyoxal side groups. The steric effect of these groups
would favor expansion of the polymer coil, which would then be able
to store much higher energy (12). The recent report (13) on a correlation of the onset wall shear stress with the intrinsic viscosity
of poly(ethylene oxide) in salt solutions, together with Clarke's
work (8), again supports the idea that expanded polymer coils enhance the observed drag reduction.

Various studies (14,15) of solvent effects in drag reduction
showed that polymers in good solvents are far more effective than

in poor solvents. Therefore, the enhancement of drag reduction caused by the glyoxal modification of polyacrylamide with a short reaction time may also be explained by the improved hydrophilic character of the polymer. The enriched hydroxyl moiety assists further hydration in proportion to the fraction of the reacted glyoxal and the observed drag reduction is increased as the polymer is rendered more soluble (a more expanded configuration) by the additional hydrophilic groups.

All the modifications on polyacrylamide reported here, i.e. by increasing the chain length, by the hydrolysis of functional groups, or by the addition of bulky hydrophilic side groups, seem to introduce a larger molecular length scale. These polymer molecules appear to assume more expanded configurations and therefore would be characterized by longer relaxation times (16). According to the correlations introduced by Elata (17) and Zakin (14), longer relaxation times would tend to promote drag reduction. These relaxation time - drag reduction correlations tend to support the interpretation of polymer structural effects as advanced in this report.

BIBLIOGRAPHY

1. B. A. Toms, Proc. 1st International Congr. Rheol. 2, 135, North Holland, Amsterdam (1948).

2. R. C. Little, J. Colloid and Interface Sci., 37, 811 (1971).

3. J. W. Hoyt and A. G. Fabula, ONR 5th Symp. on Naval Hydrodynamics, Bergen, Norway, Sept. 1964.

4. O. K. Kim, unpublished results.

5. R. C. Little and M. Wiegard, J. Appl. Polym. Sci., 14, 409 (1970).

6. E. W. Merrill, K. A. Smith, H. Shin, and H. S. Mickley, Trans. Soc. Rheol., 10, 335 (1966).

7. P. S. Virk, Sc.D. Thesis, MIT, 1966.

8. W. B. Clarke, Sc.D. Thesis, MIT, 1970.

9. R. Sakamoto and K. Imahori, Kogyokagaku Zasshi, 83, 389 (1962).

10. H. P. Gregor and D. H. Gold, J. Polym. Sci., 23, 467 (1959).

11. S. L. Vail and A. G. Pierce, Jr., J. Org. Chem., 37, 391 (1972).

12. W. O. Baker and T. H. Heiss, Bell Tel. Tech. J., 31, 306 (1952).

13. R. C. Little, J. Appl. Polym. Sci., 15, 3117 (1971).

14. H. C. Hershey and J. L. Zakin, Chem. Eng. Sci., 22, 1847 (1967).

15. P. Peyser and R. C. Little, J. Appl. Polym. Sci., 15, 2623 (1971).

16. G. E. Gadd, "Friction Reduction" in N. M. Bikales, Ed., Encyclopedia of Polymer Science and Technology, Interscience Publishers, New York, Vol. 15, 1971, p. 224.

17. C. Elata, J. Lehrer, and A. Kahanovitz, Israel J. of Tech., 4, 1 (1966).

18. W. Sholtan, Makromol. Chem., 14, 169 (1954).

19. H. Itoh et al, Kogyokagaku Zasshi, 8, 930 (1956).

THE BIOLOGIC ACTIVITY OF WATER-SOLUBLE POLYMERS

William Regelson

Division of Medical Oncology, Medical College of

Virginia, Richmond, Virginia

Polymers as structural entities are derived from natural and synthetic sources. Their role in biology within the organism relates to their development as insoluble or soluble forms in tissue and tissue fluid, some of which resemble blood proteins or nucleotides (1,2).

We are interested in presenting here those polymers that enter into biologic function through distribution or entry throughout the host. In this case, we are dealing with polymers that can be likened to blood proteins which distribute themselves via blood or lymphatic circulation, or, through cellular transport, with the cooperation of mobile phagocytic cells or adsorbing cell surfaces. Here solubility is a function of molecular weight and charge interacting with the natural proteins or lipoproteins of the tissue fluid or cell surface.

Water solubility is a relative term, and we are dealing here with hydrophilic or hydrophobic reactions depending on a number of interrelated factors (3). The problem is very much akin to that in soil conditioning where ratios of molecular weight, charge, salt concentration, and pH interreact to disperse or aggregate colloidal suspensions. The problem of systemic administration, for whatever effect, can relate to the colloidal suspension of the particulates themselves. In this regard, the physician knows from experience to rarely (or hardly ever) knowingly admit to injecting suspensions into blood vessels. The clinical results of this practice have been called "colloido-clasmic shock" or the "macromolecular syndrome" (4) and are no longer as clinically evident as they once were, as techniques in the fabrication and purification of parenteral medication or delivery systems have improved. However, when one is dealing with polymers, this problem is still with us as interreaction with blood proteins or tissue constituents can change the colloidal stability of blood or blood viscosity with resulting effects on platelet or white-cell distribution, blood clotting, and/or circulation. The effects of polymers can resemble those produced by bacterial endotoxins or nucleic acids. In contrast, therapeutically, aggregated albumin is used for diagnostic purposes and certain polymers such as dextran and polyvinylpyrolidone (PVP) can be substituted for blood proteins in the treatment of shock. Thus, for many biologic polymers, what one needs clinically depends on the similarity of the polymers to blood proteins (3).

To be more specific, it is well known that injury or infection alters the sedimentation rate of blood. This relates to an increase in fibrinogen and glycoproteins produced by the liver. The biologic role of injury-related increases for these proteins is still not known. However, glycoproteins alter the colloidal stability of proteins in urine and are now shown to increase resistance to bacterial infection, viral infection, and tumor growth.

The similarity of synthetic polymers to proteins, glycoproteins, or polynucleotides permits them to modulate a variety of biologic responses related to host-defense reactions which include resistance to viral, bacterial, protozoal, and fungal infection. Immunologic and hormonal responses and inflammation, wound repair, blood clotting, and tissue damage is also subject to the action of these macromolecules. In some ways, the polymers mimic the action of infecting organisms and thus modify their action on the host. This, in a sense, historically is a homeopathic concept: "to treat one disease with a less damaging one," thus stimulating reparative or resisting processes. This concept requires that one pay close attention to the timing and route of polymer administration. The reason for this is that timing of dosage and route of administration can be critical to the tissue distribution or half-life of these agents which relates to the effects seen.

In regard to the above, polymers can also interact through blocking or activating enzymes, and hormones, or combining with toxic materials, and in this manner their role in biology is of increasing importance.

However, whatever benefits are to be gained in the biologic use of polymers must be balanced by their acute or delayed toxicity. For example, it is well known that polymers as tissue implanted films are notoriously carcinogenic (5). The safety or carcinogenicity of water-soluble polymers is debatable as animal data do not necessarily reflect on clinical response and soluble polymers have had relatively recent entry into clinical trial.

However, polyvinylpyrolidone (PVP) which has a long history of
clinical use can produce tumor-like tissue or macrophagic reactions
that resemble tumors (6,7). Similar problems have been observed
with anionic dyes which on prolonged tissue storage induce sarcomas
(8).

To solve these and other related problems of prolonged storage,
two approaches are reasonable. One is to select polymers of a
molecular size that can be readily cleared by kidney or biliary
excretion; the other is to fabricate biodegradable polymers that
do their job and are then converted to harmless by-products. This
concept of biodegradability has been applied successfully to ab-
sorbable sutures (9), and in another area has been applied effectively to the development of degradable coatings, for timed release
of oral medication and topical gels (10-12). Important to this is
the recent work of Beasley (13), who has developed water-degradable
polymers for the timed release of fertilizers and insecticides.
These polymers have obvious potential usefulness for self-destruc-
tive packaging in the control of environmental pollution but their
systemic place in animals is not yet known.

We have worked primarily with polyanions and the bulk of our
discussion will deal with these agents.

Polyanions and polycations are physiologically important not
only to functional and structural integrity, but may control such
biologic processes as enzyme activity, the regulation of cell
division, nuclear informational transfer, intravascular coagulation,
and, as mentioned previously, host resistance to bacterial, protozoal, fungal, and viral infection. This is particularly true
for polyanions which stimulate or blockade the phagocytic reticuloendothelial system (RES) and alter immunologic response. Polyanions are neutralizers of peptic ulceration and are also useful
as antidiarrheal agents which can detoxify agents as diverse as
bacterial toxins, bacteria, viruses, magnesium sulfate, and castor
oil (14). The action of polyanions as mitotic inhibitors and their
functional role in growth processes has recently been reviewed by
us in several recent papers (15-17), as has the role of polynucleotides in immunology and virus resistance (18). Polycations have
had biologic success as topical or systemic antibiotics (e.g.
polymyxin, colymycin) and most recently, polyethylenimine and related polycations have been shown to have effective antitumor activity against transplanted tumors.

Ethylene - maleic anhydride copolymers have also been used as
drug coating agents or as antidiarrheals (10-12,14,19) where their
hydrophilic character can be of benefit. They also have potential
use as topicals with gel formation and complexing with cationic
antibiotics of benefit in controlling tissue contact and drug release.

We will only touch briefly on "neutral" or blood protein-like polymers which are important in vitro as antifreezes in tissue and blood preservation. Depending on molecular weight these polymers (polyvinylpyrrolidone, polyethylene glycols, dextrans, and mannitol) are important in vivo in blood flow, tissue oxygenation, tissue edema, and renal function, or as vehicles for drug administration and detoxification.

POLYANIONS

Heparin, a native acid mucopolysaccharide, while primarily developed for its anticoagulant and lipolytic activity (important to the prevention of atherosclerosis), can also inhibit inflammatory and immunologic processes. The anticoagulant effect of heparin and its possible relation to calcium binding was also related to inhibition of tumor growth.

Historically, more significant to the development of rationale chemotherapy, was the systematic study of anionic dyes by Paul Ehrlich and his co-workers. Trypanocidal dyes, such as isamine blue, showed clinical antitumor activity which was initially related to stimulation of phagocytic "pyrrhol" cells.

Attempts to develop synthetic or naturally derived substitutes for native heparin have produced a number of polyanions which possess antimitotic activity (heparinoids). Clinically, these compounds, which were to serve as lipolytic agents or anticoagulants, produced alopecia, ulceration of the gastro-intestinal tract, and osteoporosis leading to pathologic fractures. Polysulfonates, pyran copolymer and related polycarboxylic maleic anhydride copolymers are such compounds.

Testing of these heparinoids for antitumor activity has led to systematic clinical trial of two agents: polyethylenesulfonate and a polycarboxylic pyran copolymer derived from divinyl ether and maleic anhydride (NSC 46015). Pyran is still under clinical investigation (20). Pyran induces low levels of the antiviral substance interferon and causes a biphasic response in the functional activity of the reticuloendothelial system. It also enhances immunologic response, increases resistance to a wide variety of microbes (S. aureus, D. pneumoniae, C. neoformans, and T. duttoni) (15,21-23). Additionally, pyran possesses anticoagulant properties by inhibiting the conversion of fibrinogen to fibrin (24) and enhances vaccine effectiveness (Richmond and Campbell, 1972). See also the chapter by G. B. Butler and C. Wu.

In addition to the above, there has also been related but independent research on the inhibitory effect of heparin and other anticoagulants on the growth and spread of tumor metastasis (16, 25).

Apart from effects on living cells, heparin was used in the crystallization and isolation of tobacco mosaic virus and a number of polyanions have been shown to inhibit viruses in vitro and in vivo. However, systematic searches for effective antiviral polyanions utilizing selective enzyme inhibition have had little success. What seems to be important is an effect on cell surface. In addition, renewed interest has developed from observations starting in our laboratory regarding interferon induction by polyanions. [Interferon is an antiviral protein induced by virus infection which blocks virus replication within the cell]. There is also the little understood but very interesting area regarding the inhibition or enhancement of the virulence of viruses by polyanions vis-a-vis polycations. Most recently, the prolonged in vivo inhibiting action of synthetic polyanions, when given prior to virus inoculation, has provided impetus for assaying the fundamental role for polyanions in controlling host resistance to a variety of pathophysiology.

The following is a summary of the biologic activity of polyanions.

Embryogenesis

The function of the sperm is essential not only for transfer of genetic information but may also be involved in egg surface and cytoplasmic alterations essential for cell division. It has been shown that the jelly coat which surrounds the egg is a mucopolysaccharide esterified with sulfate. A release of sulfate occurs on fertilization and the jelly coat substance can inhibit fertilization in similar fashion to that obtained with related polyanions. Sulfated mucopolysaccharides may be critical to the phenomenon of normal development as is seen in effects on cell to cell adhesiveness associated with the necessity of sulfate ions for the development of sea urchin egg and larvae.

Growth Control

While increases in polyphosphates have not been associated with mitosis in animals, in bacteria the appearance of high levels of metachromatic polyphosphates synchronize with logarithmic growth (26). Similarly, tumor growth and normal regeneration is associated with changes in polyanionic stromal or surface polysaccharides. It is therefore not surprising that heparinoids, synthetic or derived from biological materials, show clinical side effects which may represent inhibition of rapidly growing epithelium.

Polyethylenesulfonate and polystyrenesulfonate have also been shown to displace histone bound to nucleic acid resulting in stimulation of protein synthesis (27,28) and RNA polymerase activity (29). This may be important to gene regulation of protein synthesis related to the phenomena of derepression which is fundamental to ordered growth and function.

Synthetic Polymers. We have discussed the antitumor action for polyanions in great detail in recent reviews (15,16). We will confine our comments to more recent pertinent findings and that which has bearing on the action of synthetic polynucleotides, which are synthetic cogeners of ribonucleic acid.

The current experience with poly(riboinosinic-cytidylic acid) (Poly-IC), and related thiophosphate cogeners resembles in some ways what has been seen with bacterial endotoxins or polysaccharides which possess heparin-like in vitro activity. For example, in similar fasion to poly-IC and bacterial endotoxins, lignosulfonates produced hemorrhagic necrosis of sensitive tumors.

Apart from direct effects on cells, we have shown that a number of heparinoids can inhibit transplanted mouse tumors when the drug is given intraperitoneally to mice bearing subcutaneous tumors. The most effective polyanion in this series, which included dextran sulfate, laminarin sulfate, sulfated chitosan, alginic acid, and sulfated cellulose esters as well as polyphosphate glasses and phosphorylated hesperidin, was polyethylenesulfonate. This heparinoid, of mw 13,000, which was developed for its lipolytic activity showed minimal in vitro anticoagulant properties as compared to heparin, and was tested against human tumors in forty advanced patients. Three patients of thirty-five who received an adequate trial of drug showed objective tumor regression. The intravenous toxicity of polyethylenesulfonate precluded further clinical testing.

Significant tumor inhibition of solid mouse tumors on intraperitoneal injection of poly(xenyl phosphate) was obtained by Muehlbaecher et al (30). It was demonstrated that P_{32}-tagged poly(xenyl phosphate) concentrated in the tumor and drug concentration was greater than one might expect for a predominantly reticuloendothelial pattern of distribution which had been seen for other polyanions.

Maleic Anhydride Copolymers. The search for synthetic heparinoids led us to evaluate synthetic polymers derived from ethylene and maleic anhydride (E/MA) copolymers or poly(acrylic acid). These E/MA polymers had shown lipolytic activity when given parenterally. Polymers of mw ranging from 1200 and 120,000 or higher were tested against a variety of subcutaneous rodent tumors. In each case intraperitoneal administration was associated with significant tumor regression. The higher the molecular weight the more toxic the E/MA copolymer to the mouse and dog. Optimum activity was obtained in the series where carboxamide and ionizable carboxyl groups were interdispersed on the polymer backbone. When all carboxyl groups were converted to carboxamides, significant tumor inhibition was lost. Related monomeric fragments to these polycarboxylic copolymers were completely without activity. Adequate treatment of mice prior to tumor inoculation resulted in inhibition of tumor growth evident as long as one week following drug injection. Additional activity was found against a virus-induced leukemia.

Preclinical pharmacologic evaluation of E/MA fractions of mw 1200-30,000 showed them to possess toxicity that precluded clinical trial. However, subsequent work has shown that toxicity depends on the polymer studied and a good therapeutic index has been found for vinyl ether - maleic anhydride copolymers. One fraction, a polycarboxylic pyran copolymer hydrolyzate (NSC 46015: Pyran 3,4-dicarboxylic anhydride; tetrahydro-2-methyl-6-(tetrahydro-2, 5-dioxo-3-furyl) polymer; Hercules) of transplanted tumors, with minimal bone-marrow depression. Hepatic and renal damage were limiting toxicity at high dosage. This polyanion (pyran copolymer) which is of average weight, 18-30,000, is currently in clinical trial in man. Dosage in man is limited by its thrombocytopenic effect in similar fashion to polyethylenesulfonate. Cytoplasmic inclusions have been found in all the formed elements of the blood as well as in RES cells of liver and spleen. Newer pyran samples of improved therapeutic index have been developed and are in laboratory evaluation.

Clinically, pyran copolymer has been given to more than sixty patients. At higher dosage, 12 mg/kg/day, pyran induces fever and blocks the conversion of fibrinogen to fibrin but interestingly no patient has suffered hemorrhage despite significant prolongation of clotting time.

Pyran has potent interferon-inducing capacity and pretreatment prior to inoculation with Friend virus leukemia. Raucher leukemia inhibits subsequent tumor growth induced by these viruses. Similar inhibition of tumors has been found for acrylic acid and styrene - maleic anhydride copolymers.

In addition to the inhibitory action of divinyl ether - maleic
anhydride copolymers, such as pyran copolymer, a number of polycarboxylic polyanions have been synthesized by Butler which show
activity in screens of transplanted tumors. Similar compounds
(citraconic AEMA derivatives) with antitumor effect have also been
made by R. Guile.

Polyanions have been linked to antitumor agents and antibiotics to enhance therapeutic index but this will not be discussed
here.

Antibacterial Activity

The clinical usefulness of anionic dyes or polyanions for
systemic antifacterial activity has been rather limited. Clinically the action may reflect host response rather than any direct action of polyanions on bacteria itself, although antibacterial activity has been found to reside in acrylic acid which may be pertinent to the development of the polyacrylics as interferon inducers
in recent years.

Olefin - maleic anhydride copolymers have been found to possess antidiarrheal activity when given by mouth (19). It was found
that these polymers crosslinked with vinyl crotonate (14,19) possess the ability to remove water and adsorb endotoxin and entero-
toxin produced by organisms from the gastrointestinal tract. Detoxification of the toxins of Staphylococcus aureus, Clostridium
botulinum and salmonella was obtained in vitro and in vivo, as
well as the adsorption of polio virus Type I and in vitro bacteri-
cidal action against E. coli, B. substilis, Salmonella typhimurium,
several enterocci, and an antibiotic-resistant strain of staphylo-
cocci. These polyanions are virtually without oral toxicity and,
in addition, show inhibition of ECHO IX and other viruses. Their
impact on clinical medicine is awaited with interest and study of
their action in water purification systems is underway. We have
found in vivo activity for pyran copolymer which can completely
protect mice from pneumococcal, staphylococcal, and cryptococcal
infection on prior administration. These effects are host mediated
and permit the rational development of agents which can nonspecif-
ically protect against a variety of pathogens.

In similar fashion to synthetic polyanions, naturally or syn-
thetically derived polynucleotides possess both direct and indirect
action on bacteria as well as effects on host resistance. These
effects on bacterial growth by oligonucleotides may be via stimulation of nucleotide kinases involved in the synthesis of nucleic
acid.

In addition, polyanions can bind with polypeptides to effect
nutrilites, such as vitamin B-12, which renders them unavailable
for bacterial growth. However, the relationship of polyanions to
this or their related interreaction with growth-promoting polyamines
has not been adequately explored. There has been no systematic
search for polyanions in regard to their ineraction with active
polybasic polyamine antibiotics, even though heparin has been shown
to decrease the toxicity of polyene antibiotics, such as polymixin,
without affecting antibacterial action.

Antiviral Activity

In 1942, Cohen was able to concentrate tobacco mosaic virus
with anionic hydrophilic colloids such as heparin (31). Subsequently, Takemoto et al showed that the growth of influenza virus could
be affected by polymerized salts or substituted benzoic sulfonic
acids (32). A number of workers have explored this area and tobacco mosaic virus infection has been prevented by polyglutamate,
polyacrylate and polypectate polymers.

It is obvious that there are several actions for polyanions
in regard to their effect on virus growth and multiplication within
cells. Inhibition can represent physico-chemical interreaction or
a cellular effect.

The antiviral action of polyanions in some cases is a function
of electrostatic binding between the virus and the polyanion macromolecule. In other cases, which may relate to the above but are
more pertinent to the biologic antiviral activity of polyanions,
it blocks viral adsorption and can interfere with viral recycling
as long as it is present in the medium. The need for cells to be
present for the antiviral action to be manifested is consistent
with the observations in regard to interferon production. The
action of polyanions may also be mediated through inhibition of
enzymes important for viral adsorption. However, attempts to correlate antiviral action with inhibition of enzymes have not been
too successful. Of pertinence to this is the observation that
polyethylenesulfonate inhibits deoxyribonucleotide transferase in
vaccinia-infected cells.

Enzyme Inhibition and Activation

The concept of the mitotic inhibitory role of polyanions was
suggested by Heilbrunn (33) to be analogous to their role in anticoagulation. In this chapter there is no place for a discussion
of anticoagulant action other than to stress the fact that pyran
copolymer blocks in vivo and in vitro fibrin formation. Other

reviews have ably discussed the action of polyanions as anticoagulants and inhibitors of enzyme action (34,35).

Enzyme inhibition has been shown for acid phosphatase, alkaline phosphatase, β-glucuronidase, α-amylase, hyaluronidase, lysozyme, ribonuclease, deoxyribonuclease, trypsin, chymotrypsin, pepsin, lipoprotein lipase, fumarase, glyceraldehyde phosphate dehydrogenase, catalase, adenylic deaminase, elastase, and fibrinolysin. Additional enzymes include: phosphoprotein phosphatase, lecithinase, arginine esterase, and deoxyribonucleotide transferase, as well as alcohol dehydrogenase, pyruvate kinase, glutamine dehydrogenase, pyruvate kinase, glutamine dehydrogenase, glutathione reductase and a variety of NAD- and NADH-dependent enzymes.

Serum complement is critical to immune cytocidal activity. The anticomplement (involved in immunologic cell damage) activity of heparin and related polyanions which include pyran copolymer has been described which may reflect polyanionic inhibition of the esterases in this enzyme complex.

Surface and Enzyme Activity

The formation of solid enzyme complexes which possess different kinetic qualities from the parent dissociated enzymes has bearing on what may be occurring in cell membranes. These solid enzymes fixed to specific sites by bonding to polyanions might have specific function at given locations within the cell or on the cell surface. Displacement from binding would alter functional capacity of the cell and, possibly, produce cell death by acting beyond the confines of mormal regulation, as has been proposed for lysozymal enzymes. Of particular interest to polyanion enzyme interaction has been the synthesis of an ethylene - maleic anhydride copolymer - trypsin preparations. Trypsin was covalently bound to a water-insoluble polyelectrolyte gel (36,37) which has led to the commercial development of EMA-proteases (derived from B. subtilis) in the laundry-detergent market. The in vivo biologic potential for these agents has not yet been explored. However, if insoluble enzymes formed in vivo following the administration of AEMA or vinyl ether - maleic anhydride antitumor or antiviral copolymers, it may explain their biologic activity.

In regard to "solid enzymes" such as trypsin-EMA complexes, the charged carrier protein affects the distribution of charged low-molecular-weight substrates between the gel phase and the external solution. The pH activity profile of enzyme - polyanion complex can be displaced 2.5 pH units toward more alkaline pH values at low ionic strength, while at higher ionic strength pH optima shifted toward more acid values as compared to the parent enzyme. This was felt to result from the effect of the electrostatic potential of the polyelectrolyte carrier on the local concentration of hydrogen ions.

The work of E. Katchalsky's group at the Weizmann Institute and others who have formed solid enzyme complexes is, therefore, of great interest. The effect of polyanions on the enzyme complexes may shed light on the mechanism of action for the biologic action of polyanions. In this regard, polymers react much faster with small molecules than do their lower-molecular-weight monomeric analogs. For example, carboxyls from polycarboxylic acids displace bromine from α-bromoacetamide more quickly than do carboxyls from mono- or dicarboxylic acids (39). The effect of one neighboring reactive group on another is influenced by a polymeric configuration, and this action of polymers is similar to factors governing enzyme substrate affinity (40).

Polyethylenesulfonate can effect characteristic hydrolytic cleavage patterns for peptides and proteins. This may be due to the charge density of the polymer which maintains the position of the charged groups in relation to the substrate in the presence of a mineral acid.

This observation is supported in Siegel's review on the macromolecular environment as a factor in the control of chemical processes (41) wherein he examines the action of a wide variety of proteins and macromolecules, e.g. globulin, albumin, polysaccharides and polymers that enhance a number of enzymatic reactions. This is not restricted to water-dispersible macromolecules; for example, keratin, fibrin and cellulose all exhibit appreciable activating effects of a peroxidase-catalase reaction. The colloidal character of an environment can also enhance enzyme activity and electron transfer is facilitated by the ordered structure of synthetic poly(glutamic acid).

In regard to factors governing hydrolysis reactions, the stereochemical structure of polyacids is fundamental and difference in the internal structure of copolymers can govern this reactivity. In addition, equilibrium measurements have shown that isotactic poly(methacrylic acids) bind Ca^{++} ions much more strongly than the atactic form. This means that differences in the shape and internal organization of polymers can play an important role in directing the rates of particular reactions (42). Further work needs to be done in this area.

Polymers of critical size interacting with cell surfaces may provoke lysis in similar fashion to antigen antibody activation or complement lysis of the red cell. This has been seen for polylysine agglutination and hemolysis of red cells and described by Charache

et al (43) for silicate polymers both with and without associated
red cell agglutination. Surface alterations are seen in that synthetic polypeptides which contain 50% more glutamic acid over lysine
residues increase the rate at which red cells take up and release
oxygen.

In addition to effects on enzyme reaction, polyanions can enhance antigenicity if linked to protein, or block the action of
antibody depending on its position in relation to the active sites
of antibody or antigenic determinants. Pyran copolymer has been
found to potentiate the action of a hoof-and-mouth vaccine when
given in combination with the vaccine. This protective effect is
at concentrations of pyran copolymer that are ineffective by themselves in modifying host resistance. This may be a major breakthrough in our understanding and utilization of polyanions as host
defense stimulators.

The complexing of proteins with polyelectrolytes has been regarded as an extension of the use of polyvalent ions for protein
precipitation. The stability of the complexes formed may be related to the flexibility of the polymer chain. This property permits
the use of synthetic polyelectrolytes for protein separation. Other
physical effects of polyanions that are of biologic importance
might depend on the presence of dipoles in polyanionic macromolecules
which could cause directed movements perpendicular to biological
membranes. A. Katchalsky (44) has suggested that these dipoles may
act as organizing elements which could bring biocolloids together
in orderly array. In support of this, significant electrical potential is developed by the displacement of hyaluronic acid and
piezo and photoelectric effects are obtained from bone crystal and
electric current has been found to direct the orientation of organization of bone growth.

By proper use of high polymer systems, A. Katchalsky (44)
demonstrated that chemical energy can be transferred into mechanical energy and Kuhn (45) has shown that filaments made of poly-
(acrylic acid) and poly(vinyl alcohol) can produce mechanical energy equal to that of muscle fibers. Contraction occurs on the
addition of acid and relaxation on the addition of alkali. This
could be pertinent to changes in the physical state of protoplasm
and, if polyanions could adhere to reactive groups on the cell
surface or enter within the cell and these be subjected to mechanical changes, this could result in physical alteration or disruption of the cell or cellular elements. Furthermore, the exclusion
of polymer groups at a cell surface could impede the entrance of
essential nutrients or block viral fixation to or release from the
cell surface.

Other activities for polyanions relate to their effects on ATP, their calcium-binding activity and their effects on clotting proteins. It is important to stress that polyanions have remarkable effects on the cell surface and alter the streaming of cellular protoplasm which may be important to the entry and replication of virus within the cells.

Colloidal Effects

Hydrolyzed ethylene - maleic anhydride copolymers and related polyacrylic polycarboxylic antitumor and antiviral agents have been found to improve the quality of clay and silty loam soils ("soil conditioning"). This type of effect occurring with rapidity in the circulation could accound for changes in the distribution of formed elements in the blood accompaning the administration of certain macromolecules which has been observed in vivo for the action of polyanions as relates to the production of shock and tumor necrosis.

There is evidence that tumor tissue utilizes whole protein molecules and alteration in the isoelectric point of nutritional protein by polyanions has been thought to be inhibitory to cell growth. In regard to the uptake of protein by cells, both polyanions and polycations, protamine and basic amino acids can increase phagocytosis and this relationship between polyanions and polycations may control incorporation of particulates, especially virus particles, within the cell.

POLYCATIONS

Polycations represent a similar group of charged polymers with amazingly similar properties in regard to biologic effect and toxicity as compared with the polyanions, but they have more potent neurotoxic and renal toxic effects.

Positively charged polypeptides inhibit or activate a number of enzyme systems, including pepsin, ribonuclease, and polynucleotide phosphorylase (46). These agents have antibacterial and bacteriostatic properties against both gram negative and gram positive organisms (e.g. polymixin, colymycin). They inhibit virus-replicating activity in vitro possibly through aggregating activity. One major in vivo problem relates to red cell agglutinating properties with associated vascular occlusive effects. Paradoxically they can neutralize the anticoagulant activity of heparin-like polyanions and can on their own through effects on clotting proteins produce anticoagulation.

In some cases, polycations may be of importance to bacterial and viral systems but their exact functional role is not clear. They may have growth-regulating activity and related polyamines such as arginine, spermine, spermidine, and putracine, while of biologic interest in relation to viral and bacterial effects, are too toxic to be of clinical interest. Protamine and histone proteins while cationic are of importance to sperm and gene function. Of interest in this regard, polylysine was developed in the laboratory before it was found in the chicken oviduct where it may play a role in preserving sperm function for sustained periods. Certain polycations are potent histamine-releasing agents.

Most recently, polyethylenimine, polyvinylamine and polypropylenimine fractions have been developed which inhibit transplanted tumors in vivo and in vitro. Effects are seen even against established transplanted tumors (47).

Polycations, such as polylysine or DEAE-dextran, can enhance the antiviral activity of polynucleotide interferon inducers by protecting them from the degradative action of ribonucleases or effecting their entry into the cell.

NEUTRAL POLYMERS

Polymers designed specifically to resemble albumin or globulin are in this group. The role of polymers such as polyvinylpyrrolidone, dextran and mannitol which are used clinically as plasma expanders, drug vehicles, and antitoxins has been thoroughly reviewed with particular reference to polyvinylpyrrolidone. The reader is referred to the annotated bibliographies of General Aniline & Film Co. of 1951, and of 1951-1966. Some of these references deal also with dextran and other related polymers. Polyvinylpyrrolidone and polyethylene glycols have been used as antifreeze stabilizing agents in the preservation of tissue, whole blood and blood fractions.

The potential for these high-molecular-weight polymers as friction-reducing agents in circulatory disease has not been adequately explored although mannitol and different dextran fractions can improve blood flow and tissue oxygenation and provide an osmotic stimulus to kidney function and effectively treat cerebral edema. Certain neutral polymers have antiinflammatory activity, and also possess some measure of antiviral effect. However, their main pharmacologic role has been as suspending agents for drug dispersion.

SUMMATION

While we cannot cover all the biologic activity for water-soluble polymers in this chapter, we hope what is discussed here will stimulate the interest of those working with polymers to exploit their experience with these compounds in other areas.

The author wishes to acknowledge the help of A. E. Munson, Ph.D. and P. Morahan, Ph.D. in relation to the laboratory work reported and B. Boorjian and S. Caldwell for their assistance in preparing this manuscript. This work was accomplished with the help of NCI 10537 and 10572.

BIBLIOGRAPHY

1. W. Braun and W. Fershein, Bacteriol. Rev., 31, 83 (1967).

2. W. Braun, W. Regelson, Y. Yajima, and M. Ishizuka, Proc. Soc. Exp. Biol. Med., 133, 171-175 (1970).

3. J. L. Brash and D. J. Lyman, J. Biomed. Mater. Res., 3, 175-189 (1969).

4. P. Bernfeld, J. S. Nisselbaum, B. J. Borkley, and R. W. Hanson, J. Biol. Chem., 235, 2852 (1960).

5. W. C. Hueper, Cancer, 10, 8-18 (1957).

6. C. Couinaud, Herve, Biotois, and Gioan, La Presse Medicale, 78, 1839-1841 (1970).

7. M. J. Ashwood-Smith, Nature, 1, 1304 (1971).

8. T. Gillman, Acta Haematol., 15, 364 (1956).

9. H. Takita, Polyglycolic Acid Suture in Thoracic Surgery, N. Y. State J. of Med., 70, 2991-2992 (1970).

10. A. Heyd, J. Pharmaceutical Sciences, 59, 1526-1527 (1970).

11. A. Heyd, J. Pharmaceutical Sciences, 60, 1343-1345 (1971).

12. A. Heyd, D. O. Kildsig, and G. S. Banker, J. Pharm. Sci., 59, 947-949 (1970).

13. M. L. Beasley, Personal Communication.

14. R. H. Tust and T. M. Lin, Proc. Soc. Exp. Biol. & Med., 135, 72-76 (1970).

15. W. Regelson, L'Interferon, 6, 353-379 (1970).

16. W. Regelson, Advances in Cancer Res., 11, 223-287 (1968).

17. W. Regelson, Hematologic Reviews, 1, 193 (1968).

18. R. F. Beers and W. Braun, Biologic Effects of Polynucleotides, Springer-Verlag, New York, 1971.

19. J. F. Nash and T. M. Lin (to Eli Lilly & Company), U. S. Pat. 3,224,941 (1965).

20. W. Regelson, B. I. Shnider, J. Colsky, and K. B. Olson, Initial Clinical Study of Pyran Copolymers (NSC 46015), in press.

21. W. Regelson, Adv. Exp. Med. Biol., 1, 315-332 (1967).

22. W. Regelson, A. E. Munson, W. R. Wooles, W. Lawrence, Jr., and H. Levy, L'Interferon, 6, 381-395 (1970).

23. A. E. Munson and W. Regelson, Proc. Soc. Exp. Biol. & Med., 137, (2), 553-557 (1971).

24. Y. Shamash and B. Alexander, Biochim. Biophys. Acta, 194, 449-461 (1969).

25. E. G. Elias, F. Sepulveda, and I. B. Mink, in press.

26. T. Sall, S. Mudd, and A. Takagi, J. Bacteriol., 76, 640 (1958).

27. G. Miller, L. Berlowitz, and W. Regelson, Chromasoma, 32, 251-261 (1971).

28. G. Miller, L. Berlowitz, and W. Regelson, Exp. Cell Res., 71, 409-421 (1972).

29. G. Miller, Personal Communication.

30. C. Muehlbaecher, J. Straumfjord, Jr., J. F. Hummel, and W. Regelson, Cancer Res., 19, 907 (1959).

31. S. S. Cohen, J. Biol. Chem., 144, 353 (1942).

32. K. K. Takemoto, M. L. Robbins, and P. K. Smith, J. Immunology, 72, 139 (1954).

33. L. V. Heilbrunn and W. L. Wilson, Proc. Soc. Exp., Biol. Med., 70, 179 (1948).

34. H. Engelberg, Heparin: Metabolism, Physiology, and Clinical Application, Thomas, Springfield, Illinois, 1963, p. 111.

35. P. Bernfeld, in R. M. Hochster and J. H. Quastel, Eds., Metabolic Inhibitors, Academic Press, New York, Vol. 2, 1963, p. 437.

36. Y. Levin, M. Pecht, L. Goldstein, and E. Katchalsky, Biochemistry, 3, 1905 (1964).

37. L. Goldstein, Y. Levin, and E. Katchalsky, Biochemistry, 3, 1913 (1964).

38. E. Katchalsky, Rept. Sci. Activities, Biophys. Dept., Weizmann Institute, Rehovoth, Israel, 1962-1963.

39. H. Ladenheim, E. M. Loebl, and H. Morawetz, J. Am. Chem. Soc., 81, 20 (1959).

40. H. Morawetz and E. W. Westhead, Jr., J. Polymer Sci., 16, 273 (1955).

41. M. Siegel, Ann. Histochim, 8, 133 (1963).

42. G. Smets, Angew. Chem., 1, 306 (1962).

43. P. Charache, C. M. Macleod, and P. White, J. Gen. Physiol., 45, 1117 (1962).

44. A. Katchalsky, Biophys. J., 4, 9 (1964).

45. W. Kuhn, Z. Angew. Phys., 4, 108 (1952).

46. M. I. Dolin, J. Biol. Chem., 237, 1626-1633 (1962).

47. H. Moroson, 1972.

NEW DEVELOPMENTS IN WATER-SOLUBLE POLYSACCHARIDES

Roy L. Whistler

Department of Biochemistry, Purdue University

Lafayette, Indiana

Natural water-soluble polysaccharides usually occur as food reserves, as components of plant fluids, as exudate gums or, infrequently, as components of hemicelluloses (1,2). A great many naturally occurring polysaccharides are dispersible in water and some dispersions produce gels, either directly or indirectly. Often these gels fulfill a significant function in their biological source. For example, polysaccharide gels are important in ocular fluids, in synovial fluids, in protecting microorganisms or injured plant surfaces from desiccation and, as mucilaginous gels, in acting as emollients on membrane surfaces of animals. A lesser number, but the bulk of polysaccharides, are water insoluble and are not dispersible. The principal one of these is cellulose, but the most abundant group are the plant hemicelluloses. The water-soluble and water-dispersible polysaccharides are used in various food and nonfood products. Because of the growing industrial need for water-soluble and water-dispersible polysaccharides, their number and variety are increased by chemically modifying natural polysaccharides that are less soluble or even insoluble; or, water-soluble polysaccharides are created through the biosynthetic use of microorganisms or enzymes. A new group of water-soluble molecules which may find application in the future are manmade glycoproteins.

NATURE OF POLYSACCHARIDE SOLUBILITY

Water solubility of polysaccharides is due to structural characteristics. Water solubility in naturally occurring polysaccharides is due, usually, to the presence of highly ionized anionic groups, such as sulfate or carboxyl, the presence of 1→6

179

linkages, or the presence of a branched structure. Another solubilizing feature seen in some natural polysaccharides is nonuniformity in a linear polymer produced by alteration of types of
glycosidic linkages or by the presence of different kinds of sugar
units, as in a heteroglycan. Thus, any kind of nonuniformity
which lowers the possibility of fit between chains, lessens the
energy with which molecular segments may bind to each other and
contributes to solubility and solution stability. Homogeneous
uniform linear polysaccharides are insoluble except for those 1→6
linked glycans that are exceptionally flexible by the presence of
the torsional C-5—C-6 bond and also by the greater distance that
can exist between sugar rings, which lessens torsional steric
effects (see Figure 1).

Branching reduces the intermolecular association of polymers.
However, there is a distinct difference in the effect of degree
of branching and in the length of branches on properties other
than solubility. A linear polysaccharide always produces a more
viscous solution than a branched molecule of equal molecular
weight and a molecule's contribution to solution viscosity can
decrease exponentially as the degree of branching increases. At
the same time other physical properties change with degree of
branching. Since contribution to viscosity is almost always one
useful property of an industrial polysaccharide, reduction in viscosity contribution is not a desirable effect of branching. Con-

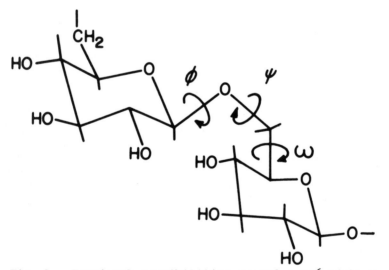

Fig. 1. Torsional possibilities around a 1→6 linkage.

sequently, the most valuable water-soluble polysaccharides are
linear polysaccharides on which only sufficient side chains are
present to effectively reduce intermolecular fit. Natural guaran
from guar seeds, or natural locust bean gum are examples of such
molecules. Both gums are industrially important although guar gum
usage is growing more rapidly because guar is an annual crop that
is easily handled by agriculturists.

NEW INDUSTRIAL POLYSACCHARIDES

The broad list of polysaccharides now used in food and nonfood applications along with a detailed survey of their manufacture, prices, chemistry and applications has been recently published (1). It is not necessary to review these polysaccharides here
but rather to present some thoughts on the direction of current
efforts to produce new gums and on the source of new gums that may
be a part of industrial production in future years.

Because of the low cost of starch ($0.06-0.09 per pound) and
cellulose ($0.09-0.14 per pound), these polysaccharides will con-
tinue to serve as basic structures for modification to produce
water-soluble or water-dispersable gums. In addition, the increase
in guar production may permit its price to decrease and lead to a
greater use of guar derivatives.

A variety of polysaccharides can be made by microorganisms,
and high-yielding mutant strains provided with low-cost energy
sources could produce fermentation gums at low costs. A host of
fermentation gums are available and more will be found. It is my
expectation that fermentation gums will eventually develop to fulfill many industrial needs and costs will decrease as volume of
production rises.

Incidental to the development of fermentation gums will appear
techniques for producing gums by microbial enzyme systems. Initially homoglycans consisting of but one type of sugar unit will
be made, as exemplified in the production of dextran (3). However,
enzyme-catalyzed syntheses of polysaccharides will eventuate. This
latter technique of enzyme modification of existing low-cost polysaccharides is rapidly maturing.

Some current efforts in the synthesis and utilization of microbial polysaccharides are described in A. Jeanes' chapter in
this book.

Trends in Polysaccharide Development and Modification

Environmental cleanliness requires purer industrial effluent
water and places a demand for better low-cost coagulants and pre-
cipitants (see the chapter by H. G. Flock and E. G. Rausch). One
of the newer reagents for coagulation and precipitation is cationic starch with a high degree of substitution (DS). Whereas cationic starch of low DS has been extensively used in the paper industry, the higher derivatized starch is finding wider application
for treatment of a range of waste waters. The starch may be made
cationic by substitution with groups containing amines, but the
most effective derivatives have the amines quaternized (see also
the chapter by G. F. Fanta, R. C. Burr, W. M. Doane, and C. R.
Russell).

Anionic starches in the form of starch monophosphates or
half-esters of dicarboxylic acids, such as succinic, have found a
growing use both because of their ionic nature and because of their
viscous properties. In these instances the DS is low, and the
products are excellent examples of the large alteration in properties caused by introduction of a low level of ionic substitution
into an otherwise soluble but neutral polysaccharide. Various
means for introducing acidic groups into starch have been examined.
It is perhaps unfortunate that, as yet, the preferential oxidation
of the C-6 carbon atom to produce a uronic acid group has not been
accomplished without extensive, undesirable depolymerization. A
product of considerable economic importance would result if this
oxidation could be appropriately controlled. Carboxymethylcellulose, although not a new derivative,has been greatly improved in
recent years and as a consequence is finding increasing applications in both the food and nonfood areas. Its use in oil recovery
is described in the chapter by R. H. Friedman.

Emulsification with polymeric materials offers advantages
over the use of low-molecular-weight emulsifiers. Hence polysaccharides capable of producing needed surface effects are in growing demand. Carrageenan and hydroxypropyl alginate have emulsifying properties, and starch and cellulose derivatives are often
applicable; but work is needed to improve further the emulsifying
action of polysaccharides. Starch esters and perhaps starch ethers
have potential in this area.

As already indicated, guar has grown extensively in use and
its amenability to standard agricultural production as a normal
annual crop will surely lead to its becoming a permanent commodity.
Furthermore, since no extensive genetic work has been directed to
improving its yield per acre and its adaptation to temperate cli-
mates, very significant yield increases can be expected if appropriate agronomic effort is applied. Larger yields would result

in a lower price for guar gum, or at least would prevent price increases.

The guaran molecule (Figure 2) has many excellent properties. It may be viewed as a linear chain, of 1→4-linked β-D-mannopyranosyl units, which is derivatized uniformly in that every second chain unit has a single α-D-galactopyranosyl unit attached at the C-6 position. The otherwise insoluble mannan chain is made soluble by substitutent D-galactosyl units because they reduce the possibility of close fit between chain segments. As a consequi, the polysaccharide is soluble and forms stable solutions. Morever, the molecule is still long and, hence, its solutions are highly viscous. To these valuable characteristics must be added the neutral nature of the polymer. This feature permits guar gum to be compatible with a variety of other compounds which it may come in contact in its applications. It has the ability to bind to cellulose fibers, making it an excellent wet-end paper additive, and its adherence to certain crystal surfaces gives it usefulness in the flotation separation of certain ores and minerals. A further advantage is that, being neutral, its properties can be significantly altered by relatively minor affixation of charges. Thus, either anionic or cationic substituents at relative low DS produce large changes in physical characteristics, which is similar to the effects observed when starch is derivatized. This possibility of controlling properties by charge affixation presages the development of new guar derivatives. The D-galactosyl side chains also offer sites for specific modification that are not possessed by most other commercial polysaccharides. For example, the enzyme galactose oxidase converts the primary carbon at C-6 of the D-galactosyl units to an aldehyde. The polymer then becomes a good agent for improving the wet strength of paper owing to cross-linking with formation of acetal linkages. Other specific modifications are under investigation which should further enhance the industrial applications of guar gum.

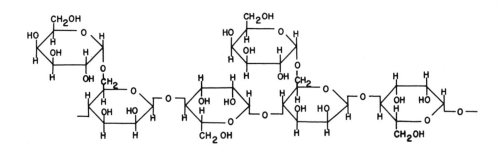

Fig. 2. Structure of guaran.

Modification of cellulose by neutral substituents, such as
methyl and more particularly hydroxyethyl and hydroxypropyl, renders the polymer water soluble; the viscosity depends on the chain
length of the resulting cellulose derivative. The newest members
of the family are hydropropylcellulose and the mixed hydroxy-
ethylhydroxypropyl derivatives. These have greatly expanded the
range of properties and hence the application of neutral water-soluble cellulose derivatives.

Biosynthetic Polysaccharides

Biosynthetic polysaccharides are just attaining industrial
growth. The first of these is xanthan gum obtained as an extracellular polysaccharide from Xanthomomas campestris. This gum is
attaining wide industrial application and has been given govern-
mental clearance for use in foods. One can be certain that other
fermentation gums will be produced commercially. Scleroglucan
from Sclerotium glucanicum has been examined extensively by the
Pillsbury Company. This gum consists of a linear chain of 1→3-link-
ed β-D-glucopyranosyl units with a β-D-glucopyranosyl side-unit
joined 1→6 to every third main chain unit.

Fermentation polysaccharides have an increasing industrial
usefulness because of the variety of structures that can be produced and because large-scale production can be achieved at relatively low cost.

Together with the development of fermentation polysaccharides
there is a growing interest in the direct enzymatic production of
polysaccharides with specific structures. The initial phases of
development make use of transferase enzyme systems. Later it is
conceivable that polymers will be made enzymatically by stepwise
conversion of sugars to 1-phosphates, extending this intermediate
to such energy-rich molecules as uridine diphosphosugars, followed
by transferase action to place the sugar unit on a growing polysaccharide acceptor chain.

Finally, this discussion of new developments must not be concluded without suggesting that a beginning has been made in the
development of an entirely new field of water-soluble polymers -
the glycoproteins. The glycosidic joining of sugar or oligosaccharide units to a protein chain offers the possibility of producing water-soluble products with a host of new properties. It
is certain that the future will see the development of such polymers for new end uses.

BIBLIOGRAPHY

1. R. L. Whistler, Industrial Gums, Academic Press, New York, 1973.

2. R. L. Whistler, "Polysaccharides" in N. M. Bikales, Ed., Encyclopedia of Polymer Science and Technology, Interscience Publishers, New York, Vol. 11, 1969, pp. 396-424.

3. A. Jeanes, "Dextran" in op cit, Vol. 4, 1966, pp. 805-824.

WATER-BASED FINISHES IN RESPONSE TO ENVIRONMENTAL CONSTRAINTS

Thomas J. Miranda

Whirlpool Corporation

Benton Harbor, Michigan

This chapter is concerned with the application of water-soluble polymers in the coating of appliances in response to environmental constraints. Coatings requirements for appliances range from the severe environments of the washer and diswasher to less severe conditions of the dryer and the refrigerator. As a result, the coating industry has developed finishes like the epoxies for primers and the thermosetting acrylics for topcoats (1). These coatings have proved their value in the appliance industry in providing durability, color stability, corrosion and detergent resistance as well as long-term decorative effects.

With the current emphasis on ecological problems, the coating process has come under close scrutiny, particularly where solvent is released to the atmosphere during the application and baking operations. In this chapter, methods for reducing pollution through a materials approach and the role of water-soluble coating polymers will be reviewed. The preparation and properties of water-soluble polymers has been adequately described in the literature (2,3) and, therefore, will not be extensively covered.

TYPICAL POLYMERS

Typical polymers used in water-based coatings may be either addition or condensation types. The polymers are modified in such a way that hydrophilic sites are introduced in the backbones or side chains to impart water-soluble character. The following examples are representative.

Addition Type (4). A polymer is prepared by addition of the following monomer - catalyst mixture to 55.5 parts of refluxing isopropanol:

Acrylic acid	3.55 parts
N-Methylolacrylamide	7.53
Ethyl acrylate	16.6
Methyl methacrylate	16.6
Caprylyl peroxide	0.22

After sixteen hours the mixture is thinned with water containing ammonium hydroxide and the solution azeotroped until a 50% solids isopropanol-water solution of the polymer results. The hydrophilic monomers acrylic acid and N-methylolacrylamide impart water solubility and serve as crosslinking sites for the polymer. The ethyl and methyl ester groups, on the other hand, are hydrophobic.

Condensation Type (5). An air-drying or baking condensation polymer is prepared by adding to a flask and refluxing for one hour the following:

Trimethylolpropane	49.4 parts
Succinic anhydride	36.8
Toluene	100

Then the following are added and heated, and water is removed until an acid number of 54.6 is reached:

Phthalic anhydride	110.5 parts
Trimethylolethane	71.0
Dehydrated castor fatty acids	138.8

The mixture is then reduced to 51% solids using the following mixture:

Butoxyethanol	38.5 parts
n-Propanol	115.5
Ammonium hydroxide	15.5
Potassium hydroxide	14.2
Water	542.0

Water solubility is imparted by the intermediate which also serves as a crosslinking site.

$$H_3C-CH_2-\underset{\underset{CH_2OH}{|}}{\overset{\overset{CH_2OH}{|}}{C}}-CH_2-O-\overset{\overset{O}{\|}}{C}-CH_2CH_2-COOH$$

Hydrophilic

BACKGROUND OF DEVELOPMENT

Several years ago, Whirlpool Corporation's management anticipated the developing ecological situation and took steps to reduce
pollution so that ecological standards could be met or exceeded.
Of immediate concern was that of producing a number of different
appliances in several locations using facilities and processes ger-
mane to each appliance. For example, the washing machine and dish-
washer are subjected to severe environments of hot water, detergents,
oxidizing agents, and humidity. These appliances demand primers
that have excellent detergent resistance. To satisfy these needs,
durable epoxy-modified thermosetting acrylics are employed, being
applied by the flow-coat processes. Topcoats are thermosetting
acrylics applied by electrostatic spray using bells or discs. Other
appliances such as refrigerators, dryers, freezers, and furnaces
employ coatings having varying degrees of performance characteristics commensurate with each appliance.

Because of the variety of application methods, types and age
of equipment, cost of replacement, energy availability, local,
state and federal codes and differences in plant location, no single
set of criteria would be appropriate for all our facilities. The
immediate goal, therefore, was to reduce the amount of organic solvent emission to 80% within two years of initiation of the program.
A target figure was obtained after determining the amount of solvent
emission per facility per day. The plan then called for reducing
emissions beyond the 80% level as the program advanced. A typical
primer solvent make-up may be as follows:

Solvent	%
Mineral spirits	81
Butanol	5.69
Xylene	5.61
Heavy aromatics	4.04
SC-100	3.09
Isopropanol	0.57

Another consideration was the possibility of local or state
enforcement. This would require a need to have a system ready for
immediate uses, which at that time was not available.

Compliance Program

As a first step, a close relationship was developed between
our research center and the research centers of leading coating
suppliers as well as between our engineering groups and divisional
personnel (Figure 1). Options were then reviewed and analyzed in

Fig. 1. Compliance scheme.

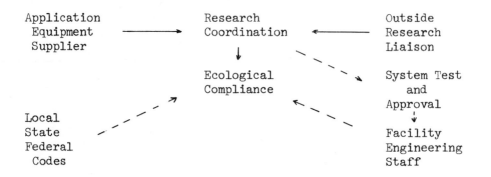

order to take advantage of available methods for reducing solvent emissions. Obviously, a number of approaches would be suitable and these were evaluated in detail.

In the first analysis suppliers suggested a number of potential solutions. The best thinking was heavily aligned toward water-based coatings but these were not sufficiently developed, particularly with regard to application techniques and detergent resistance. However, the water-based approach appears to provide the best immediate and long-term solution to the ecological problem.

Analysis of Pollution-Complaince Approaches

In Table I a summary is given of the processes considered and of their advantages and disadvantages. These will be considered in detail.

Incineration. This method for reducing effluents has been used to reduce hydrocarbon levels by combusion. The initial investment in this equipment (up to $500,000) would be high, require an increase in fuel consumption, and involve a maintenance function of the order of $30,000 per year. While a portion of the energy can be recovered for preheating purposes and permit the use of present finishes, this method was not deemed acceptable as a long-term solution.

Exempt Solvents. Exempt solvents have been used extensively in California as a means of reducing photochemical oxidants under Rule 66 legislation. Exempt solvents are defined as solvents that are not photochemically reactive. Compounds such as aromatic hydrocarbons, branched aldehydes, ketones and branched alcohols, and certain unsaturated and halogenated hydrocarbons are considered nonexempt. Xylene is a nonexempt solvent but can be used in coating

TABLE I. Potential Pollution-Reduction Methods

Process	Advantages	Disadvantages
Incineration	Reduce emission; use convential solvents	High capital cost; energy availability; maintenance cost
Exempt Solvents	Comply with Rule 66	Pollution by exempt solvents; reformulation required; increased cost; performance may be affected
100% Solids	Little or no emission; application knowledge valuable for long-term needs	Application and performance criteria in developmental stage
Powder Coatings	Low pollution, no solvent	Cost of powder; color match and shading; U. S. Technology behind Europe
Radiation Curing; Electron-Beam Curing UV Curing	Low solvent emission; room-temperature cure	High cost of equipment (E.B.); flat surface requirement
Nonaqueous Dispersions	Exempt solvents; high performance potential	Pollution by exempt solvent
Precoat Steel	No pollution	Requires modification of present mfg. procedure; scrap problem; touch-up problem
Aqueous Coatings	Low solvent emission; flow and leveling of solvent systems; high latitude of polymer design; presently used in electrocoating	Application

TABLE II.   Comparison of Solvent Content of Exempt and Nonexempt
            Appliance Topcoat

Nonexempt %	Solvent	Exempt %
41.34	Xylene	8.0
18.32	Butanol	16.2
12.96	Hi-Sol 85 aromatic	--
7.82	Cellosolve acetate	7.2
6.70	Methyl isobutyl ketone	--
5.32	Methyl ethyl ketone	--
0.84	Toluene	11.2
--	Butyl acetate	12.5
--	Isopropanol	2.9
	Methylamyl acetate	39.7
	Butyl Cellosolve acetate	2.3

formulations up to 8%. A comparison of a typical solvent make-up
of a conventional appliance finish topcoat and an exempt-solvent
system is shown in Table II.

Exempt solvents can still be a source of pollution even though
the photochemical oxidant content can be greatly reduced. However,
the cost of reformulation and retesting of the exempt system made
this approach unacceptable for our purposes.

100%-Solids Systems. The concept embraced by 100%-solids
systems refers to liquid systems that are devoid of solvent and can
be completely converted evolving little or no byproduct in the curing step.

The recognition of the need for solventless coatings had been
anticipated earlier by those active in acrolein chemistry (6). An
interesting series of compounds were described by Hochberg (7) in
which dioxolanes formed by reacting vicinal diols and acrolein
would yield a polymer having a reactive vinyl dioxolane which could
polymerize by autoxidation of the unsaturated function, i.e.,

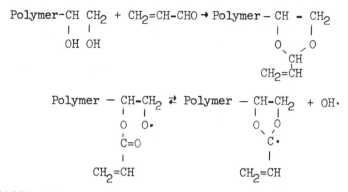

The difficulty involved with this system was the unpredictability of initiating the allyl-type double bond.

The advantage for this system is that little or no emission would be experienced. (Some emission may occur if volatile by-products are eliminated during crosslinking, i.e., CH_2O, H_2O, $RCHO$, etc.) Another advantage is that knowledge gained in the application of such a system would be of good value to other technologies, such as radiation curing or powder coating.

The principal disadvantage lies in the application problems, particularly in terms of using presently installed equipment with little or no modifications. In developing approaches such as this, a close working relationship must be developed between the research centers of the coating supplier, user, and equipment manufacturer as well as the finish engineers. The total systems approach must be considered to bring about successful online application of 100%-solids coatings. In this work 85% solids were permitted in order to work out the engineering problems while evaluating the performance parameters of the polymer system used in the finish. Solids contents were then raised to 95% with equally successful results. Complete removal of the solvent is an anticipated future development.

Powder Coatings. A considerable amount of attention has been devoted to powder coating as a means of reducing pollution. While powder coating is not new, its application is being treated as though it were the latest breakthrough in the coating industry. This is evident by the increase in articles and conferences dealing with the subject (8,9). When the emotionalism subsides, powders should find their place in the coating industry but will by no means become a panacea. The potential for powder applications in the appliance industry has been reviewed recently (10).

Powder coating does present a route to low-emission coatings and as a potential replacement for porcelain enamels in applications such as crisper pans or washer tops, or to replace plating on wire racks. The present emphasis on powder coating is toward the epoxy types for durability, single-coat application, and corrosion resistance. One disadvantage of the epoxy resin is its inherent tendency to chalk; because of this, coating suppliers are developing polyesters and thermosetting acrylics. Several high-performance polyesters are now available (11). For application in appliances all three types mentioned above will be used, the epoxies in shelving and interior parts and the acrylics and polyesters on certain external parts.

Two factors limit the general application of powders to the exterior of appliances, namely color matching and shading. Because of poor flow, attempts to shade appliances with powders have been unsuccessful to date. With the solution of the flow and leveling problem of powders, shading may become an established practice.

Radiation Curing. The introduction of room-temperature curing by radiation has attracted considerable interest in the coating industry in recent years. Radiation curing affords a 100%-solids approach in that the monomeric solvents become an integral part of the coating upon polymerization. Two approaches are suggested; electron beam (12) and ultraviolet curing (13).

Polymers used for electron-beam curing have been described by Burlant (14), Miranda (15), and Hoffman (16), Nordstrom (17) and others and consist of polyesters, acrylics, polyurethanes, polymeric thiourethanes, and modified epoxy polymers. The key to the polymerization is the presence of a conjugated double bond derived from a maleate or an acrylic monomer.

A difference between electron-beam and ultraviolet curing is that the latter requires an activator such as benzophenone. The advantages and disadvantages of radiation curing have been adequately described in the literature (12). Electron-beam curing does not appear to have a prominent position in the coating of appliances. The high capital cost, flat surface requirement,'solvent emission', ozone generation, and nitrogen blanket requirement would preclude its application at this time.

Ultraviolet curing is not in an advanced state which would permit its broad application in appliance finishes to reduce pollution. As a matter of fact, the 'solvents' used in ultraviolet and electron-beam curing themselves present a problem due to evaporation and would require some type of removal. Another problem with ultraviolet curing is the difficulty of using pigmented systems such that cure of thick films up to 2 mils could be achieved.

It should be pointed out, however, that ultraviolet curing does have potential in appliance-finish applications and is currently being used in black coatings for compressors and for inks in the appliance industry.

Nonaqueous Dispersions (NAD). A special class of acrylic polymers similar to aqueous emulsions but having an organic nonsolvent as the continuous phase are the nonaqueous dispersions (18). These polymers exhibit low viscosities, having a milky appearance and can be prepared at fairly high solids contents. A comparison of the properties of a typical emulsion and a nonaqueous dispersion is shown below:

	Emulsion	NAD
Solids, %	50.55	50-55
Particle size, μ	0.1	0.1
Appearance	White	White
Viscosity, cP	1000	40
Continuous phase	Water	VM & P Naphtha

Film formation occurs by air drying or by baking to yield coatings having excellent hardness and gloss. Prior to the strong emphasis on environmental pollution, nonaqueous dispersions were probably one of the strong contenders for new finishes. In localities where Rule 66 type legislation is acceptable, they would provide a good finish system. However, for the ultimate in solvent-emission control, they also suffer from the same disadvantages cited for exempt solvents.

Precoated Steel. This approach leaves the pollution problem with the steel supplier and could be of some value. Several problems negate the use of precoated steel. These include manufacturing technique, scrap steel disposal, methods of fastening wrapper, edge corrosion, and repair.

Water-Based Finishes. Water-based finishes were selected as having high potential in meeting the goal for pollution compliance. Successful experiences have been achieved using anodic electrocoating for washer cabinets and for refrigerator compressors. The knowledge gained here would have good carryover to other priming operations. Of particular advantage is the low cosolvent content of electrocoat systems as well as the low emissions level of the deposited finish. Furthermore, water-based systems afford the opportunity for a broad base of polymer types including acrylic, epoxies, polyesters, polyurethanes, vinyls, silicones, etc.

Water-based systems for electrocoating may be the solution
type, epoxy, alkyd, polyester, acrylic or Diels-Alder adduct types.
In all cases, the carboxyl-terminated systems are base neutralized
and solubilized with cosolvents. Emulsion types are also available
and have been used particularly where good salt-spray resistance is
required. Care, however, must be exercised to assure dilution and
mechanical stability of these types of water-dispersed polymers.

Electrocoating, therefore, is seen as a key process for obtaining pollution compliance. The advantages and disadvantages of this
process have adequately been described in the literature (19,20)
and are discussed further in the chapter by G. E. F. Brewer. For
the long term, efforts should be directed toward reducing or completely eliminating the cosolvent and removal of other contaminants
to the water shed (amines). Ultrafiltration (21) has greatly improved the efficiency of the electrocoating process.

A commercial development that was recently introduced to the
appliance industry is cationic electrodeposition (22,23). Improved
salt-spray resistance (up to 1,400 hrs.), simple bath control,
elimination of dissolution of phosphate and iron on the substrate,
and single-coat potential are some of the advantages.

Other applications of cosolvent-solubilized water-soluble
coatings are in the development of nonfoaming flow-coat primers.
These coatings contain 20% or less of organic solvent and exhibit
good flow leveling and corrosion resistance. Figure 2 illustrates
the salt-spray resistance of a flow-coat primer. The coating on
the left shows 3/32 inch rust creepage on the surface while that
on the right almost no creepage after 500 hours of salt spray. The
central panel is the panel before testing. An early concern was
the ability to achieve adequate detergent resistance for washing
machine primers. The coatings cited here are more than adequate
in detergent resistance. Coupled with this development is the
need for water-based topcoats that will fulfill the various speci-
fication grades for appliances. Success in achieving this objective through cooperative research to research contacts with leading
coating suppliers has been realized.

The principal difficulty with water-soluble coatings is in
the application step. Electrostatic spraying is a preferred route
for obtaining the high-quality finishes on appliances. Water-based
coatings present a problem in that the electrostatic bell system
produces finishes with unsatisfactory appearances. Modifications
on the electrostatic disc have been achieved such that excellent
results have been obtained. Electrogasdynamics is another solution
to successful application of aqueous finishes. As with systems
containing 100% solids, close cooperation between suppliers, users
and equipment manufacturers is mandatory to implement the total
system.

Fig. 2. Water-reducible flowcoat primer.
Left: after 500-hr salt spray;
        baked 20 min at 400°F.
Center: original before salt spray.
Right: after 500-hr salt spray;
        baked 20 min at 450°F.

 In spite of the success of soluble-type aqueous finishes, the cosolvent may in the future become a source of concern as regulations demand lower concentrations of organic effluents. Here lies the challenge to coating chemists!

 Emulsion Types. In considering emulsion-type aqueous coatings, those systems that vary in particle size between the true emulsion and the colloidal state are included. With these types, reduction of the organic solvent to 1% or less can be achieved. Coatings based on these systems which exhibit high gloss and passes all laundry type specifications are now available. The principal difficulty lies in electrostatic application. This presents a good challenge to equipment suppliers to help overcome troubles that may prevent the successful implementation of an otherwise good finish. Some properties of a completely water-reducible topcoat are shown in Table III.

TABLE III. Typical Properties of a Water-Based Finish

Wet

Solids	70-80%
Viscosity (seconds)	90-110
Flash point	None
Solvent	Water
Bake schedule	15 min at 350°F

Dry

Density	2.2
Pencil hardness (range)	F-4H
Impact (in-lb), direct reverse	80-140
Taber abrasion	50
Salt spray (500 hr)	Pass
Gloss (range)	Flat to 93
Detergent (250 hr at 165°F)	Pass

CONCLUSIONS

Water-based finishes offer high potential to satisfy the constraints imposed by pollution regulations. Initially there was some reluctance by coating manufacturers to use these systems. Arguments against water include rust, high heat of vaporization, slow evaporation, poor chemical resistance, and stability; these all have been shown to be myths. The low cost of water, the high latitude for polymer design and finish formulation it permits, the ability to use conventional equipment with little or no modification and the flow and leveling characteristics inherent in a 'solvent'-based system all favor water-based finishes as a key route for the finish of the future.

The author is indebted to Erwin Kapalko of PPG Industries who provided the salt-spray data.

BIBLIOGRAPHY

1. T. J. Miranda, J. Paint Technology, $\underline{38}$ (499) 469 (1966).

2. T. J. Miranda, Official Digest, $\underline{37}$ (489) 62 (1965).

3. D. H. Solomon, The Chemistry of Organic Film Formers, John
    Wiley & Sons, New York, 1967.

4. H. J. Essig (to B. F. Goodrich Company), Canadian Pat. 620,439
    (1961).

5. T. J. Miranda (to O'Brien Corporation), U. S. Pat. 3,329,635
    (July 4, 1967).

6. C. W. Smith, Ed., Acrolein, John Wiley & Sons, New York, 1962.

7. S. J. Hochberg, Oil Color Chemists Assoc., $\underline{48}$, 1043 (1965).

8. First International Powder Coating Conference, Toronto,
    Canada, February, 1971.

9. Powder Coating Conference, Society of Manufacturing Engineers,
    Cincinnati, March 28-30, 1972.

10. T. J. Miranda, Powder Coating Potential in the Appliance
    Industry, Society of Manufacturing Engineers Midwest Symposium,
    FC71-845, November 10, 1971.

11. Eastman Technical Bulletin TBP.5, Eastman Kodak Company,
    Rochester, New York.

12. T. J. Miranda and T. F. Huemmer, J. Paint Technology, 41 (529)
    118 (1969).

13. S. H. Schroeter, Organic Coatings and Plastics Chemistry
    Preprints, $\underline{32}$ (2) 401 (1972).

14. W. J. Burlant (to Ford Motor Company), U. S. Pat. 3,247,012
    (April 19, 1966).

15. T. J. Miranda (to O'Brien Corporation), U. S. Pat. 3,600,359
    (August 17, 1971).

16. A. S. Hoffman, Proceedings IAEA Meeting, Seoul, Korea,
    Sept. 28, 1970.

17. J. D. Nordstrom and J. E. Hinsch, Ind. Eng. Chem. Product
    R & D, $\underline{9}$ (2) 155 (1970).

18. C. J. Schmidle and G. L. Brown (to Rohm and Haas Company) U. S. Pat. 3,397,166 (August 13, 1968).

19. R. L. Yeates, Electro Painting, Teddington, England, Drapper Ltd., 1966.

20. M. W. Ranney, Electrodeposition and Radiation Curing, Park Ridge, N. J., Noyes Data Corporation, 1970.

21. L. R. LeBras and R. M. Christenson, J. Paint Technology, 44 (566) 63 (1972).

22. M. Wismer and J. F. Bosso, Chem. Engr., 78 (13) 114 (1971).

23. J. F. Bosso and G. R. Gacesa (to PPG Industries) U. S. Pat. 3,619,398 (November 9, 1971).

ELECTRODEPOSITION OF WATER-DISPERSIBLE POLYMERS

George E. F. Brewer[a]

Ford Motor Company

Detroit, Michigan

The interest in the formulation and manufacture of water-dispersible polymers has increased during the last ten years due to the awareness for environmental requirements. (See also the chapter by T. J. Miranda.) Much work has been done, aiming at reduced use - or elimination - of organic solvents as carriers of coating materials. One of the outstanding innovations in the exclusive use of water as carrier of coating materials is the electrocoating process (1).

The electrocoating process is based on the fact that dispersed, ionized polymeric materials carry electric charges of one polarity and are deposited on the electrode of the opposite polarity in an electrolytic cell (Figure 1). The impact of the electrocoating process is illustrated by the fact that in the last ten years approximately 150 major electrocoating installations have been installed in the United States, another 350 overseas, and more are being built at an accelerated rate. Several hundred scientific papers report on various facets of the electrocoating process and are accessible through compilations (2-6).

Advantages. The advantages of the electrocoating process are as follows:

Use of water as practically the only carrier fluid virtually eliminates stream and air pollution, fire hazards, and materially reduces the cost of facilities for controlling these conditions.

- - - -

[a]Present affiliation: Coating Consultant, Brighton, Michigan

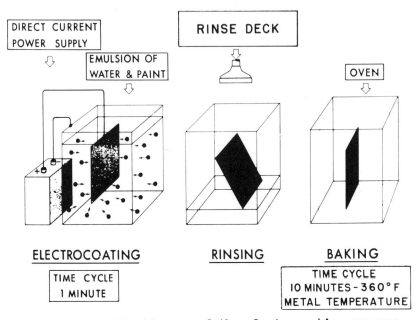

Fig. 1.   Schematic diagram of the electrocoating process.

Freshly deposited coats emerging from the coating tank are practically free from volatile materials. Thus, the freshly painted coat allows some handling and shows the lowest known tendency to sag or "wash" during a subsequent bake.

A second coat, or color coat, of water-borne or solvent-borne spray paint can be applied directly over the uncured electrocoat.

Low paint-bath viscosity (approximately equal to water) results in ease of agitation and pumping, and allows fast entry or drainage.

Extremely uniform film thickness on edges, flanges, cavities, box sections, etc., result in very much improved corrosion protection.

The formation of uniform, thin protective films brings savings of up to 50% in paint consumption.

Overall savings, considering materials, labor, facilities, electric power, etc., are reported to reach 20-30%, when compared with spray, electrostatic spray, or dip-coat painting.

DESCRIPTION OF PROCESS

Practically all types of paint vehicles such as epoxies, acrylics, polyesters, etc., can be formulated to carry ionizable groups, such as -COOH or -NH$_2$. When reacted with a base, the -COOH group will ionize to give a -COO$^-$ anion, while the -NH$_2$ group with an acid will form a -NH$_3^+$ cation. If we symbolize the resinous portion by R, we can symbolize an anion-forming resin by R(COOH)$_n$. On reaction with a base we obtain:

$$R(COOH)_n + m\ YOH \rightarrow R(COO^-)_m(COOH)_{n-m} + m\ Y^+ + m\ H_2O$$

A typical electrocoating bath is formed from such a resin, neutralized to about 30 or 60%, resulting in an apparent <u>electrical equivalent weight</u> of 1,000 to 2,000. The apparent electric equivalent weight depends upon the degree of neutralization. Assume that the carboxylic acids (RCOOH) are practically un-ionized, while the ionization of their salts (RCOO$^-$ + Y$^+$) approaches 100%. Then, let us assign weights and numbers to the symbol R(COOH)$_n$. Suppose that the molecular weight is 16,000 and n is equal to 20. If the degree of neutralization is 40%, then 8 of the 20 carboxyl groups are ionized (m is 8), and we can symbolize the molecule as R(COOH)$_{12}$-(COO$^-$)$_8$. The electrical equivalent weight will then be 16,000 divided by 8, i.e., 2,000. At 50% neutralization, 10 carboxyl groups will be ionized, and the equivalent weight will be 1,600.

Thus the maintenance of a constant solubilizer concentration is important in several respects, notably with regard to current consumption and the available cooling capacity of the equipment, since practically all of the current is converted into heat.

The properties of the final, polymerized film will largely depend upon the chemical nature of the resin backbone, R. Thus, if R carries many epoxy groups, the final film will have essentially the properties of epoxy spray or dip primers; if R carries many acrylic groups, the electrodeposited film will have conventional acrylic coat properties. Additional latitude in the choice of resins is given through the use of electrodepositable resins as emulsifying agents for water-indispersible resins. Many pigments will electrodeposit together with the macroions.

Resins for electrocoating paints need not only ionizability, but have to meet other special requirements:

Lower viscosity when compared with conventional paint resins, since in absence of solvents, the resin itself has to provide "flow out."

Low redispersibility of freshly deposited resin, to prevent
film damage through bath droplets, which are lifted from the coating
tank with the merchandise.

High pumping stability of the electrocoating bath, which contains enough paint solids for ten days or even six months of production, depending on the shape of the merchandise. Paint additions
are of course made daily, but the turnover of the paint inventory
in the tank is slow. At the end of the turnover period, a substantial
part of the tank content is still of the original paint fill, as can
be predicted from the law of probability.

The solubilizing counterion is chosen not to interfere with
continuous use of the counterelectrode. Alkali metal ions, ammonia,
and particularly organic amines are used.

Finally, the most important requirement, the ability of the
deposited resin to form a continuous and highly electrically resistant film.

Electrodeposits start on the section, or edge of the merchandise which is nearest to the counterelectrode. If the deposit
blocks further through-flow of electricity, the current will find
the nearest still available path, namely bare metal, until all the
metal surface is covered. Indeed, when the outer surface is completely coated, deposition of resinous films is extended even into
very remote recesses of highly formed pieces. Figure 2 shows the
gradual growth of an electrocoat on the inner surfaces of a can,
starting in the vicinity of an access hole. The ability to extend
uniform films into recessed areas of merchandise is called "throwing power" and is one of the outstanding advantages offered by the
electrocoating process.

Much research has been done to define the factors which influence the throwing power. There seem to be five electrical parameters involved, which can be used to direct more current flow into
recessed areas and, consequently, increase throwing power:

1. High electric resistance of the first-deposited film blocks off
the flow of electricity to already coated areas and directs it to
uncoated areas.

2. Low minimum deposition voltage of the paint composition increases
throwing power. Low minimum deposition voltage (usually 5 to 15
volt) allows film deposition in recessed areas where only small
voltages are available due to the electric resistance of the coating
bath in confined, already coated areas.

3. High electric conductivity of the coating bath is desirable to
allow the transport of electricity into recessed areas.

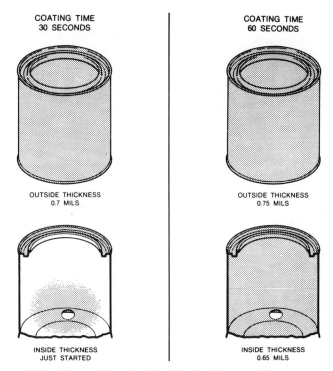

COATING TIME
30 SECONDS

COATING TIME
60 SECONDS

OUTSIDE THICKNESS
0.7 MILS

OUTSIDE THICKNESS
0.75 MILS

INSIDE THICKNESS
JUST STARTED

INSIDE THICKNESS
0.65 MILS

Fig. 2.   Film formation in Ford electrocoating process.

4.   Large electrical equivalent weight of the deposited material results in deposition of more material through the small quantities of electricity available in recessed areas.

Two of the above parameters - film resistance and minimum deposition voltage - can only be built into the paint by the paint manufacturer. The next two parameters - electrical conductivity of bath, and electrical equivalent weight - while originally predetermined by the paint manufacturer, are influenced by the user's care in the maintenance of the solubilizer concentration.

5.   The coating voltage is entirely in the hands of the user: the dial setting on the power source.   There is, however, a very definite upper limit for the coating voltage dictated by a phenomenon called "film rupture."   Figure 3 shows two test panels coated in the same bath.   The panel on the left was coated on application of 250 volt, while the panel on the right was coated at 300 volt.   The "roughness" of that panel is due to film rupture, which lifted the freshly deposited film from many minute areas.   Electricity then flowed through these created voids in the film, and deposited more paint there.

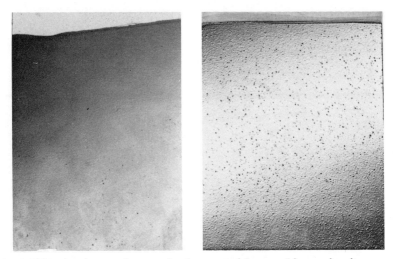

Fig. 3. Film rupture observed when coating voltage is increased
          from 250 volts (left) to 300 volts (right).

Figure 4, graphs of volts vs. time and amperes vs. time, was
recorded during the coating of the two panels shown in Figure 3.
The graph of volts vs. time shows that the voltage was raised from
zero to 250 volts in about 45 seconds and then held at 250 volts.
The simultaneously recorded plot of amperes vs. time shows a peak
at about 15 seconds. The electrical resistance of the just forming
film depresses the current flow to approach the zero ampere line
asymptotically. The very small area under the curve between the
60 and 240 second marks indicates the small number of ampere-sec-
onds (coulombs), corresponding to an extremely small weight of
electrodeposited material. In other words, the electrodeposition
was completed in about 60 seconds.

The right half of Figure 4 presents the data taken during the
electrodeposition of the ruptured panel shown in Figure 3. Again,
the voltage was raised at a constant rate and again an amperage
peak is shown. When the voltage increased beyond the 280 volt mark,
film rupture occurred as indicated by the increased flow of current
in the diagram.

OPERATION OF COATING TANKS

If the dispersed electrodepositable resin is symbolized as
$RCOO^- + Y^+$, it is implied that the macroanion $RCOO^-$ will be deposited on the merchandise and will leave the bath in form of the

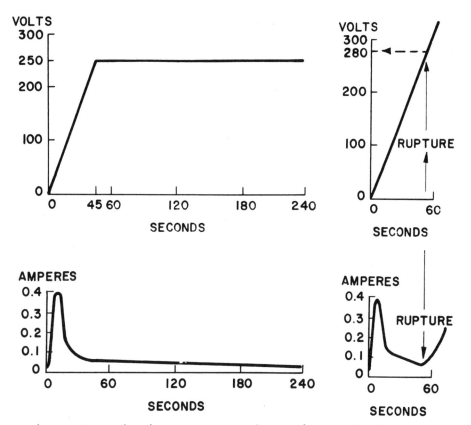

Fig. 4.   Voltage (top) and amperage (bottom) as a function of time
in the electrocoating of the panels shown on Figure 3.
Left, unruptured panel; right, ruptured panel.

coat. The solubilizing ion Y^+ will remain in the bath, where it
will accumulate and will eventually interfere with the coating
operation. Two methods have been developed to avoid the accumulation of solubilizer in the bath:

1.   Solubilizer removal through electrodialysis, ion exchange, or
dialysis (Figure 5).

2.   Solubilizer reuse, accomplished through feeding solubilizer
deficient feed (Figure 6).

The solubilizer reuse method is based on the principle that an
electrocoating bath will remain operable as long as all materials
are replaced which have left the tank, be it as electrodeposit,

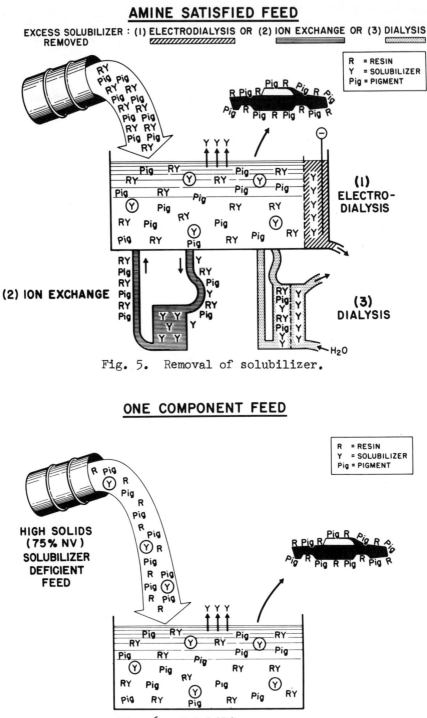

AMINE SATISFIED FEED

EXCESS SOLUBILIZER : (1) ELECTRODIALYSIS OR (2) ION EXCHANGE OR (3) DIALYSIS
REMOVED

R = RESIN
Y = SOLUBILIZER
Pig = PIGMENT

(1) ELECTRO-DIALYSIS

(2) ION EXCHANGE

(3) DIALYSIS

Fig. 5.   Removal of solubilizer.

ONE COMPONENT FEED

R = RESIN
Y = SOLUBILIZER
Pig = PIGMENT

HIGH SOLIDS
(75% NV)
SOLUBILIZER
DEFICIENT
FEED

Fig. 6.   Solubilizer reuse.

through evaporation, filtration, etc. This principle is valid regarding any paint component. It is known that the electrodeposited coat is richer in pigment than the bath it was deposited from. Thus, a typical bath may, eg, contain 22 g of inorganic pigment combined with 78 g of resin. The film deposited from this bath may contain 26 g of pigment with 74 g of resin. Consequently, the replenishment material of this bath has to contain 26 g of pigment for every 74 g of resin. Similar differences may be determined regarding other bath components.

As a general rule, the original bath is formulated to give the desired film properties, while the replacement material contains the materials needed to return to the original bath composition.

Operational Characteristics

The important operational characteristics of the electrocoating process are the following:

Current consumption ranges from 2 to 4 amperes per square foot of coated merchandise.

Coating time varies from a few seconds to about 3 minutes.

Power sources supply direct current at less than 10% ripple factor, usually variable from 50 to 500 volt.

Bath temperature is commonly held between 65 and 90°F, through the use of cooling equipment.

Agitation of the bath is accomplished through pumps, draft tubes, ejector-nozzle systems, etc., capable of continuously moving from one sixth to one thirtieth of the total bath volume per minute, to prevent paint settling.

Bath replenishment is accomplished through mixing the paint supply and some of the bath in a reservoir tank and returning the mixture to the main bath.

Bath filtration for the removal of lint and other particulate impurities, is accomplished by filters of 5 to 75 μ pore size, capable of filtering the entire bath volume continuously in 30 to 120 min.

A rinse consisting of a water shower removes bath droplets adhering to the freshly coated merchandize as it emerges from the tank. Pollution prevention and economy make it desirable to recover the rinsed-off paint solids. This is accomplished through a process involving ultrafiltration.

Bake or cure of the deposited film requires from 5 to 25 min at metal temperatures ranging from 250 to 400°F.

BIBLIOGRAPHY

1. H. N. Bogart, G. L. Burnside, and G. E. F. Brewer, "Electro-coating Reaches Those Inaccessible Surfaces," J. Soc. Auto-motive Engineers, 73, No. 8, 81-33 (Aug. 1965).

2. R. L. Yeates, Electropainting, Draper Ltd., Teddington, England, 1966.

3. R. H. Chandler, Advances in Electropainting, 42 Grays Inn Road, London, 1969.

4. M. W. Ranney, Electrodeposition and Radiation Curing, Noyes Data Corp., Park Ridge, N. J., 1970.

5. W. Brushwell, "Electrocoating," Am. Paint Journal, pp. 71-77 (July 13, 1970) and pp. 64-70 (July 27, 1970).

6. G. E. F. Brewer, "Electrodeposition of Coatings," in N. M. Bikales, Ed., Encylopedia of Polymer Science and Technology, Interscience Publishers, New York, Vol. 15, 1971, pp. 178-191.

II. SYNTHESIS

PREPARATION OF ACRYLAMIDE POLYMERS

Norbert M. Bikales

Consulting Chemist

Livingston, New Jersey

Acrylamide polymers have been available commercially since the early 1950s. The first important applications were as flocculants in the treatment of acid-leached uranium ores and as paper additives to increase dry strength. Numerous other applications have since been developed which can be summarized as follows:

Paper Chemicals: dry strength agents, retention aids, drainage aids;

Water Treatment: clarification and treatment of municipal and industrial effluents, potable water, foods, etc;

Oil-Well Stimulation: fracturing, flooding;

Mineral Processing: flocculation of ores, tailings, coal, etc;

Miscellaneous: friction reduction, thickening, soil stabilization, gel chromatography and electrophoresis, photography, fog dissipation, breaking of oil-in-water emulsions, etc.

Many of these applications are discussed in detail in other chapters of this book. Those of greatest current interest involve water treatment, oil-well stimulation, and friction reduction but acrylamide polymers continue to enjoy substantial markets as paper chemicals and in mineral processing. It is estimated that 1972 consumption of acrylamide polymers was about 35 million lb in the United States and it is known to be increasing rapidly.

Although there are at least nine producers of acrylamide polymers in the United States, virtually nothing has been published by them about the methods they use to manufacture this important class

of water-soluble polymers. The patent literature is, however,
quite extensive and this chapter is an attempt to assess current
industrial practice. No attempt is made to present a comprehensive
review of all pertinent literature, and only those aspects believed
to represent typical practice or to be of special interest will be
described.

POLYMERIZATION MECHANISMS

Acrylamide is a crystalline, relatively stable monomer that
is soluble in water and many organic solvents. It readily undergoes polymerization by conventional free-radical methods, but can
also be polymerized photochemically, in the solid state with ionizing radiation, and anionically (1-3). In the latter case, the
polymer obtained is poly(β-alanine), a polyamide. All current industrial production is believed to be by free-radical polymerization.

$$n\ CH_2 = CH\text{-}CONH_2 \quad \nearrow \quad -\!\!\left[CH_2CH\right]_n^{} \quad\quad \text{free radical} \quad (1)$$
$$\underset{CONH_2}{}$$

$$\searrow \quad -\!\!\left[NHCH_2CH_2CO\right]_n^{} \quad\quad \text{anionic} \quad (2)$$

Of greatest interest have been the redox initiators, but
studies have also been reported with single initiators (4,5),
ultraviolet radiation (6), gamma rays (7), oxygen carriers (8),
electrochemical initiation (9), ultrasonic radiation (10), high
pressure and shear (11), as well as without added initiators (12,13).
If the rate of polymerization is expressed by Equation 3,

$$\text{rate} = k\ [M]^a\ [I]^b \tag{3}$$

then the order with respect to monomer has been reported variously
as 1 (4,14), 1.25 (5), and 1.5 (15), but the order with respect to
initiator is believed to be 0.5 (4,5,14,15). Termination by dis-
proportionation (4) and by coupling (15) have been postulated.

The pH of the reaction medium is also important, since hydrolysis of amide groups can take place at high pH, whereas imidization
(Equation 4) is favored at low pH and high temperatures.

$$\begin{array}{cc}
-CH_2\text{-}CH- & -CH_2\text{-}CH- \\
\quad | & \quad | \\
\quad CONH_2 & \quad CO \\
 & \quad \backslash \\
 & \quad NH \\
 & \quad / \\
\quad CONH_2 & \quad CO \\
\quad | & \quad | \\
-CH_2\text{-}CH- & -CH_2\text{-}CH-
\end{array} \qquad \rightarrow \qquad\qquad (4)$$

The heat released on free-radical polymerization is substantial, 19.8 kcal/mole (1).

INDUSTRIAL PROCESSES

Industrial processes utilize free-radical methods of polymerization to produce a variety of polymers differing in molecular weight (from ca. 200,000 to ca. 10,000,000) and composition. By far the greatest current interest is in those polymers having very high molecular weights (>5,000,000). This requires very low concentrations of initiators and the absence of oxygen and of other interfering substances. A high purity of the monomer is, therefore, essential, particularly with respect to traces of metal. Commercial grades of acrylamide having, eg, less than 10 ppm of iron are satisfactory and can be used without further purification. Since acrylamide is neurotoxic (1), pellets of monomer may be advantageous in that they reduce exposure of workers to dust.

Initiators

Free-radical initiators of the redox type are normally used. Concentrations range from 0.005-1% based on acrylamide - the lower concentrations being preferred for high-mw polymers. Typical initiators are shown in Table I. One of the components of the redox system can be advantageously dissolved in the reactor together with the monomer; the other component, eg, the reducing agent, is then added gradually to control the rate of polymerization and prevent the temperature from becoming excessive.

Metal ions are frequently used as cocatalysts, particularly when complexed with salts of ethylenediaminetetraacetic acid or salicylic acid. Copper is preferred because, unlike ferric ion, cupric ion will not act as a chain terminator.

Reaction Media

The choice of the reaction medium is dictated primarily by the need to dissipate the strong exotherm of polymerization and by considerations involving the ease with which the finished polymer can be recovered from the medium. Cost, flammability, and ability to dissolve the reactants are additional factors to be considered.

Reaction in Aqueous Solution. Water meets the requirements of a satisfactory medium, with the exception of ease of recovery of the polymer (to be discussed below). Acrylamide is dissolved

TABLE I. Initiators for Acrylamide Polymerization

ammonium persulfate - 3,3',3"-nitrilotrispropionamide	U. S. Pats. 3,002,960; 3,0176,740 (Cyanamid)
sodium bromate - sodium sulfite	U. S. Pats. 3,076,740; 3,090,761 (Cyanamid)
sodium bromate - persulfate - sodium (bi)sulfite	U. S. Pats. 3,102,548; 3,215,680 (Cyanamid)
potassium persulfate - sodium metabisulfite	U. S. Pats. 3,278,506; 3,374,143; 3,450,680 (Nalco)
persulfate - Fe^{++} - metabisulfite	U. S. Pats. 3,234,076; 3,316,181 (Nalco)
ammonium persulfate - Cu^{++} - sodium bisulfite	U. S. Pat. 3,332,922 (Calgon)
ammonium persulfate - sodium metabisulfite - EDTA	U. S. Pat. 3,442,803 (Calgon)
hydrogen peroxide - Cu^{++} - thiourea	U. S. Pat. 3,255,072 (Dow)

to a concentration of 8-30%, the pH is adjusted to 3-6, and the reaction vessel and medium are thoroughly deaerated and blanketed with an inert gas such as nitrogen or carbon dioxide. A typical recipe is as follows (16):

	Parts
Acrylamide	32
Deionized water	268
Sodium bromate	0.00302
Sodium sulfite	0.00075
Dry Ice (CO_2)	5.3
Sulfuric acid, 1 $\underline{N}$	0.165

The temperature is kept below about 30°C or, alternatively, allowed to rise adiabatically from room temperature to a maximum of about 60°C. The aqueous solution becomes first very viscous and eventually a tough, rubbery gel, thus making heat transfer difficult. For this reason, acrylamide concentrations must be kept low in comparison with its solubility in water. Reaction is about 98% complete in 3-6 hours.

Reaction in Inverse Emulsion. In this method, the acrylamide is present in concentrated aqueous solution dispersed in an organic medium in the form of small droplets (17). A surface-active stabilizer must be used to prevent coagulation of the emulsion. The reaction is conducted at temperatures up to 70°C in an inert atmosphere. A typical recipe is as follows (18):

Part 1

Acrylamide	14.6
Methacrylic acid	1.1
Maleic anhydride	0.13
Water	12.6
Sodium hydroxide	1.2

Part 2

Toluene	67.3
Sorbitan monooleate	2.3

Initiator

Sodium metabisulfite (1% aq.)	0.56
Potassium persulfate (1% aq.)	0.14

Unlike simple solution polymerization, this process does not increase the viscosity of the medium significantly. It has certain analogies to conventional emulsion polymerization (19), but kinetically it probably resembles solution polymerization. Reaction is completed in about an hour.

Reaction with Precipitation. In this variation, the monomer is soluble in the reaction medium but the resulting polymer is not. The medium thus never gets very viscous and the polymer is relatively easy to isolate and dry.

The most important medium of this type is a 40-55% solution
of t-butyl alcohol in water (20) preferably in the presence of a
salt. Essentially anhydrous media have also been described: 98%
t-butyl alcohol (21), 98% butyl acetate (22), and t-butyl alcohol
with lithium chloride (23). Alternatively, an essentially aqueous
medium can be used if the polyacrylamide is highly hydrolyzed and
the reaction is at low pH (24), since poly(acrylic acid) is insoluble in acid.

Removal of Residual Monomer

For many industrial applications, such as in the treatment of
municipal waste, the flooding of oil wells, or in fire fighting,
the small amounts of residual monomer (ca. 2%) remaining after
polymerization are acceptable. However, for those applications
falling under food and drug regulations, such as in the clarification of potable water or cane sugar juice, or in the manufacture
of paper and paperboard that may come in contact with foods, a
maximum level of 0.05% residual acrylamide has been established
in the United States by the Food and Drug Administration and the
Environmental Protection Agency.

There are various ways of reducing the concentration of monomer in acrylamide polymers. One method is to force the polymerization reaction to completion by addition of a different initiator;
thus, an azo catalyst can be added after polymerization has been
completed with a redox catalyst (25,26). Another method is to add
a compound that adds readily to the double bond of acrylamide (1,3),
thus converting the latter to a derivative that is presumably less
toxic. Among the suitable compounds are sodium bisulfite (27) and
ammonia (28).

Isolation and Drying

The methods of isolation and drying chosen depend on the medium in which the polymerization was effected. If the polymer
precipitates during reaction, filtration followed by drying will
be sufficient. If an inverse emulsion is used, then azeotropic
removal of the water followed by filtration will yield polymer
granules.

The greatest difficulties arise when polymerization is carried out in aqueous solution. The solution becomes so viscous
that it is difficult to handle it. Furthermore, there is typically a great amount of water to remove, since concentrations of polymer
are below 30%. Drum drying has proven most effective; however,
at the temperatures used (150-170°C) there is considerable degra-

dation of the polymer, resulting on the one hand in a decrease in
molecular weight and on the other in some crosslinking by imidization. The residence time on the dryer must, therefore, be kept as
short as possible. In some cases, it has been found advantageous
to dilute the polyacrylamide solution with large quantities of inorganic salts, such as sodium sulfate (29) or sodium silicate, in
order to increase the rate of drying.

A further method consists of extruding the aqueous gel-like
solution into a rapidly moving stream of a water-soluble organic
precipitant, such as methanol. Apparatus for such a technique has
been described (30) and the method appears to be used in cases where
thermal degradation cannot be tolerated.

Because drying is the most difficult and probably the most expensive of the steps involved in polyacrylamide preparation, there
is increasing interest in selling aqueous polymer solutions. This
approach, which decreases the manufacturing costs of the polymer
but increases shipping costs, is especially suitable if the plant
is located near the customer.

Stabilization

Polyacrylamide solutions decrease in viscosity on standing for
several days; eg, η_{sp}/c decreases from about 14 to about 11 in nine
days (31). Several explanations have been offered for this phenomenon, but at least one of the causes appears to be degradation caused
by residues of peroxide initiators (32). Empirically, it has been
found that certain additives protect polyacrylamide from degradation;
these include alkali metal nitrites or iodides (33), thiourea or
sodium thiocyanate (34), and cyanamide, urea, or guanidine (35).
These stabilizers are necessary when aqueous solutions rather than
dry polymer granules are to be sold.

Treatments to Improve Solubility

Acrylamide polymers redissolve with a great deal of difficulty
even when not crosslinked, often forming gelatinous insoluble masses
called "fish eyes." The rate of solution is dependent on the molecular weight and the method of manufacture, and especially on the
method of drying. Drum-dried polymer produced in aqueous solution
tends to dissolve more rapidly than inverse-emulsion-polymerized
polymer, particularly if the polymer has been diluted with an inorganic salt.

A number of methods have been described for increasing the
solubility. These include compaction (36) and pretreatments with

toluene (37), carbitols (38), chromic chloride (39), polyamines
(40), surfactants, and various organic solvents.

INTRODUCTION OF IONIC GROUPS

Polyacrylamide is essentially nonionic, except for a small
number of carboxyl groups introduced inadvertently by hydrolysis.
Many applications require, however, polyelectrolytic character,
eg, to impart dry strength to paper or to flocculate municipal
wastes. The amount and nature of ionic character can be varied over
a broad range by postreactions of polyacrylamide or by incorporation
of a comonomer with the required functionality.

Anionic Acrylamide Polymers

Anionic character can be most readily introduced into polyacrylamide by partial hydrolysis of amide groups as shown in Equation 5 (41).

$$- CH_2CH - CH_2CH - \quad \xrightarrow{\text{NaOH}} \quad - CH_2CH - CH_2CH - \qquad (5)$$
$$\underset{CONH_2}{|} \quad \underset{CONH_2}{|} \qquad\qquad\qquad \underset{COONa}{|} \quad \underset{CONH_2}{|}$$

Another alternative is to polymerize and hydrolyze simultaneously by carrying out the polymerization in strongly alkaline medium
(42). A more controlled method consists of copolymerizing acrylamide with an unsaturated acid, such as acrylic acid, methacrylic acid,
or ethylenesulfonic acid, or an acid precursor, such as maleic anhydride. Both posthydrolyzed and copolymerized types are produced
commercially.

Cationic Acrylamide Polymers

Cationic character can, likewise, be imparted either by copolymerization or by postreaction, and both approaches are being
used at this time. This aspect of acrylamide polymer preparation
has recently been thoroughly reviewed (43). Table II lists some
of the comonomers believed to be of commercial interest. At least
one product on the market is a polyampholyte, i.e., it has both
anionic and cationic groups in the same molecule (44).

The Mannich reaction (Equation 6) is an excellent way of introducing amino groups into polyacrylamide.

$$-CH_2-CH- \quad \xrightarrow[(CH_3)_2NH]{CH_2O} \quad -CH_2-CH- \qquad (6)$$
$$\underset{CONH_2}{\qquad} \qquad \underset{CONHCH_2N(CH_3)_2}{\qquad}$$

This reaction is used by at least one major manufacturer (45). Reaction with polyamines is also of interest (Equation 7),

$$-CH_2-CH- \quad \xrightarrow{R'NHRNH_2} \quad -CH_2-CH- \qquad (7)$$
$$\underset{CONH_2}{\qquad} \qquad \underset{CONHRNHR'}{\qquad}$$

as is the classical Hofmann degradation (46,47) shown in Equation 8.

$$-CH_2-CH- \quad \xrightarrow{NaOCl} \quad -CH_2-CH- \qquad (8)$$
$$\underset{CONH_2}{\qquad} \qquad \underset{NH_2}{\qquad}$$

The amino derivatives can then be quaternized if desired. Sulfonium copolymers can also be prepared and one company had this type on the market a few years ago.

The sales of cationic acrylamide polymers are increasing very rapidly, especially in paper and flocculation applications.

TABLE II.   Some Cationic Comonomers for Acrylamide

$HN(CH_2CH=CH_2)_2$

$(CH_3)_2\overset{\oplus}{N}(CH_2CH=CH_2)_2 \ Cl^-$

$(C_2H_5)_2\overset{\oplus}{N}(CH_2CH=CH_2)_2 \ Cl^-$

$\underset{CH_2=CCOOCH_2CH_2NH_2}{\overset{CH_3}{|}}$

$\underset{CH_2=CCOOCH_2CH_2N(CH_3)_2}{\overset{CH_3}{|}}$

$\underset{CH_2=C-COOCH_2CH_2\overset{\oplus}{N}(CH_3)_3 \ CH_3OSO_3^-}{\overset{CH_3}{|}}$

$\underset{CH_2=C-COOCH_2CH_2\overset{\oplus}{N}(C_2H_5)_2(CH_3) \ CH_3OSO_3^-}{\overset{CH_3}{|}}$

$\underset{CH_2=C-COOCH_2\overset{OH}{\underset{|}{C}}HCH_2\overset{\oplus}{N}(CH_3)_3 \ Cl^-}{\overset{CH_3}{|}}$

$CH_2=CH-COOC_2H_4\overset{\oplus}{S}(CH_3)_2 \ CH_3OSO_3^-$

MISCELLANEOUS POLYMERIZATION PROCEDURES

There is increasing interest in polymerizing acrylamide by methods other than solution or inverse emulsion with redox catalysts. Acrylamide is very susceptible to polymerization with ionizing radiation, and one U. S. company has recently introduced high-mw products made by cobalt-60 irradiation. Photopolymerization is another potentially useful method, but it has not in the past offered sufficient advantages over chemical initiation. There are indications, however, that new methods will attract increasingly greater attention. Several circumstances account for this, including the facts that markets for acrylamide polymers are increasing rapidly, that new higher-capacity production facilities must be built, that severe competition is forcing producers to reduce costs, and that several potential producers would like to enter this market with products made by a superior process. Most of the attention will be directed toward speeding up the process, so that it can be made continuous, and toward reducing the amount of water or other diluent that must be removed.

Among the new processes that have been recently disclosed are the following:

spray drying of catalyzed monomer solutions to give dry polymer directly (48);

free-radical polymerization at high concentrations on heated trays (49) or continuously in an organic medium (50);

Co-60 irradiation of concentrated acrylamide solutions (51);

photopolymerization at high concentrations, either batchwise (52) or on a moving polytetrafluoroethylene-coated belt (53).

To become practical, these methods must overcome the tendency of polyacrylamide to become crosslinked when polymerization is forced by high temperatures or high concentrations of initiators or monomer. Nevertheless, it is probably that future processes will utilize at least some of these features.

BIBLIOGRAPHY

1. N. M. Bikales, "Acrylamide," in A. Standen, Ed., Encyclopedia of Chemical Technology, 2nd ed., Interscience Publishers, New York, Vol. 1, 1963, pp. 274-284.

2. W. M. Thomas, "Acrylamide Polymers," in N. M. Bikales, Ed., Encyclopedia of Polymer Science and Technology, Interscience Publishers, New York, Vol. 1, 1964, pp. 177-197.

3. N. M. Bikales, "Acrylamide," in E. C. Leonard, Ed., _Vinyl and Diene Monomers_, Part I, Wiley-Interscience, New York, 1970, pp. 81-104.

4. E. A. S. Cavell, Makromol. Chem., _54_, 70 (1962).

5. J. P. Riggs and F. Rodriguez, J. Polymer Sci., Part A-1, _5_, 3151 (1967).

6. G. K. Oster, G. Oster, and G. Prati, J. Am. Chem. Soc., _79_, 595 (1957).

7. G. Adler, D. Ballentine, and B. Baysal, J. Polymer Sci., _48_, 195 (1960).

8. N.-L. Yang and G. Oster, Polymer Letters, _7_, 861 (1969).

9. M. S. Tsvetkov and E. P. Koval'chuk, Vysokomol. Soedin., Ser. B, _11_, 42 (1969).

10. A. Henglein, Makromol. Chem., _14_, 15 (1954).

11. A. G. Kazakevich et al, Dokl. Akad. Nauk SSSR, _186_, 1348 (1969).

12. A. Nakano and Y. Minoura, Polymer, _10_, 1 (1969).

13. N. M. Bikales, unpublished results.

14. G. S. Kachalova et al, Isv. Vyssh. Ucheb. Zaved., Khim. Tekhnol., _14_, 427 (1971).

15. I. Geczy and H. I. Nasr, Acta Chim. (Budapest), _70_, 319 (1971).

16. G. R. Backlund and J. F. Terenzi (to American Cyanamid), U. S. Pat. 3,090,761 (21 May 1963).

17. J. W. Vanderhoff and R. M. Wiley (to Dow Chemical), U. S. Pat. 3,284,393 (8 November 1966).

18. M. J. Jursich and G. T. Randich (to Nalco Chemical), U. S. Pat. 3,450,680 (17 June 1969).

19. E. W. Duck in N. M. Bikales, Ed., _Encyclopedia of Polymer Science and Technology_, Interscience Publishers, New York, Vol. 5, 1966, pp. 801-859.

20. D. J. Monagle and W. P. Shyluk (to Hercules Incorporated),
    U. S. Pat. 3,336,269 (15 August 1967).

21. H. v. Brachel and F. Engelhardt (to Cassella Farbwerke),
    Brit. Pat. 1,102,708 (7 February 1968).

22. Y. Matsunaga, M. Matsuura, and H. Mitsuhashi (to Mitsubishi
    Chemical), Japan Pat. 69-3,832 (17 February 1969).

23. D. K. Ray-Chaudhuri and J. E. Schoenberg (to National Starch),
    Ger. Offen. 2,135,742 (20 January 1972).

24. H. Volk and P. J. Hamlin (to Dow Chemical), Brit. Pat.
    1,216,105 (16 December 1970).

25. E. A. Gill, Fr. Pat. 1,530,821 (28 June 1968).

26. R. B. Thompson and M. J. Jursich (to Nalco Chemical), U. S.
    Pat. 3,414,547 (3 December 1968).

27. D. J. Pye (to Dow Chemical), U. S. Pat. 2,960,486
    (15 November 1960).

28. G. D. Jones (to Dow Chemical), U. S. Pat. 2,831,841
    (22 April 1958).

29. E. R. Kolodny (to American Cyanamid), U. S. Pat. 3,215,680
    (2 November 1965).

30. J. F. Terenzi (to American Cyanamid), Ger. Pat. 1,116,402
    (2 November 1961); U. S. Pat. 3,208,829 (28 September 1965).

31. N. Narkis and M. Rebhun, Polymer, 7, 507 (1966).

32. H. C. Haas and R. L. MacDonald, Polymer Letters, 10, 461
    (1972).

33. Dow Chemical Co., Brit. Pat. 950,022.

34. G. F. Schurz and K. R. McKennon (to Dow Chemical), U. S.
    Pats. 3,234,163 (8 February 1966); 3,235,523 (15 February
    1966).

35. M. N. D. O'Connor (to American Cyanamid), Fr. Pat.
    1,577,800 (8 August 1969).

36. J. Carlin (to American Cyanamid), U. S. Pat. 3,053,819
    (11 September 1962).

37. R. E. Friedrich and R. G. Martin (to Dow Chemical), U. S.
     Pat. 3,282,874 (1 November 1966).

38. A. B. Savage (to Dow Chemical), U. S. Pat. 3,350,338
     (31 October 1967).

39. A. J. Gentile (to American Cyanamid), U. S. Pat. 3,251,814
     (17 May 1966).

40. P. Economou (to American Cyanamid), U. S. Pat. 3,637,564
     (25 January 1972).

41. K. Nagase and K. Sakaguchi, J. Polymer Sci., Part A, 3,
     2475 (1965).

42. C. S. Scanley (to American Cyanamid), U. S. Pat. 3,414,552
     (3 December 1968).

43. M. F. Hoover, J. Macromol. Sci. - Chem., A4, 1327 (1970).

44. F. S. Varveri, R. J. Jula, and M. F. Hoover (to Calgon),
     U. S. Pat. 3,639,208 (1 February 1972).

45. R. L. Wisner (to Dow Chemical), U. S. Pat. 3,539,535
     (10 November 1970).

46. G. D. Jones, J. Zomlefer, and K. Hawkins, J. Org. Chem., 9,
     500 (1944).

47. A. M. Schiller and T. J. Suen, Ind. Eng. Chem., 48, 2132
     (1956).

48. A. J. Frisque and R. Bernot (to Nalco Chemical), U. S. Pat.
     3,644,305 (22 February 1972).

49. H. I. Patzelt, L. J. Connelly, E. G. Ballweber, D. B.
     Korzenski, and K. L. Slepicka (to Nalco Chemical), S. African
     Pat. 69-1,065 (19 August 1969).

50. K. Schmitt, F. Reichel, M. Marx, and G. Storck (to BASF),
     Ger. Offen. 2,043,663 (16 March 1972).

51. A. D. Abkin (to Karpov Institute), Brit. Pat. 1,139,917
     (15 January 1969).

52. H. Willersinn and C. H. Krauch, Ger. Offen. 2,009,748
     (16 September 1971).

53. J. P. Communal, J. Fritz, and B. Papillon (to Progil), Ger.
     Offen. 2,050,988 (13 May 1971).

EXTRACELLULAR MICROBIAL POLYSACCHARIDES: NEW HYDROCOLLOIDS HAVING BOTH FUNDAMENTAL AND PRACTICAL IMPORT

Allene Jeanes

Northern Regional Research Laboratory
U. S. Department of Agriculture
Peoria, Illinois

Polysaccharides that occur free in the culture fluids of many nonpathogenic microorganisms show great diversity as well as novelty in composition, structure and properties. Only within approximately the last thirty years, however, has the practical potential of these macromolecular biopolymers been given serious consideration. Thus far, only a beginning has been made in discovering microbial sources, selecting products suitably constituted for specific applications, and developing the new science and technology of biopolymer fermentation. This activity was stimulated initially by wartime need for a blood plasma substitute and, more recently, by forecasts of impending shortages of natural plant hydrocolloids, by the advisability of national independence from foreign imports, and by demands not adequately met by currently available substances. See also the chapter by R. L. Whistler.

EXTRACELLULAR MICROBIAL POLYSACCHARIDES OF PRACTICAL SIGNIFICANCE

Extracellular microbial polysaccharides that have been investigated most systematically and have achieved at least potential practical significance are grouped according to compositional type and class for consideration here. The constituent sugar moieties and position and anomeric form of the glycosidic linkages are emphasized since they, primarily, determine the shape, conformation and specific properties of the macromolecules.

Several classes of neutral, extracellular homopolysaccharides are shown in Table I. Dextrans were the first microbial polysaccharides to be produced industrially; the one now in use has 95% $1 \rightarrow 6$- and 5% $1 \rightarrow 3$-linkages (1) and is completely metabolized by man

227

TABLE I
Neutral, Extracellular Microbial Homopolysaccharides*

Polysaccharide Class	Microbial Source	Monomeric Unit	Linkages**	Mol. Wt.
Dextran	Certain *Leuconostocs* and *Streptococci*	α-D-Glucopyranosyl	$1 \to 6$, $1 \to 3$, $1 \to 4$, $1 \to 2$***	$n \times 10^6$
Levan	*Aerobacter levanicum*	β-D-Fructofuranosyl	$2 \to 6$, $2 \to 1$	$n \times 10^6$
Pullulan	*Pullularia* sp.	α-D-Glucopyranosyl	$1 \to 4$, $1 \to 6$	750×10^3
β-Glucosylglucan	*Sclerotium* sp.	β-D-Glucopyranosyl	$1 \to 3$, $1 \to 6$	
β-Glucan	*Agrobacterium* sp.	β-D-Glucopyranosyl	$1 \to 2$	
α-Mannan	*Hansenula* sp.	α-D-Mannopyranosyl	$1 \to 2$, $1 \to 3$, $1 \to 6$	
β-Mannan	*Rhodotorula* sp.	β-D-Mannopyranosyl		

* Restricted to those of established or proposed practical applicability.
** Linkage present in higher proportion is italicized.
*** All non-1,6-linkages are not necessarily present in any one dextran.

when ingested or infused. Its use as the source of solutions for intravenous infusion and of the Sephadex gels for separation procedures results from broad basic research in Sweden. The dextrans themselves have contributed extensively to fundamental research in carbohydrate and immunochemistry, biochemistry, medical chemistry, and enzymology.

The dextran class illustrates several pertinent principles applicable to all microbial polysaccharides. The classes have many members; more than a hundred structurally different dextrans have been characterized which differ in the identity and proportion of non-1→6-linkages. Properties vary accordingly, such as solubility or insolubility in water, sparsely or highly branched, and all gradations between (1). Characteristics are determined genetically by the specific dextran-producing strain. Thus, in choosing a polysaccharide to serve a specific purpose, a strain must be selected that produces a desirable product.

Levans, like dextrans from Leuconostocs and Streptococci, can be produced only from the disaccharide, sucrose. All other classes of polysaccharides discussed here usually are produced from the monosaccharide glucose as the carbon source in the production medium, although other monosaccharides or sucrose might be used. The technology is well developed for producing dextrans and levans enzymatically rather than in growing cultures; improved efficiency of operation and quality of product result (1). Pullulan (2), the only characterized α-glucan from the diversely constituted genus Aureobasidium (Pullularia), probably represents only one of the structural types that further research may be expected to disclose from these widely distributed black yeast-like fungi (3). Uncharacterized polysaccharides from unidentified strains are effective flocculants in recovering uranium and other ores (4).

TABLE II
Extracellular Microbial Polysaccharides Having Novel Anionic Components*

Polysaccharide Class	Microbial Source	Composition of Polysaccharide		
		Neutral Glycose (Gly)	Acidic Group (A)	Molar Ratio Gly:A
Phosphomannan	Hansenula holstii NRRL Y-2448	α-D-Mannopyranose	$M^1-O-\overset{\overset{O^-}{\mid}}{\underset{\underset{O}{\parallel}}{P}}-O-^6M$	5:1
Succinoglucan**	Alcaligenes faecalis var. myxogenes 10C3	β-D-Glucopyranose β-D-Galactopyranose	Half-ester succinic	7:1:1.5
Galactoglucan	Arthrobacter stabilis NRRL B-3225	D-Glucopyranose D-Galactopyranose	Half-ester succinic Pyruvic acid ketal	2:1:–:–***
Glycosaminoglycuronan	Unidentified NRRL Y-6272	n-Acetylglucosamine	n-Acetylglucosaminuronic	2:1

* Work done at Northern Regional Research Laboratory unless indicated otherwise
** T. Harada et al, Osaka University.   *** Also present, 5% O-acetyl

The predominantly β-1→2-linked glucan from Agrobacterium tume-
faciens and the α- and β-mannans are included as exampled of polysaccharides available only from microbial sources. They have not yet
been developed industrially. Neutral microbial polysaccharides composed predominantly of galactopyranose units are not yet known.

Biopolymers of the classes already mentioned may be water soluble. In contrast, the branched β-glucosylglucans and the linear
β-1→3-glucans are essentially insoluble, but swell greatly in water
and form gels. These substances, which have numerous microbial
sources, are unusually stable to acid, but labile to alkali (2).

Not included here are the glucans and other polysaccharides
produced by certain microorganisms from n-alkanes of petroleum.

Several classes of unique extracellular microbial polysaccharides
having novel anionic constituents are shown in Table II. All are
produced from glucose as carbohydrate substrate and isolated as
neutral salts. Phosphomannans, produced by members of Hansenula
and related yeast genera, consist mainly of sequences of α-D-man-
nopyranosyl residues polymerized through diester orthophosphate
groups. Ratios of mannose to phosphate are known in the range
from 2.5 to 28 (5); the phosphomannan having a ratio of 5, from
NRRL strain Y-2448, is discussed later (6). Fundamental research
interest in these polyelectrolyte biopolymers and their subsequently discovered counterparts in yeast cell walls and membranes, has
been high. One aspect of practical potential is their exceptional
resistance to spontaneous contamination by individual microorganisms
but biodegradability under natural conditions by mixed cultures.

A glucan partially esterified by succinic acid, discovered by
Harada and co-workers (Table II), shows high viscosity which in-

Extracellular Microbial Anionic Heteropolysaccharides

Polysaccharide from Strain NRRL·	Microorganism	Composition of Polysaccharide		Linkages
		Components	Molar Ratio	
B-1459	*Xanthomonas campestris*	D-Mannose	3.0	GA $(\beta,1\rightarrow 2)$ M
		D-Glucose	3.0	M $(1\rightarrow 4)$ G
		D-Glucuronic acid (K salt)	2.0	G $(1\rightarrow 4)$ GA
		O-Acetyl	1.7	G $(\beta,1\rightarrow 4)$ G
		Pyruvic acid	0.63	
B-1973	*Arthrobacter viscosus*	D-Glucose	1	MA $(\beta,1\rightarrow 4)$ G
		D-Galactose	1	G $(\beta,1\rightarrow 4)$ Gal
		D-Mannuronic acid (K salt)	1	Gal $(\beta,1\rightarrow 4)$ MA
		O-Acetyl	4	
Y-1401	*Cryptococcus laurentii* var. *flavescens*	D-Mannose	4	M $(\quad)$ M
		D-Xylose	1	X $(1\rightarrow)$
		D-Glucuronic acid (K salt)	1	GA $(\beta,1\rightarrow 2)$ M
		O-Acetyl	1.7	
(No. 271)*	*Bacillus polymyxa*	D-Glucose	3	Predominantly
		D-Mannose	3	α, 1→3
		D-Galactose	1	
		D-Glucuronic acid (salt)	2	
B-1828		••	—	

* E. Ninomiya *et al*, Meiji Sugar Mfg. Co., Ltd.
** NRRL results not yet published

creases in the presence of salts even when heated (7). The galac-toglucan from strain NRRL B-3225 has two anionic substituents: half-ester succinic acid and pyruvic acid ketal (8). Treatment with mild alkali removes the half-ester but not the pyruvic acid; some unusual properties of the product are discussed later. The polymer of N-acetyl glucosamine and N-acetyl glucosaminuronic acid (Table II), discovered in our research (3), is the first extracellular member of a new and novel class and has unusual fundamental significance (3). Microbial sources for polycationic glycosamino-glycans having nonacetylated amino groups are known and under investigation.

Several anionic heteropolysaccharides discovered in our research (2), which constitute the main subjects for further consideration here, are shown in Table III. Each has a hexuronic acid component; PS B-1459[a] has, in addition, a pyruvic acid ketal substituent. Three of these biopolymers have O-acetyl substituents which are easily removed by dilute alkali; the deacetylated products have modified properties, especially that from strain B-1973. PS B-1459 (9) is now in industrial production and use under the names xanthan, Kelzan[R] and Biopolymer XB-23[R]; a food grade, Keltrol[R], is used in the food industry under FDA approval (10). PS B-1973 (11) has two unusual compositional features: D-mannuronic acid and 25%

- - - - - -

[a]When used with a NRRL strain number, PS stands for polysaccharide.

by weight of O-acetyl, which is 50% of that theoretically possible.
PS Y-1401 (12) differs from the other polysaccharides included here
in having a pentose constituent, which occurs as a single unit side
chain. It functions well to suspend laundry soil (13) and stabilize
emulsion paints (14). Two strains of Bacillus polymyxa, discovered
independently in Japan (15) and by us, appear to give essentially
the same product. The outstanding property of the aqueous solutions
is soft gelatin when heated or in the presence of alcohol or salts
and exceptional high waterbinding capacity.

PRODUCTION OF EXTRACELLULAR MICROBIAL POLYSACCHARIDES

Batch-type fermentation is used (2), although a continuous-flow
process has been established for PS B-1459 (16). This process
effects significant reduction in cost, and the principles are generally applicable to other biopolymers. Media contain corn sugar,
low-cost sources of nitrogen and growth factors, and essential inorganic elements. During the approximately three-day batch fermentation, cultures are aerated, stirred, and the pH is controlled.
The quality of finished product determines which of the following
steps are taken for purification and recovery: centrifugation to
remove cells; precipitation of polysaccharide by a nonsolvent or a
quaternary substance to separate it from soluble components of the
medium; dissolution and reprecipitation; and dehydration by drum- or
spray-drying or by use of a nonsolvent, usually an alcohol. For
some purposes, dehydration may be omitted. Yields, based on glucose,
are around 70-75%. Commercial products may contain about 1% nitrogen
(proteinaceous) and variable amounts of inorganic ash. Polysaccharide products used in our research are essentially free of both
nitrogen and extraneous inorganic salts.

SELECTED ANIONIC MICROBIAL POLYSACCHARIDES

Structure. An outstanding structural feature of the phosphomannan Y-2448 macromolecule (Table II) is the presence of diester
orthophosphate groups which not only join together, apparently in
linear chains, mannose oligosaccharide sub-units linked mainly
α-1$\rightarrow$3-, but which also attach these chains to a high-molecular-weight
mannan core (5). Each phosphate group carries two types of ester
bonds: the mannosyl-1-phospho-, an acid-labile, alkali-stable hemi-
acetal type which permits depolymerization of the macromolecule by
mild acid; and the -phospho-6-mannose-, which involves the primary
mannose C-6 hydroxyl position and is stable to both acid and alkali.
Nonreducing end groups have not been detected.

Phosphomannan Y-2448 was found to be innocuous in acute toxicity, skin and feeding tests (17). It rates high for organoleptic

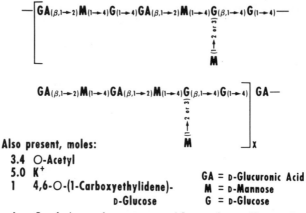

Also present, moles:
3.4 O-Acetyl
5.0 K⁺
1   4,6-O-(1-Carboxyethylidene)-
           D-Glucose

GA = D-Glucuronic Acid
M = D-Mannose
G = D-Glucose

Fig. 1.   Block of sixteen hexose residues from the polysaccharide
produced by <u>Xanthomonas</u> <u>campestris</u> NRRL B-1459.

quality (18), stabilizing beer foam (19), and suspending laundry
soil (13).   Phosphomonoesters resulting from autohydrolysis of de-
cationized phosphomannans are excellent dispersants (20).

The constituent sugars and substituents of PS B-1459, the molar
ratio in which they occur, the position and anomeric type of the
linkages and the sequence of certain sugar residues are detailed in
Table III.   Variations observed in composition of experimental and
commercial products usually have been within the ranges: glucuronic
acid 18-19%, O-acetyl 4.5-4.8% and pyruvic acid 3.0-3.5%.   Apparent
repeating units established by various structural techniques (21)
(Figure 1) and also by precise methylation structure analysis (22),
have minor differences.   The size is that required to include one
pyruvic acid residue, which occurs as 4,6-O-(1-carboxyethylidene)-
D-glucose in a terminal side-chain position (21).   Both structures
show frequent branching.   For the block of sixteen sugar residues
depicted, there appears to be seven protruding groups, namely: two
side-chain mannose residues, the 4,6-O-(1-carboxyethylidene)-D-glu-
cose (which is assumed to constitute a single-unit side chain) and
the four mannose residues in the main chain which, by virtue of
their linkage at C-1 and C-2, project away from the axis of the main
chain to produce a bulge approximately equivalent in size and steric
effect to that of a single-unit non-reducing side-chain residue (21).

PS B-1973 has a linear structure consisting of a trisaccharide
repeating unit (Table III and Figure 2) (23,24).   The O-acetyl groups
appear to occur mainly on the glucose and galactose residues (23).

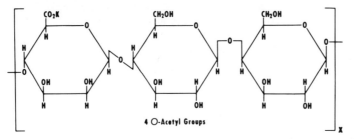

Fig. 2. Repeating trisaccharide unit in the polysaccharide produced by <u>Arthrobacter viscosus</u> NRRL B-1973.

<u>Hydrodynamic Data.</u> These reveal fundamental properties of
PS Y-2448 (25) and B-1459 (26) (Table IV). In gel-like dispersions,
molecular entities are not monodisperse.

<u>Solvation Behavior.</u> Dry particles absorb water rapidly at
normal temperatures and swell; slow addition to well agitated solvent assures uniform dispersal. Dispersal into salt solutions may
be slower. Dispersions for research were equilibrated overnight
before use. Good solvents are, for PS B-1459: ethylene glycol or
glycerol - the concentrated (warmed) or aqueous, especially 10%
di- or triol; and for PS B-1973: aqueous ethanol - up to 33% ethanol.

TABLE IV

Microbial Polysaccharides: Hydrodynamic* and Related Data

Property	Polysaccharide	
	Y-2448	B-1459
Aqueous dispersions	Gel-like	Gel-like
	In 1M KCl	In 4M urea or 0.01M NH$_4$Ac
S_{25}^0	90S	5S
$\overline{M}_{SD}$	15 × 10^6	(1.4 ± 0.3) × 10^6
$\overline{M}_{LS}$	33-39 × 10^6	(3.6 ± 0.7) × 10^6
Particle character	Strong conc. dependence; more like a flexible random coil than CMC.	Strong conc. dependence; relatively rigid, strongly interacting, high-molecular weight entities.

* F.R. Dintzis, G.E. Babcock and R. Tobin

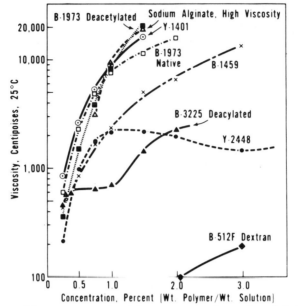

Fig. 3. Viscosity - concentration curves for microbial polysac-
         charides and an alginate.

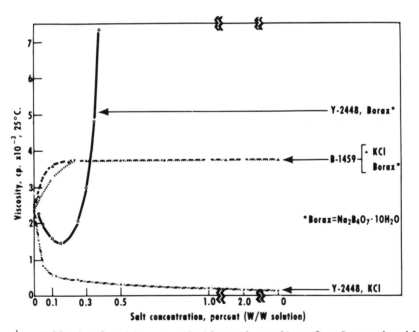

Fig. 4. Effect of salts on solution viscosity of polysaccharides
         from Hansenula holstii NRRL Y-2448 (2% conc) and Xantho-
         monas campestris NRRL B-1459 (1% conc).

Viscosity-Concentration Relationship. The various microbial anionic polysaccharides are compared with each other, with a neutral dextran ($\bar{M}_w$ 90 x 10^6), and with a commercial alginate in Figure 3. The viscosity measurements were made with a Brookfield LVT viscometer at 30 rpm and 25°C.

Most of the polysaccharides (including native B-3225, which is omitted) show similar behavior and very high viscosity even at low concentrations. An explanation has not been established for the behavior of deacylated B-3225, which is novel in having pyruvic acid ketal (K-salt) (6.3%) as the only anion (8). Phosphomannan Y-2448 characteristically shows a viscosity maximum at 1.5% concentration and a minimum at 3%. Partial depolymerization or the presence of salt result in the curve being linear (6).

Effects of Salts on Viscosity. These were investigated by adding salt incrementally to aqueous polysaccharide dispersions. The typical polyelectrolyte phosphomannan Y-2448 (6), is compared with polyanionic PS B-1459 (9), the viscosity of which is increased by all salts (Figure 4). At concentrations ≤ 0.25%, viscosity of PS B-1459 is diminished somewhat by salt, but much less than that of anionic synthetic and plant hydrocolloids. The strong complexing of phosphomannan Y-2448 with borax indicates the presence of the C-2 - C-3 cis-diol in some D-mannose residues.

For PS B-1459, representative examples are shown (Figure 5) of enhancement of viscosity by inorganic salts and of tolerance for high salt concentrations. There is similarity of behavior between the phosphates of dibasic ammonium and mon- and dibasic potassium; the acetates of sodium and calcium; and the chlorides of mono- and divalent cations in the neutral pH range. Eventual phase separation is indicated only for sodium acetate. At about pH ≥ 9, di- and trivalent cations insolubilize the polysaccharide.

Similar data are shown for native (N) and deacetylated (D) PS B-1973 (11) (Figure 6). N, having 25% O-acetyl groups, is more tolerant of high salt concentrations than is D. Ammonium acetate solution is a better solvent than water for N. A film cast from a brilliantly clear, unheated solution containing 15% ammonium acetate was clear, tough, flexible and stable when dry. Borax complexes strongly with D, but weakly with N unless partial deacetylation has resulted from heating or aging the solution; the flow properties then change from short, gelatinous to very long and cohesive.

Film Properties. The film properties of polysaccharides B-1459 and B-1973, cast from aqueous solutions (27), compare favorably with those of amylose triacetate, which are generally regarded as excellent (Table V). The native polysaccharides do not form coherent films unless the aqueous solutions are heated or autoclaved - presumably solvation or dispersion is improved or conformation is

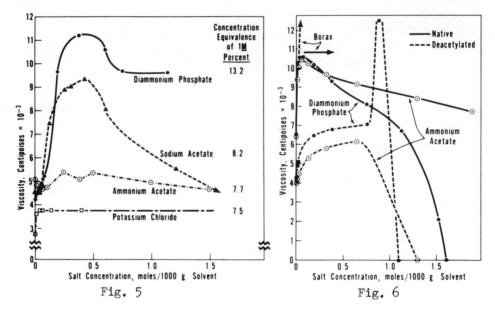

Fig. 5.

Fig. 6.

Fig. 5.   Native polysaccharide from NRRL B-1459, 1% conc:
          viscosity vs. molal salt concentration.

Fig. 6.   Native and deacetylated polysaccharides from NRRL B-1973,
          1% conc: viscosity vs. molal salt concentration.

TABLE V

Microbial Polysaccharides: Average Test Values of Films

Polysaccharide		Glycerol	Tensile Strength*	Elongation*	Schopper Double Folds
		%	kg./mm.²	%	Number
B-1459	Native	0	5.8	3	6
	Heated	0	8.0	5	70
	Deacetylated	0	7.0	4	45
	Native	30	3.6	6	1,500
	Heated	30	4.0	11	2,000
	Deacetylated	30	5.0	8	3,600
B-1973	Autoclaved	0	7.5	7	10,000
	Deacetylated	0	9.5	8	3,500
	Autoclaved	30	4.8	9	10,000+
	Deacetylated	15	7.7**	10	10,000+
Corn amylose		0	7.2	13	900
		15	5.4	20	280

* Scott IP2 incline plane serigraph, jaw setting 40 mm.
** 7.7 kg./mm.² = 10,950 lb./in.²

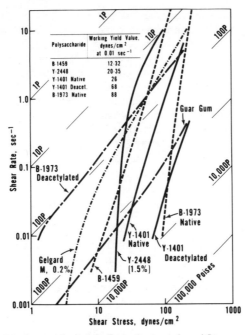

Polysaccharide	Working Yield Value, dynes/cm² at 0.01 sec⁻¹
B-1459	12-32
Y-2448	20-35
Y-1401 Native	26
Y-1401 Deacet.	68
B-1973 Native	88

Fig. 7.   Extracellular microbial polysaccharides and controls, 1%
conc: shear stress - shear rate - viscosity relations,
25°C.

changed.   Films from PS B-1973 are outstanding in flexibility.
After drying, the films are water soluble.

Rheology.   Viscosity and rheological relationship were established (28) by the method of T. C. Patton (29) using the Wells-Brook-
field Micro Cone and Plate Viscometer, Model RVT.   Shear stress vs.
shear rate relations are plotted on log-log scales; viscosity may
be read where any point along the flow curve coincides with or inter-
sects constant lines along which viscosity of Newtonian solutions
would fall (Figure 7).

Grossly, the flow curves differ in the rapidity of decrease in
flow rate with decrease in shear stress, that is, in the resistance
the solutions present to the shearing force.   Two general types of
curves are differentiated by criteria as follows: pseudoplastic -
concavity to the shear-rate scale and tending asymptotically to par-
allel the viscosity lines of constant Newtonian flow that run at
45° angles to the perpendicular; plastic - convexity to
the shear-rate scale and tending asymptotically to become perpendicular to the shear stress axis at ultra-low shear rates (29).   Pseudoplastic dispersions appear to be continuously deformable under shear

stress; the corresponding macromolecules are believed to possess approximately linear backbone structure and to be flexible, and their solution configuration may be that of a random coil. Plastic dispersions do not continue to flow at or below a definite shear stress, and the rate of shear decreases rapidly towards zero. The corresponding macromolecules are believed to be relatively stiff or involved in crosslinked colloidal networks or other structured aggregates that resist shearing stress. The rheological classification of a substance, however, is influenced by concentration and, with sufficient decreases in concentration, may change from plastic through pseudoplastic to Newtonian (29).

The flow curves for deacetylated PS B-1973 and guar gum accord with the general criteria for pseudoplastic substances; those for Gelgard M (a plastic-type synthetic from Dow Chemical Co.) and microbial polysaccharides Y-2448, Y-1401, native B-1973, and native B-3225 accord with the criteria for plastic substances. The flow curve for PS B-1459 seems to be a composite. Its slight concavity identifies it with the pseudoplastic substances, but its exceptionally sharp decrease in viscosity with shear and its almost perpendicular approach to the shear stress axis are characteristic of plastic-type substances. The striking behavior of phosphomannan Y-2448 appears to result from electrostatic charge effects since, in the presence of 0.1% KCl, viscosity is diminished below the range of measurement.

The conformation and solution configurations of the polysaccharide macromolecules accord with those expected for their respective flow types.

Patton defines a practical or "working" yield value as the shear stress required to produce a shear rate of 0.01 sec^{-1} (29). Microbial polysaccharides having significantly high working yield values are listed in Figure 7. The value for native PS B-3225, 26, is not included. With the exception of PS B-1459, all of these are plastic-type substances. This is a further exhibition of plastic characteristics by PS B-1459.

Removal of O-acetyl groups from PS B-1973 (11) and both O-acetyl and half-ester succinic acid substituents from B-3225 (8) changed the rheological class from plastic to pseudoplastic and decreased the apparent working yield values to 1.6 and 2.5, respectively. Deacetylated PS B-1459, also, is pseudoplastic, with working yield value 20.5. Deacetylation of PS Y-1401 (12), however, increased the working yield value, but did not change the type (Figure 7).

In the presence of 0.5% KCl, the respective working yield values for polysaccharide B-1459, B-1973, and B-3225 are increased five-, two-, and fourfold. After being heated to 80°C and tested at 25°C,

the value for B-1459 increases if 0.5% KCl is present during heating but decreases if it is absent. Working yield values of polysaccharides B-1459 and B-3225 are increased significantly in the presence of locust bean gum and, to a lesser extent, guar gum.

All of these biopolymers having high yield values are excellent
suspending agents in keeping with the established relation between
suspending power and yield values. The demonstrated superiority of
PS B-1459 for suspending heavy particles is the basis for numerous
patented applications (2,30) in petroleum-drilling fluids, paints,
glazes, abrasives, explosives, and cements, as well as for applications in the food (31) and metal beneficiation (32) industries.

Patton (29) obtained a measure of thixotropy from the ratio of
working yield values determined before and after shearing the sample.
The microbial polysaccharide dispersions were sheared in the viscometer for 5 min. at 100 rpm (384 sec^{-1}); the thixotropic index
has been no more than 1.1 to 1.15. Thus, although all the microbial
polysaccharides with the exception of deacetylated B-1973 show rapid
decrease in viscosity with increase in shear stress (Figure 7), all
regain the initial viscosity without lag when shear ceases. This
ease of thinning while under shear is cited frequently as an unusual property of xanthan (18,30-33) as well as of PS B-1973 (34).
Reversible shear-thinning is a valuable property of the biopolymers
in operations that involve movement of dispersions such as flow of
foodstuffs (31) or paints (29), flow through pipes (30,31) or orifices
(30,33,34), and centrifugal separations (32). See also the chapters
by J. W. Hoyt and R. H. Wade and by D. C. MacWilliams, J. H. Rogers,
and T. J. West.

Nature of Molecular Entities in Aqueous Dispersions

Observations with an ordinary polarizing microscope reveal
that polysaccharides B-1459, B-1973, and B-3225 are unique in readily assuming elongated birefringent forms which tend to aggregate
along the long axis (35). Thus, when salt-free dispersions are
spread on a microscope slide, patterns are evident that assume somewhat more order with time and are stable when dry. These remarkably
ordered patterns are distinctive for each polysaccharide. Dispersions of these same polysaccharides show high retardation of polar-
ized light (birefringence) at low or zero rates of shear. Apparent
orientation occurs with ease and the magnitude of birefringence does
not depend entirely on "orientation." Salt weakens or prevents both
the ordered microscopic growth and the nonflowing birefringence.
Salt-free aqueous dispersions of phosphamannan Y-2448, which appear
to be three-dimensionally crosslinked, show birefringence effects
comparable to "liquid crystals" (35).

Birefringence effects are well-known indications of ordered structural arrangements among large molecules. Thus, observations from birefringence, hydrodynamic and microscopic procedures, and from the dependence of good film formation on the method of polysaccharide dispersal, are consistent in indicating the presence in some of the microbial polysaccharides of supramolecular entities which require special attention for dispersal. Electrostatic charges have a role in the existence of these entities; however, the range and extent of differences found suggest that additional factors may be involved. Composition, structure, and conformation of certain of these polyelectrolytes may be especially conducive to molecular interaction and ordering through electrostatic effects. Other evidence that PS B-1459 may assume an ordered conformation, as indicated in part by gel formation when the polysaccharide is admixed with locust bean gum (10), has been established (36).

SUMMARY

Now established among the hydrocolloids of commerce are several macromolecular polysaccharides produced by nonpathogenic microorganisms cultured on glucose or sucrose as carbon sources. The distinctive properties of these biopolymers, based on composition and structure different from other hydrocolloids, determine their uses in the food, pharmaceutical, petroleum and various other industries. Selection and development of other types may be expected from among the myriads of natural sources. Fundamental research, upon which vigorous practical developments are dependent, will be advanced by elucidating basic information on these biocolloids which deviate from the conventional in both chemical constitution and physicochemical behavior.

Microbial polysaccharide research at the Northern Regional Research Laboratory has been cooperative among bioengineers, carbohydrate chemists and microbiologists.

Mention of a company and/or product by the U. S. Department of Agriculture does not imply approval or recommendation of the company or product to the exclusion of others which may also be suitable.

BIBLIOGRAPHY

1. A. Jeanes, "Dextrans" in N. M. Bikales, Ed., Encyclopedia of Polymer Science and Technology, Interscience Publishers, New York, Vol. 4, 1966, pp. 805-824.

2. A. Jeanes, "Microbial Polysaccharides" in op cit, Vol. 8, 1968, pp. 693-711; and references to original publications cited therein.

3. A. Jeanes, K. A. Burton, M. C. Cadmus, C. A. Knutson, G. L. Rowin, and P. A. Sandford, Nature New Biol., 233, 259 (1971).

4. J. E. Zajic (to Kerr-McGee Oil Industries, Inc.), U.S. Pat. 3,320,136 (May 16, 1967).

5. M. E. Slodki, "Yeast Phosphohexosans," in Develop. Ind. Microbiol., Amer. Inst. of Biological Sciences, Vol. 11, 1970, p. 86.

6. A. Jeanes, J. E. Pittsley, P. R. Watson, and R. J. Dimler, Arch. Biochem. Biophys., 92, 343 (1961).

7. T. Harada and T. Yoshimura, Agr. Biol. Chem., 29, 1027 (1965).

8. C. A. Knutson, J. E. Pittsley, and A. Jeanes, Abstracts of Papers 161st ACS Meeting, CARB 28 (1971).

9. A. Jeanes, J. E. Pittsley, and F. R. Senti, J. Appl. Polym. Sci., 5, 519 (1961).

10. Federal Register 34 (53), 5376 (March 19, 1969); 36 (96), 9010 (May 18, 1971).

11. A. Jeanes, C. A. Knutson, J. E. Pittsley, and P. R. Watson, J. Appl. Polym. Sci., 9, 627 (1965).

12. A. Jeanes, J. E. Pittsley, and P. R. Watson, J. Appl. Polym. Sci. 8, 2775 (1964).

13. A. Jeanes, R. G. Bistline, and A. J. Stirton, JAOCS, in press.

14. Heyden Newport Corp., French Pat. 1,395,294 (March 1, 1965).

15. E. Ninomiya and T. Kizaki, Angew. Makromol. Chem., 6, 179 (1969).

16. S. P. Rogovin, U.S. Pat. 3,485,719 (Dec. 23, 1969).

17. A. N. Booth, A. P. Hendrickson, and F. DeEds, Toxicol. Appl. Pharmacol., 5, 478 (1963).

18. A. S. Szczesniak and E. Farkas, J. Food Sci., 27, 381 (1962).

19. E. Segel (to J. E. Siebel Son's Co., Inc.), U.S. Pat. 2,943,942 (July 5, 1960).

20. M. E. Slodki, U.S. Pat. 3,084,105 (April 2, 1963).

21. J. H. Sloneker, D. G. Orentas, and A. Jeanes, Can. J. Chem., 42, 1261 (1964).

22. I. R. Siddiqui, Carbohyd. Res., 4, 284 (1967).

23. J. H. Sloneker, D. G. Orentas, C. A. Knutson, P. A. Watson, and A. Jeanes, Can. J. Res., 46, 3353 (1968).

24. I. R. Siddiqui, Carbohyd. Res., 4, 277 (1967).

25. F. R. Dintzis, G. E. Babcock, and R. Tobin, in Solution Properties of Natural Polymers, Chemical Society (London) Special Publication No. 23, 1968, pp. 195-206.

26. F. R. Dintzis, G. E. Babcock, and R. Tobin, Carbohyd. Res., 13, 257 (1970).

27. P. R. Watson, A. Jeanes, and C. E. Rist, J. Appl. Polym. Sci., 6, S12 (1962).

28. A. Jeanes and J. E. Pittsley, J. Appl. Polym. Sci., in press.

29. T. C. Patton, J. Paint Technol., 38, 656 (1966); Cereal Sci. Today, 14, 178 (1969).

30. W. H. McNeeley, "Biosynthetic Polysaccharides," in H. J. Peppler, Ed., Microbial Technology, Reinhold Publishing Corp., New York, 1967.

31. J. K. Rocks, Food Technol., 25, 476 (1971).

32. L. Valentyik, Trans. Soc. Mining Eng. AIME, 252, 99 (1972).

33. F. H. Deily, G. P. Lindblom, J. T. Patton, and W. E. Holman, Oil Gas J., 62 (June 26, 1967).

34. J. D. Floyd and M. J. O'Connor (to Hercules Inc.), U.S. Pat. 3,484,299 (Dec. 16, 1969).

35. J. E. Pittsley, J. H. Sloneker, and A. Jeanes, Abstracts of Papers 160th ACS Meeting, CARB 21 (Sept. 1970).

36. D. A. Rees, Biochem. J., 126, 257 (1972).

SOME PROPERTIES OF CARBAMOYLATED POLYETHYLENIMINE AND ITS HYDROLYSIS

PRODUCT

Norio Ise and Tsuneo Okubo

Kyoto University

Kyoto, Japan

In a previous paper, we reported on the catalytic influence of polyelectrolytes in the ammonium cyanate - urea conversion (1). In the course of the study, it was found that polymers having primary and secondary amino groups, such as polyethylenimine hydrochloride (PEI.HCl) and poly($\underline{L}$-lysine) hydrobromide, easily react with cyanate ions to form carbamoylated polymers. In the present chapter, we report the details of the preparation of carbamoylated polyethylenimine (CPEI) and its hydrolysis product (HCPEI), their structure, viscosity behavior and the results of an exploratory study of their flocculation properties. A preliminary report was published earlier (2).

PREPARATION

Silver cyanate was prepared from silver nitrate (Merck) and urea (Merck) by the method of Warner and Warrick (3). The polyethylenimine (PEI) used was produced by Nihon Shokubai Chemical Co. or by Dow Chemical Co. A 50% solution of PEI was diluted and purified by ion-exchange resins. The concentration of polyethylenimine was determined by conductometric titration. The molecular weights of PEI (Nihon) and PEI (Dow) are claimed to be 10^4 and 10^5, respectively.

To a PEI.HCl solution, an excess of silver cyanate was added and shaken; the mixture was kept at room temperature for twenty-four hours and the silver chloride produced was separated by centrifugation. The filtrate was condensed under reduced pressure and freeze-dried. Carbamoylated polyethylenimine with high degrees of car-

243

bamoylation was usually obtained as a highly hygroscopic white
powder.

Instead of silver cyanate, sodium cyanate can also be used.
The sodium salt gives, however, sodium chloride as a product of the
carbamoylation reaction which cannot be separated unless a dialysis
technique is applied.

The degree of carbamoylation can be adjusted by changing the
amount of the cyanates and can be determined by conductometric
titration.

Hydrolysis products were obtained from the CPEI (Dow) by treating 13% aqueous solutions with NaOH, the final concentration being
about 4 $\underline{N}$. These solutions were heated at 80°C for eight hours.
The hydrolyzed polymer salted out and was dried at 50°C under reduced pressure. The degree of hydrolysis was determined by conductometric titration to be practically unity. The hydrolysis products (hygroscopic white powders) designated as HCPEI (Dow) (50:100)
and HCPEI (Dow) (100:100) were obtained at yields of 90 and 83%,
respectively. The two figures in the parentheses denote the degree
of carbamoylation and the degree of hydrolysis.

Reaction Kinetics

The carbamoylation reaction was followed by two methods, namely
1. the Volhard method and 2. electric conductivity measurements.
In the first, the amount of unreacted cyanate ions was titrated indirectly by ammonium thiocyanate (3). The conductivity measurements
utilize the fact that the polycations and cyanate anions form neutral
substance as a reaction product. PEI.HCl was mixed with silver
cyanate, silver chloride was filtered off, and the conductivity of
the filtrate measured in a Jones-Ballinger type cell (4) (cell constant, 7.64 cm^{-1}) with an autobalance precision conductivity bridge
of Wayne-Kerr (Model 331) at 1591.55 Hz with an accuracy of 0.01%.
The conductivity method furnishes kinetic data of higher precision
than the Volhard method.

Figure 1 gives time - conversion curves, which were obtained
by the Volhard method. With silver nitrate the conversion is 99%
at 40 min. For silver cyanate and for sodium cyanate, the reaction
slowed down after twenty minutes. The conversion for sodium cyanate
was lower than that for silver cyanate and leveled off at 80%. This
is due to the primary salt effect (5) of sodium chloride, produced
as a by-product, in the reaction between ionic species of unlike
charges.

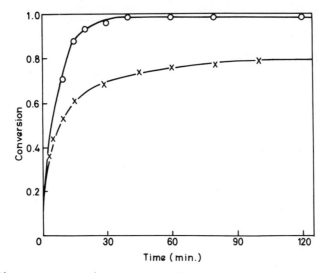

Fig. 1.   Time - conversion curves of the carbamoylation of polyethylenimine hydrochloride at 30°C.

-O-: PEI(Nihon).HCl-AgOCN, [reactant] = 0.2 mole/l
-X-: PEI(Nihon).HCl-NaOCN, [reactant] = 0.1 mole/l

TABLE I.   Kinetic Data for the Carbamoylation Reaction at 30°C
and 40°C

[reactant](mole/l)	k_2 (1/mole-min) 30°C	40°C
	$[PEI^+ - OCN^-]$	
0.00766	1.37	
0.00957	1.15	1.64
0.0139	0.536	1.31
0.0184	0.572	0.889
0.0232	0.457	0.904
0.0312	0.442	0.882
0.0411	0.315	0.582
0.0822	0.196	
	$[NH_4^+ - OCN^-]$ (6)	
0.0376	0.00429	0.0148

The second-order rate constants for PEI (at a degree of neutralization of 0.92) were calculated in a conversion range between 1% and 5%. The second-order plot was linear in this range. The results are shown in Table I, together with those for the NH_4^+ - OCN^- reaction. Clearly at 30°C, the PEI^+ - OCN^- reaction is faster by a factor of about 10^2 than the NH_4^+ - OCN^- reaction. This is quite understandable in view of the strong attractive forces between PEI cations and OCN anions. Table I shows furthermore that the rate constant decreases with increasing concentration. This is a feature of the primary salt effect.

Table II gives the thermodynamic quantities for the carbamoylation reaction. The activation enthalpy (ΔH^*) for PEI^+ is smaller than that for NH_4^+. This is due to the fact that the electrostatic attractive forces between cations and anions become stronger with increasing valency. The activation entropy (ΔS^*) for the polycation reaction is much smaller than that for the NH_4^+ - OCN^- reaction. According to our previous study on the partial molal volume (7), the PEI cation is hydrated with electrostriction by two water molecules per repeating unit at the full degree of neutralization. On the other hand the ammonium is reported to be hydrated by three water molecules, from compressibility measurements (8). If these values are correct, the NH_4^+ - OCN^- reaction should have a larger ΔS^* than the polycation reaction. In addition to the dehydration factor, the local accumulation of OCN^- in the vicinity of the polycation by strong electrostatic attractive forces would result in small ΔS^* values for the PEI^+ - OCN^- reaction.

In Table III, the extent of the hydrolysis of CPEI (Dow), determined by conductometric titration with hydrochloric acid, is shown as a function of time. It is clear that the hydrolysis rate was rather high, and it was assumed that the hydrolysis was complete within several hours in 4 M NaOH aqueous solution.

TABLE II.   Thermodynamic Parameters for the Carbamoylation Reaction at 40°C[a]

Reactant	ΔG^* (kcal/mole)	ΔH^* (kcal/mole)	ΔS^* (cal/deg-mole)
NH_4^+ + OCN^- (1)	23.6	22.9	-2 (+4)
PEI^+ + OCN^-	20.6	8.0	-40.2

[a][reactant] = 0.0513 mole/l.

TABLE III. Extent of Hydrolysis of Carbamoylated Polyethylenimine
(Dow) at 85°C

Time, hr	Hydrolysis, %[a]
0	0
0.5	0.06
2	0.16
5.5	0.34
12.5	0.82

------

[a] based on the primary and secondary amino groups.

CHARACTERIZATION

Elemental Analyses. 100% Carbamoylated PEI (Nihon) was
analyzed with the following results: calcd, H, 6.7%; C, 41.9%,
N, 32.6%; O, 18.7%; Cl, 0%; obsd, H, 8.77%; C, 45.6%; N, 24.4%;
O, 19.2%, Cl, 2.1%. The calculated values were obtained assuming
that the parent PEI contains no primary amine, 75% secondary amine
and 25% tertiary amine, that all of the secondary amine was carbamoylated, and that carbamoylated PEI has the structure to be
mentioned below. The fraction of tertiary amine was determined by
an acetylation method (9). The above results show that almost of
all the secondary amino groups were carbamoylated.

Infrared Spectra. The IR spectra of PEI.HCl and CPEI were
compared using an IR spectrophotometer, model IR-27 of Shimazu
Manufacturing Co., Kyoto. The peaks of carbamoyl groups at
3050-3200 cm^{-1} (ν_{NH}), 1650 cm^{-1} ($\nu_{C=O}$), and 1640 cm^{-1} (δ_{NH})
were observed for CPEI, whereas they did not exist in PEI. Further-
more, the absorptions of protonated secondary amines at 1575-1600
cm^{-1}, 2000 cm^{-1}, and 2250-2700 cm^{-1}, which were observed for the
parent PEI, became weak by carbamoylation. Therefore, it was concluded that partially carbamoylated PEI has the following structure:

```
      H   H         H   H   H
      |   |         |   |   |
    —C - C - N - C - C - N —
      |   |   |     |   |   |   +
      H   H   |     H   H   H
            C=O
             |
            NH₂
```

The IR spectra of HCPEI (Dow) showed absorptions at 1610-1550 cm^{-1} and 1400 cm^{-1} due to carboxylate group. The structure of hydrolyzed carbamoylated polyethylenimine is:

$$
\begin{array}{ccccccc}
H & H & & H & H & \\
| & | & & | & | & \\
-C & -C & -N & -C & -C & -N- \\
| & | & | & | & | & | \\
H & H & C{=}O & H & H & H \\
& & | & & & \\
& & O^- & & & \\
& & Na^+ & & &
\end{array}
$$

Viscosity

The carbamoylation reaction changes cationic amino groups into neutral ones. Thus, as the degree of carbamoylation becomes larger, the polyelectrolyte behavior in the viscosity should become less and less distinct. For example, the intrinsic viscosity, $[\eta]$, decreases with increasing extent of reaction as is shown in Figure 2. The specific viscosity at finite concentrations also decreased with carbamoylation.

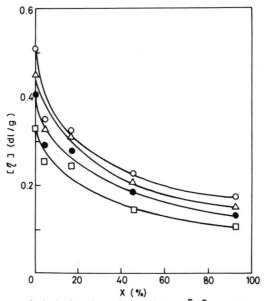

Fig. 2. Change of intrinsic viscosity, $[\eta]$, with degree of car-
        bamoylation (x);
        solvent: H$_2$O; added salt: NaCl; temp: 30°C; PEI (Nihon).
        [NaCl], mole/l -O-: 0.01   -Δ-: 0.02
                        -●-: 0.05   -□-: 0.1

FLOCCULATION PROPERTIES

As is well-known, polyacrylamide and its derivatives are use-ful polymer flocculants. Their flocculation rate is fairly high but the transparency of the supernatant fluid is not always satis-factory. On the other hand, PEI gives small rate of flocculation with high transparency. Thus, it is quite interesting to study the flocculation properties of CPEI, which has the dual structural character of PEI and polyacrylamide.

Carbamoylated polyethylenimine for the flocculation tests was prepared from the hydrochloride of Dow PEI and NaOCN. Aqueous sus-pended solutions of kaolin, bentonite, or diatomaceous earth and a CPEI solution were mixed in a measuring cylinder with stopper, and the pH of the resulting mixture was adjusted with HCl or NaOH. The minerals were supplied from Nakarai Chemicals, Kyoto, and used without further purification in the present exploratory study. The mixture was suspended completely by rapidly turning the cylinder up and down 10 times. The rate of precipitation was recorded. After 4-10 min the transparency of the supernatant liquid at 630 mμ and the precipitate volume were read. All measurements were car-ried out at 23°C. Tables IV, V, VI, and VII give the results. There are several points to be mentioned.

1. The rate of flocculation with CPEI is higher than that with Nihon PEI or Dow PEI. It is also much higher than that with cationic polyacrylamide (except for the kaolin suspension at pH=5.0) for the bentonite and diatomaceous earth suspensions at pH 5.5-10.2, where cationic polyacrylamide cannot cause flocculation during the time interval investigated. It is seen further that the rate of floc-culation with hydrolyzed carbamoylated polyethylenimine is larger than that with CPEI. HCPEI (50:100) gives a slightly higher rate than HCPEI (100:100) (see Table VII).

2. In most cases, HCPEI gives higher transparency than CPEI, and especially more than PEI. At small dosages HCPEI (50:100) gives higher transparency than HCPEI (100:100). In comparison with Dow PEI, the change of the transparency by carbamoylation appears not to have a definite tendency. However, CPEI seems to show much higher transparency than polyacrylamide.

3. The volume of precipitate is larger with HCPEI than with CPEI' or PEI. It seems that CPEI gives a larger volume of precipitate than does cationic polyacrylamide as can be observed from the sedi-mentation boundary.

4. In Table VI, the flocculation rate is seen to become higher with increasing degree of carbamoylation. The transparency becomes small at high degree of carbamoylation. The precipitate volume also decreases by carbamoylation.

TABLE IV.   Flocculation of 1% Kaolin Suspensions

Flocculant	Dosage ppm	Transparency, %	Flocculation Rate, cm/min	Precipitate Volume, ml
		pH = 5.0		
CPEI-100[a]	2.5	29.9[c]	16	15.1
PEI(Dow)	2.5	10.3[c]	-	16.6
PEI(Nihon)	2.5	3.0[c]	8	13.8
Cat. polyacrylamide[b]	2.5	45.0[c]	24	11.6
		pH = 8.0		
CPEI-100[a]	5	5.7	19.2	14.0
	10	59.7	19.2	18.5
	25	93.1	16.9	19.0
PEI(Nihon)	5	91.0	10.3	23.0
	10	59.0	19.7	22.5
Polyacrylamide[d]	20	0	f	6.5
	90	0	f	9.0
Cat. polyacrylamide[b]	5	11.4	f	14.5
	10	11.4	f	14.0
		pH = 9.0		
CPEI-100[a]	10	13.9[e]	19	13.0
PEI(Dow)	5	58.8[e]	12	15.2
Cat. polyacrylamide[b]	10-60	f	f	f
		pH = 10.2		
CPEI-100[a]	10	0.3[e]	32	6.2
PEI(Dow)	5	0.4[e]	-	9.5
	10	11.7[e]	16	13.8
Cat. polyacrylamide[b]	5-100	e	f	f

– – – – – – –

[a]100% carbamoylated Dow PEI
[b]commercially available cationic derivative
[c]after 4 minutes
[d]commercially available grade (degree of polymerization = 10^4)
[e]after 10 minutes
[f]no flocculation

TABLE V.   Flocculation of 1% Bentonite Suspensions

Flocculant[a]	Dosage ppm	Transparency, %[b]	Flocculation Rate, cm/min	Precipitate Volume, ml
		pH = 9.0		
CPEI-100	100	35.7	4	41.0
PEI(Dow)	100	94.2	0.9	63.0
		pH = 10.2		
CPEI-100	50	46.2	1.2	57.0
	100	20.2	2.5	44.0
PEI(Dow)	50	35.9	0.6	74.8
	100	58.3	1.0	45.0
PEI(Nihon)	100	51.2	0.8	66.0
Cat.Polyacrylamide	50-100	0.7	c	c

- - - -

[a]see Table IV for description
[b]after 10 min
[c]no flocculation

TABLE VI.   Flocculation of 1% Kaolin Suspensions by CPEI of Various Degrees of Carbamoylation at pH = 8.0

Degree of Carbamoylation, %	Dosage ppm	Transparency, %	Flocculation Rate, cm/min	Precipitate Volume, ml
98	5	5.7	19.2	14.0
	10	59.7	19.2	18.5
94	5	78.0	13.1	21.0
	10	86.5	12.5	20.5
58	5	81.8	10.7	22.0
	10	68.0	11.5	18.0
20	5	88.6	10.3	24.0
	10	59.5	10.7	22.5
0	5	91.0	10.3	23.0
	10	59.0	10.7	22.5

TABLE VII. Flocculation of 1% Diatomaceous Earth Suspensions at 23°C

Flocculant	Dosage ppm	Flocculation Rate, cm/min	Transparency, %[a]	Precipitate Volume, ml[b]
		pH = 5.5		
HCPEI	0.25	19	70	11.8
(100:100)	0.5	16	78	14.0
	1.5	14	82	13.8
	3.5	14	76	13.0
HCPEI	0.25	12	80	11.5
(50:100)	0.5	19	86	11.2
	1.5	16	73	12.7
CPEI-100	1.5	12	77	12.2
PEI	1.5	7	70	10.0
Cat.Polyacrylamide	1.5	no flocculation		
		pH = 9.0		
HCPEI	2.5	12	60	12.7
(100:100)	5	13	72	12.0
	15	14	68	12.0
	25	12	64	12.0
HCPEI	1	19	81	13.1
(50:100)	2.5	16	82	12.1
	5	12	71	11.2
CPEI-100	5	9	61	9.8
PEI	5	very slow	22	8.0
Cat.Polyacrylamide	5-40	no flocculation		
		pH = 10.2		
HCPEI	2.5	9	50	9.0
(100:100)	10	9	74	12.0
	25	9	53	10.7
HCPEI	1.5	14	58	10.5
(50:100)	2.5	14	70	10.7
	10	10	67	11.1
CPEI-100	5	very slow	31	5.5
PEI	10	very slow	31	6.7
Cat.Polyacrylamide	100	no flocculation		

- - - - - -

[a]measured at 660 mμ after 8 min of mixing

[b]measured after 8 min of mixing

Though further detailed study is necessary before definite conclusions can be drawn, it seems at present that CPEI shows high flocculation velocity especially at neutral and alkaline pH values at which polyacrylamide, its cationic derivatives, or polyethylenimine, which are commercially available, are not highly effective. Hydrolysis increases the rate of flocculation and increases or did not affect the transparency in comparison with carbamoylated polyethylenimine.

A part of the experimental work has been carried out by H. Moritani. The cationic derivative of polyacrylamide was donated by M. Hirooka, Sumitomo Chemicals Company, Osaka, Japan.

BIBLIOGRAPHY

1. T. Okubo and N. Ise, Proc. Roy. Soc. (London), A327, 413 (1972).

2. N. Ise, H. Moritani, and T. Okubo, Polymer, 13, 187 (1972).

3. J. C. Warner and E. L. Warrick, J. Am. Chem. Soc., 57, 1491 (1935).

4. J. Jones and M. Ballinger, J. Am. Chem. Soc., 53, 411 (1931).

5. For the primary salt effect, see textbooks of physical chemistry, e.g., W. J. Moore, Physical Chemistry, Prentice-Hall, Inc., Englewood Cliffs, N. J., 1962, p. 368.

6. W. J. Svirbely and J. C. Warner, J. Am. Chem. Soc., 57, 1883 (1935).

7. N. Ise and T. Okubo, J. Am. Chem. Soc., 90, 4527 (1968).

8. J. F. Hinton and E. S. Amis, Chem. Rev., 71, 627 (1971).

9. C. D. Wagner, J. Am. Chem. Soc., 69, 2609 (1947).

WATER-SOLUBLE POLYISOTHIOCYANATES

G. D. Jones, Carole Kleeman, and Doreen Villani-Price

The Dow Chemical Company

Midland, Michigan

Polyisothiocyanates have been made as oil-soluble polymers of vinylphenyl isothiocyanate (1) and as biological absorption materials (2). In earlier work it was reported (3) that the water-soluble copolymer of acrylamide and vinyl isothiocyanate was stored up to a week at room temperature and then gelled by the addition of hexamethylenediamine. In the present work, water-soluble polyisothiocyanates have been made from polymers of 2-chloromethylbutadiene. The preparation of this monomer and of water-soluble polymers from it has been described previously (4).

PREPARATION AND CHARACTERIZATION

2-Chloromethylbutadiene was polymerized with boron trifluoride at -10°C, with azobisisobutyronitrile(AIBN) and bromoform at room temperature and at 50°C in emulsion with 0.5% t-dodecyl mercaptan. Some 10-20% hydrolysis occurred during emulsion polymerization. In all methods of polymerization it was necessary to interrupt polymerization short of completion in order to preserve solubility. The emulsion polymer had the same elution time in gel permeation chromatography (GPC) as polystyrene of molecular weight 80,000, as shown in Figure 1, but is of smaller size. The boron trifluoride-catalyzed polymer was sticky rather than rubbery. The molecular weight peaked at 5,000 but the weight average was 67,000.

The NMR spectrum of poly(2-chloromethylbutadiene), a portion of which is shown in Figure 2, is composed of three main bands: a high field band at 2.2 ppm due to $=CCH_2$, a band consisting of partially resolved lines at 4.04 and 4.06 due to $=C-CH_2Cl$, and a

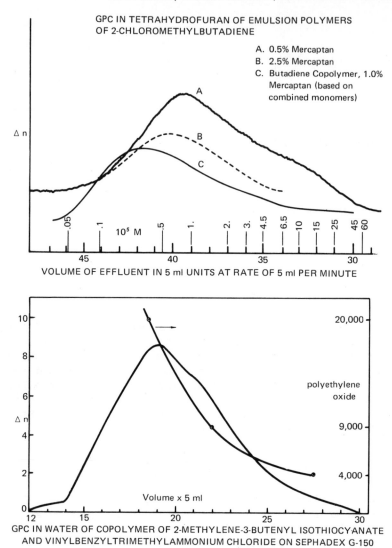

GPC IN TETRAHYDROFURAN OF EMULSION POLYMERS
OF 2-CHLOROMETHYLBUTADIENE

A. 0.5% Mercaptan
B. 2.5% Mercaptan
C. Butadiene Copolymer, 1.0% Mercaptan (based on combined monomers)

VOLUME OF EFFLUENT IN 5 ml UNITS AT RATE OF 5 ml PER MINUTE

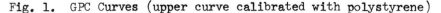

GPC IN WATER OF COPOLYMER OF 2-METHYLENE-3-BUTENYL ISOTHIOCYANATE
AND VINYLBENZYLTRIMETHYLAMMONIUM CHLORIDE ON SEPHADEX G-150

Fig. 1.   GPC Curves (upper curve calibrated with polystyrene)

low field band having two lines at 5.56 and 5.42 superimposed on an
unresolved band. From the work of Schué and Dole-Robbe (5a) the
5.56 line is believed to be due to the =CH of the cis structure and
the 5.42 line due to that of the trans structure. The BF$_3$-catalyzed
polymer is mainly trans. The amount of 1,2-structure is small in
the free-radical polymer, approximately 3% as shown by the small
line at 3.52 ppm due to nonallylic -CH$_2$Cl. The amount of 1,2 plus
3,4-structure is about 10% as shown by a small band at 1.5-1.9 ppm.
The corresponding values for the BF$_3$-catalyzed polymer are approximately 20 and 42%, respectively.

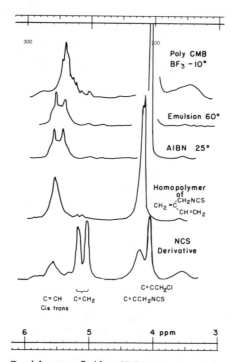

Fig. 2.   Portions of the NMR spectra at 100 MHz.

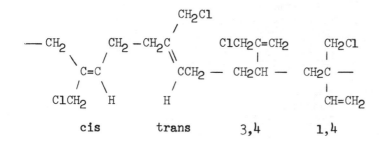

The infrared spectrum of the free-radical polymer resembled that of emulsion polyisoprene (5b) with respect to the $=CH_2$ at 10.7 μ and trans = CH at 11.5 μ.

The chlorine in poly(2-chloromethylbutadiene)is readily reacted with trimethylamine or dimethyl sulfide to produce a water-soluble derivative. It was found that the half quaternary and half sulfonium derivative would react with inorganic thiocyanate in water solution to give the isothiocyanate. Only the sulfonium groups were displaced. It was then found that it was not necessary to make the sulfonium intermediate. The water-soluble derivative,

obtained by amination of half of the chlorine, reacted with inorganic
thiocyanate to give the water-soluble isothiocyanate derivative.
Residual thiocyanate ion could be removed by dialysis against saline
solution. A solution of this derivative was readily gelled by the
addition of a diamine. It was further derivatized with p-bromoan-
iline, ammonia, and hydrazine.

Nor was it necessary to preaminate the polymer in order to
obtain a polyisothiocyanate. The dioxane solution of poly(2-chloro-
methylbutadiene) reacted with excess ammonium thiocyanate in dioxane
solution to give a polyisothiocyanate which in this case was not
water-soluble. A solution in dimethyl sulfoxide could be gelled
by hexamethylenediamine. One could also limit the reaction with
ammonium thiocyanate and then post aminate the remaining chlorine
with trimethylamine to give a water-soluble polyisothiocyanate.
This sequence of reactions, however, was accompanied by some hydrolysis of the isothiocyanate group and sometimes resulted in crosslinking.

Preparation and Polymerization of
2-Methylene-3-Butenyl Isothiocyanate

The monomeric isothiocyanate, 2-methylene-3-butenyl isothiocyanate, was made. 2-Chloromethylbutadiene reacted with ammonium
thiocyanate in t-butyl alcohol at 50°C to produce a mixture of the
thiocyanate (mainly) and some isothiocyanate. Upon distillation,
rearrangement occurred and the isothiocyanate, which boiled at 60°C
at 7 mm (lower than the thiocyanate), was separated as an odoriferous yellow liquid (n_D^{25} 1.5660; d_4^{25} 1.0174). The thiocyanate could
be concentrated by distillation at a lower temperature where rearrangement was slow. Analysis by GLC on Gas Chrome Z was consistent with NMR if the column was short (2 ft) and operated at
80°C. The infrared spectrum showed diene vibrations at 1642 and
1601 cm^{-1}, and the allyl CH band at 1442 cm^{-1} gave rise to Fermi
resonance bands at 2925 and 2861 cm^{-1}. The -NCS antisym vibration
came at 2092 cm^{-1}. The vinylidene wagging band was observed at
907 and vinyl twisting and wagging vibrations at 989 and 912 cm^{-1}.

2-Methylene-3-butenyl isothiocyanate is a diene with a strongly electron-withdrawing group in a nonconjugated position. It does
not show the high dimerization tendency that is shown in water solution by 2-methylene-3-butenyltrimethylammonium chloride or the
dimethylsulfonium chloride. (It is not water-soluble.) The downfield shift of diene hydrogens of 2-methylene-3-butenyl isothiocyanate is not as large an effect as that produced by the protonated amino group in the isoprene derivative nor that of vinyl isothiocyanate.

TABLE I.   Comparative NMR Values on
2-Chloromethylbutadiene Derivatives

Chemical shift, ppm relative to TMS in $CDCl_3$					Coupling constants		
	1	3	4	5	J3,4a	J3,4b	Ja,b
NCS	5.33,	6.40	5.18,	4.31	18.4	11	21
	5.26		5.19				

Downfield shift relative to
    isoprene

			$\Delta\delta$		δ
	solvent	1	3	4	
NCS	$CDCl_3$	0.43	0.05	0.19	4.31
SCN	$CDCl_3$				3.75
Cl	$CDCl_3$	0.33	0.02	0.28	4.17
$N(CH_3)_3Cl$	DMSO	0.88	0.19	0.52	4.41
NH_2	DMSO	0.30	0.08	0.14	3.31
ND_2DCl	D_2O	0.57	0.14	0.33	3.85

$$CH_2=C \overset{\displaystyle CH_2-}{\underset{\displaystyle C=CH_2}{}}$$

    H
    1   3   4
                5

2-Methylene-3-butenyl isothiocyanate (containing 10% of the
thiocyanate isomer) was polymerized with AIBN at 60°C in the presence of bromoform (6.6 wt %). The polymer was precipitated in
ether.   By GPC in tetrahydrofuran it had the retention of polystyrene of molecular weight 7,000. The UV absorption paralleled
the refractive index curve. Like allyl isothiocyanate, the polymer
had an absorption maximum at 250 mµ. The polymer was not soluble
in carbon tetrachloride or ethanol but readily soluble in chloroform or mixtures of chloroform with chlorinated solvents and in
tetrahydrofuran.

A portion of the NMR spectrum of the homopolymer of 2-methyl-
ene-3-butenyl isothiocyanate is shown in Figure 2, with bands at
5.53 ppm due to =CH and 4.15 ppm due to $=CCH_2NCS$. There is also a
band at 2.24 ppm due to $=CCH_2-$. The amount of 1,2 and of 1,2 plus
3,4 structure is estimated to be 5.5 and 11%. The spectrum of the
isothiocyanate derivative of emulsion poly(2-chloromethylbutadiene)

shows both unreacted =CCH$_2$Cl and =CCH$_2$NCS at 4.07 and 4.22 ppm, respectively. There remains a cis =CH band at 5.62 but the trans =CH is gone and there are two new bands at 5.22 and 5.07 which are assigned to CH$_2$= of the structure below which results from an allylic rearrangement. There is an enhanced band at 1.8-1.9 ppm, due to the nonallylic CH$_2$.

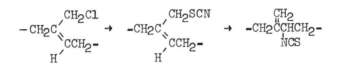

With an unheated sample the intermediate thiocyanate can be detected in the infrared spectrum by the sharp line at 4.7 μ superimposed on the broad isothiocyanate band at 4.8 μ. A warmed sample shows a reduction in allylic CH$_2$Cl wag at 7.9 μ, a loss of =CH bending band at 11.6 μ and increase in the =CH$_2$ wag band at 11.0 μ due to the allyl rearrangement. A further effect was the appearance of a conjugated C=C stretch band at 6.2 μ, indicating some double-bond shift to give the conjugated isothiocyanate. It was not for this reason, however, that the derivative was yellow; vinyl isothiocyanate is not colored (it has an absorption band at 227 mμ). There was present in the 2-chloromethylbutadiene about 2% 1-chloro-2-methyl-butadiene which gave rise to a methyl band at 7.25 μ. This band was not noticeably increased by the thiocyanate reaction. The infrared spectrum of the homopolymer differs from that of the derivative in showing strong bands for the sym-NCS stretch at 8 μ and -CH$_2$NCS wag at 7.5 μ.

Copolymerization. Copolymerization of 2-methylene-3-butenyl isothiocyanate and vinylbenzyltrimethylammonium chloride gave a water-soluble curable copolymer from a feed having a monomer ratio 1:1.73. The polymerization was carried out in t-butyl alcohol at 55° with azobisisobutyronitrile as the catalyst, the polymer precipitating as it formed. The polymer emerged from a Sephadex G-150 column at the same elution volume as a polyglycol of molecular weight 17,000 (Figure 1). A 10% water solution of the copolymer was treated with a water solution of hexamethylenediamine and a gel formed within a minute.

The half-life of the isothiocyanate group in water solution was observed with this copolymer. The shelf life was limited by a tendency to crosslink on long storage rather than by any substantial loss of isothiocyanate groups. A 10% solution containing diethyl tartrate as an internal standard was used to cast films periodically over a period of several weeks. There was no change in the ratio of infrared absorption bands of the functional groups.

Copolymerization with acrylamide involved preferential polymerization of the diene but not so greatly as with butadiene. The
initial copolymer from a monomer feed having a mole fraction of
acrylamide of 0.94 was found to have a mole fraction of contained
acrylamide of 0.79 whereas a value of 0.59 would be expected with
butadiene. This copolymer gave a turbid water solution which could
be gelled by the addition of hexamethylenediamine.

Other Reactive Halides

The dimers of 2-chloromethylbutadiene contain allylic chlorine;
both chlorines are allylic in the higher boiling dimer II. The
ring chloromethyl group has the structure corresponding to <u>trans</u>-
crotyl chloride. Dimer II gives a diisothiocyanate on reaction at
room temperature with inorganic thiocyanate. The dimerization of
2-chloromethylbutadiene gives I and II in equal amounts but the
dimerization of the aqueous sulfonium derivative of 2-chloromethyl-
butadiene give 79% II. Reaction of this mixture with inorganic

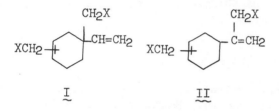

thiocyanate gave a product which acted as a curing agent for a
polyamine. It was, however, odoriferous and not water-soluble.

In contrast to the results with 2-chloromethylbutadiene and
polymers thereof, many other reactive halides give the thiocyanate
as the isolated product. These include 1,4-dichlorobutene,
α,α'-dichloro-<u>p</u>-xylene, bis-chloromethyl diphenyl ether, vinylbenzyl
chloride and poly(vinylbenzyl chloride) and include the sulfonium
derivatives thereof.

Vinylbenzyl Thiocyanate

Vinylbenzyl thiocyanate rearranged poorly upon distillation.
It was rearranged in refluxing dimethylformamide containing potassium iodide. The rearrangement rate was independent of the concentration of thiocyanate ion, as expected. The rearrangement of
poly(vinylbenzyl thiocyanate) was accompanied by crosslinking unless the iodide concentration was low. Vinylbenzyl isothiocyanate
was made by the reaction of potassium thiocyanate with vinylbenzyl

chloride in refluxing dimethylformamide containing potassium iodide. The method was reported by others (6). A mixture of isothiocyanate and thiocyanate was obtained from the reaction of vinylbenzyl nitrate and ammonium thiocyanate and from the reaction of vinylbenzyl chloride with mercuric thiocyanate.

Distinguishing the vinylbenzyl products by NMR required the use of a dilute solution in an aliphatic solvent in order to spread the chemical shifts. At 10% concentration in $CDCl_3$ the CH_2NCS of vinylbenzyl isothiocyanate comes at 4.58 ppm and the CH_2Cl of vinylbenzyl chloride at 4.53 (the CH_2SCN of vinylbenzyl thiocyanate, at 4.02). At 5% concentration the values are 4.70, 4.57 and 4.16 ppm. In dimethylformamide they are 4.94, 4.80 and 4.48 ppm.

The meta/para vinylbenzyl isothiocyanate had an UV absorption band at 250 mµ as does benzyl isothiocyanate; that of benzyl thiocyanate is at 261 mµ.

Inamuch as vinylbenzylthiourea was one of the desired derivatives of vinylbenzyl isothiocyanate, it was attempted to make it from vinylbenzylamine and thiocyanic acid. This salt is, however, resistant to rearrangement.

APPLICATIONS

The half-quaternary half-isothiocyanate derivative was tested as a binder for nonwoven sheets. Handsheets were made using 1 part paper pulp and 3 parts rayon with binder added to the beaten pulp. The binder was precipitated and cured on the pulp by the addition of a copolymer of methacrylic acid and 2-aminoethyl methacrylate. For comparison, the amino polymer was coacervated on the pulp with

TABLE II.   Beater Additive Tests with Half-Quaternary Half-Isothiocyanate Derivative

Sample	Loading, %	pH	Tensile, psi	Elong, %
Control			12	1.9
Derivative alone	22		67	4.4
Coacervated and cured	18.3	acid	137	5.2
	15.9	alkaline	154	5.7
	16.3*	alkaline	186	5.9
Coacervated only	13.1	acid	105	5.2
	10.4	alkaline	75	4.5
	10.8*	alkaline	80	5

poly(vinylbenzyltrimethylammonium chloride) and did not give as strong a sheet. In the tests marked with an asterisk the amino polymer was a terpolymer containing acrylamide and of higher vis-

cosity. Nitrogen analysis indicated that retention of the co-
acervates was complete but only 56% was retained when the derivative was used alone.

A sample of anisotropic nonwoven rayon was impregnated with
resin and the strength tested in both cross and machine directions.
The results were not sensitive to drawing rate as long as it was
greater than 0.5 in./min. Sample A was the homopolymer of 2-methyl-
ene-3-butenyl isothiocyanate which was applied from a solution in

TABLE III.   Strength of Impregnated Nonwoven Tested at 10 in./min
             in Cross and Machine Directions

Additive	Add-on %	Direction	Tensile, psi	Elong, %
A	17	CD	60	48
		MD	1180	28
B	27	CD	84	17
		MD	1923	7
Acrylic latex	17	CD	142	79
		MD	1940	8.7
Control		CD	6.2	57
		MD	243	4.5

1,1,1-trichloroethane containing 20% chloroform. It was soaked in
aqueous hexamethylenediamine (0.2 $\underline{M}$), and cured at 85°C for twenty
minutes. Sample B was the reaction product of 2-chloromethylbuta-
diene - isoprene (50 wt %) copolymer and ammonium thiocyanate. It
was applied from methylene chloride solution and cured at 100°C for
fifteen minutes. A portion was cured with ammonia.

Mylar[R] film was coated with half-quaternized poly(2-chloro-
methylbutadiene) which had been reacted with ammonium thiocyanate
and to which hexamethylenediamine had been added as crosslinking
agent. The coating weight was 20 mg-cm^{-2} and the half-life of
charge retention was 0.7 sec at 17% RH. While this charge decay
was desirably rapid, the coating swelled on immersion in water.
The lower charge density of a derivative made from a butadiene copolymer of 2-chloromethylbutadiene produced a more durable coating.

Polypropylene glycol was mixed with the half-quaternary,
half-isothiocyanate derivative and cured by heating with stannous
octoate.

A polyisothiocyanate can be used to heat-cure a polycarboxylic
acid in its free form. In a test using the model compounds,

p-phenylene diisothiocyanate and adipic acid, it was found that both carbon oxysulfide and carbon dioxide were evolved at 160°C. Both amide and thioamide linkages had formed.

ISOTHIOCYANATE VS. THIOCYANATE

It has been suggested by Fava (7) and Renson (8) that Sn1 character in halide displacement favors isothiocyanate over thiocyanate in the product. Streitwieser (9) involved a π-allyl complex formulation of the transition state in reactions of allyl chlorides. The reactivity of allyl and benzyl chlorides to nucleophiles increases greatly as the ionizing power of the solvent increases. cis-Crotyl chloride, which is more reactive than the trans isomer (10), may be considered a model for that portion of poly(2-chloromethylbutadiene) which has the trans-1,4-structure and methallyl chloride for the 3,4-structure. Crotyl chloride is reported to

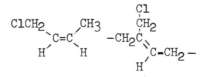

solvolyze much faster than allyl or methallyl chloride (11). It gives the thiocyanate as the initial product of reaction with inorganic thiocyanate in ethanol (12,13) but rearranges on warming. Benzyl chloride and especially p-methylbenzyl chloride solvolyze faster than allyl chloride and yet give a relatively stable thiocyanate as the main product (14). Apparently the Sn2 mechanism takes precedence in the reaction with such a strong nucleophile as thiocyanate and the thiocyanate once formed is not very susceptible to solvolysis. Sulfonium compounds react rapidly by Sn2 mechanism in nonpolar solvents.

Benzyl nitrate has been suggested to be more susceptible to solvolysis than benzyl chloride (15). It is well known that solvolysis can be promoted by silver or mercuric ion. It is not clear, however, whether the isothiocyanate is formed as a primary or rearrangement product in these cases.

The chemical shifts of the various allylic hydrogens can be correlated qualitatively with the ease of rearrangement to the isothiocyanate (Table IV).

TABLE IV.   Chemical Shift of CH_2Cl of the Electrophiles Used, ppm

	DMF	t-BuOH	CHCl$_3$	CCl$_4$	"Neat"	Benzene
2-Chloromethylbutadiene					4.17	3.95
Dimer II side chain				4.13		
Dimer II ring				4.03		
Allyl chloride			4.02			3.75
Crotyl chloride			3.92			3.9
Vinylbenzyl chloride	4.8	4.5	4.57		4.3	4.24
Vinylbenzyl nitrate		5.25				

Allyl chloride and crotyl chloride give doublets whereas 2-chloromethylbutadiene gives a single line at 60 MHz. Vinylbenzyl chloride might be expected from the downfield location of the $-CH_2Cl$ to be less prone to solvolysis than 2-chloromethylbutadiene. Low water solubility is an unfavorable factor, also, and is not entirely corrected by use of mixed solvents. Thus a low concentration of t-butyl alcohol is reported to severely retard solvolysis of benzyl chloride in aqueous solution (16).

BIBLIOGRAPHY

1.   N. A. Barba, A. M. Shier, and N. K. Zin, Izv. Vyssh. Ucheb. Zaved., Khim. Teknol., 13, 267 (1970); CA, 73, 25899.
     A. M. Shur and N. A. Barba, J. Org. Chem. (USSR), 2, 1819 (1966).
     C. G. Overberger and H. A. Friedman, J. Polymer Sci., Part A, 3, 3625 (965).

2.   G. Manecke and G. Gunzel, Naturwiss.,54, 531 (1967).
     R. Axen and J. Porath, Nature, 210, 367 (1966).
     L. H. Kent and J. H. R. Slade (to Secretary of State Defence), Brit. Pat. 993,961 (1965);CA, 63, 8130.

3.   G. D. Jones and R. L. Zimmerman (to Dow Chemical Co.) U. S. Pat. 2,757,190 (1956).

4. G. D. Jones, N. B. Tefertiller, C. F. Raley, and J. R. Runyon, J. Org. Chem., 33, 2946 (1968).
   G. D. Jones, W. C. Meyer, N. B. Tefertiller, and D. C. MacWilliams, J. Polymer Sci., Part A-1, 8, 2123 (1970).
   G. D. Jones, G. R. Geyer, and M. J. Hatch (to Dow Chemical Co.), U. S. Pat. 3,544,532, U. S. Pat. 3,494,965 (1970).

5a. F. Schué and J. P. Dole-Robbe, Bull. Soc. Chim., 975 (1963).

5b. W. S. Richardson and A. Sacher, J. Poly. Sci., 10, 353 (1947).

6. E. J. Tarlton and A. F. McKay (to Monsanto Can. Ltd.), Ger. Pat. 1,148,540 (1963); CA, 60, 2825.
   Syoneda, H. Kitano, and K. Fukui, Kogyo Kagaku Zasshi, 65, 1816 (1962); CA, 59, 2679.

7. A. Fava, A. Iliceto, and S. Bresadola, J. Am. Chem. Soc., 87, 4791 (1965).

8. Renson, Bull. Roy. Soc. (Liège), 29, 78 (1960).

9. A. Streitwieser, Jr., Solvolytic Displacement Reactions, McGraw Hill, N. Y., 1962.

10. L. J. Brubacher, L. Freindl, and R. E. Robertson, J. Am. Chem. Soc., 90, 4611 (1968).

11. R. H. De Wolfe and W. G. Young, Chem. Rev., 56, 753 (1956).

12. W. G. Young, I. D. Webb, and H. L. Goering, J. Am. Chem. Soc., 73, 1081 (1951).

13. O. Mann and H. Richter, Ber., 73B 843 (1940).

14. U. Tonellato and G. Levorato, Boll. Sci. Fac. Chem. Ind. (Bologna), 27, 261 (1969).

15. G. R. Lucas and L. P. Hammett, J. Am. Chem. Soc., 64, 1928 (1942).

16. S. J. Dickson and J. B. Hyne, Can. J. Chem., 49, 2394 (1971).

PREPARATION OF CATIONIC POLYELECTROLYTES BY SPONTANEOUS POLYMERIZATION

J. C. Salamone and E. J. Ellis

Lowell Technological Institute

Lowell, Massachusetts

In recent years the preparation and study of cationic poly-electrolytes have been of considerable interest. This field has recently been reviewed (1), and it can be noted that the syntheses of these water-soluble polyions has been concerned primarily with the preparation of ionene polymers (2) by polycondensation reactions, with the polymerization of cationic vinyl monomers in which the positively charged site is located far from and not in resonance with the reactive double bond, and with the quaternization of preformed, nucleophilic polymers. In general, there have been few studies on the preparation of cationic polyelectrolytes from monomers in which the positive charge is in resonance with the double bond. Furthermore, of the quaternary vinyl monomers that have been prepared, several have been shown not to undergo polymerization (1).

Several years ago Kabanov and Kargin reported that cationic polyelectrolytes based on poly(4-vinylpyridine) could be prepared in one step by the spontaneous polymerization of 4-vinylpyridine (I) by alkyl halides (3-5). These reactions were reported to result in the formation of poly(4-vinyl-N-alkylpyridinium salts) (II) with no formation of monomeric salts (reaction 1). Similar spontaneous polymerizations were reported to occur when strong acids were reacted with 4-vinylpyridine producing the protonated salts of poly(4-vinylpyridine) (III, reaction 2). The mechanisms of initiation of these polymerizations had been reported by Kabanov to result from the attack of the anionic counterion of the initially formed, unstable monomeric salt which yielded an anionic carbon capable of further attack (propagation) on other cationic, monomeric salts.

SPONTANEOUS POLYMERIZATION OF 4-VINYLPYRIDINE

$$-CH_2-CH-$$

(1)

II

(2)

III

I

Recently, however, it has been reported by our group (6-8) and by Ringsdorf et al (9-11) that stable monomeric salts could be prepared from 4-vinylpyridine with several alkylating agents and monomeric acids. In contrast to the suggested counterion initiation mechanism, polymerization was found to be initiated by a pyridylethylation reaction involving the attack of neutral 4-vinylpyridine upon the β-position of a 4-vinylpyridinium ion. Whereas this reaction leads to poly(4-vinyl-N-alkylpyridinium salts) (II) when 4-vinylpyridine is treated with alkylating agents at room temperature, the reaction with strong acids in dilute solution leads to the ionene structure (IV) in which the positively charged pyridinium ring is in the polymer main chain (6,7,9). It was further

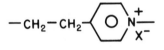

IV

shown that the acidic salts of poly(4-vinylpyridine) (III) resulted only with concentrated strong acids. The mechanism of the latter type of spontaneous polymerization appears to be quite different than that with dilute acids, and this is presently under investigation.

In this work we wish to present how spontaneous polymerization could be used to prepare cationic polyelectrolytes from one-step reactions from 2- and 4-vinylpyridine with either alkylating agents or concentrated, strong acids. Conditions under which other monomers can undergo spontaneous polymerization will be discussed.

RESULTS AND DISCUSSION

In order to obtain spontaneous polymerization by alkylating agents or by dilute, strong acids with either 2- or 4-vinylpyridine (6-8,12,13), it is necessary that the initially formed monomeric salts be in a state favorable for polymerization. In cold solvents, these alkylation and protonation reactions can lead to crystalline monomeric salts which are not attacked (initiated) by neutral vinylpyridine. Consequently, these reactions should be done at room temperature for polymer formation. As mentioned previously, spontaneous polymerization reactions with concentrated strong acids appear to involve a different mechanism of polymerization, and other conditions could be utilized to prepare cationic (protonated) polyelectrolytes. These are described below.

Spontaneous Polymerization by Alkylating Agents

In Table I are listed the results of several spontaneous polymerization reactions between 2-vinylpyridine (2VP) and 4-vinylpyridine (4VP) with either dimethyl sulfate or methyl iodide. The reactions were done in benzene at room temperature for 3 or 5 days using 1.0 ml of vinyl monomer. The ratios of alkylating agents to vinyl monomers and the amount of solvent were varied in order to determine the most favorable conditions for the formation of high-molecular-weight polymers. The water-soluble products poly(4-vinyl-N-methylpyridinium methylsulfate) (V, $X^- = CH_3SO_4^-$) and iodide (VI, $\bar{X}^- = I^-$) and poly(4-vinyl-N-methylpyridinium methylsulfate) (VII, $X^- = CH_3SO_4^-$) and iodide (VIII, $X^- = I^-$) were purified by precipitation from trifluoroethanol into either acetone or ethyl ether. All the products were hygroscopic.

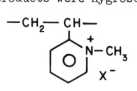

$$V, X^- = CH_3SO_4^-$$
$$VI, X^- = I^-$$

$$VII, X^- = CH_3SO_4^-$$
$$VIII, X^- = I^-$$

TABLE I.   Spontaneous Polymerization Reactions by Alkylation

Monomer	RX	$\dfrac{[RX]}{[Monomer]}$	Solvent, ml benzene	Days	Conv., %	$[\eta]^a$
2VP	$(CH_3)_2SO_4$	3.34/1.00	75	5	95	$8.0, 12.0^b$
2VP	$(CH_3)_2SO_4$	1.05/1.00	75	5	18	1.75
2VP	$(CH_3)_2SO_4$	1.05/1.00	10	3	86	6.5
4VP	$(CH_3)_2SO_4$	3.34/1.00	75	5	70	7.7
4VP	$(CH_3)_2SO_4$	1.05/1.00	75	5	25	2.5
4VP	$(CH_3)_2SO_4$	1.05/1.00	10	3	94	2.2
2VP	CH_3I	11.34/1.00	75	5	25	0.12
4VP	CH_3I	11.34/1.00	75	5	85	0.59

- - - - - - - -

^{a}obtained in 0.01 $\underline{M}$ KBr solution at 25 C; dl/g
^{b}data obtained for two reactions

It can be seen from Table I that these simple quaternization reactions can lead easily to polymer formation. The reactions with dimethyl sulfate, in particular, are quite rapid, and high-molecular-weight products apparently result. The reactions with methyl iodide are much slower, and even with a greater ratio of alkylating agent to monomer than was used with dimethyl sulfate, only relatively low-molecular-weight products result. This effect appears to be related to the amount of cationic monomer which is available for incorporation into the kinetic chain, with the faster reactions by dimethyl sulfate giving more monomer present.

Further details of these alkylation reactions as well as the preparation and study of the corresponding monomeric salts are under investigation. Other preliminary data have been published (13).

Spontaneous Polymerization by Concentrated Acid

It has previously been shown that the reactions of 2- and 4-vinylpyridine with concentrated strong acids in homogeneous and heterogeneous reactions can lead to the rapid formation of the corresponding poly(vinylpyridinium salts) (7,13), such as represented

by reaction 2. These resulting polymeric acid salts are water soluble, and the neutral (water-insoluble) polymeric precursors,
poly(2-vinylpyridine) and poly(4-vinylpyridine), can be obtained by
treatment with strong base. The homogeneous reactions can be done
with either "neat" strong acid or with concentrated acid in water.
The heterogeneous reactions can be carried out by a suspension of
strong acid in an organic solvent.

In Table II are reported the results of the spontaneous polymerizations of 2- and 4-vinylpyridine in "neat" sulfuric acid (96%
by weight) at room temperature for 5 days. The reactions result
in the formation of the protonated polymeric salts in the bisulfate
form. The intrinsic viscosities obtained are reported for the neutralized polymers. It is seen that these simple reactions, done in
the presence of air, produce high-molecular-weight polymers. The
same reactions under nitrogen produce similar results.

TABLE II. Spontaneous Polymerization by "Neat" Sulfuric Acid

Monomer	$\frac{[H_2SO_4]}{[VP]}$	Conversion, %	$[\eta]$, dl/g
2VP	3.7/1.0	14.0	3.59[a]
4VP	3.7/1.0	78.5	1.57[b]

- - - - - - - - -

[a]obtained for neutralized polymers in benzene solution at 25°C
[b]obtained for neutralized polymers in ethanol solution at 25°C

Using five-day reactions of these monomers with sulfuric acid
in water, somewhat different results were obtained. In Table III
are listed the results of the spontaneous polymerization of 2-vinyl-
pyridine as a function of varying acid concentrations. A total of
15 ml of water and 1 ml of 2-vinylpyridine was used for each reaction. The amount of sulfuric acid used is reported in a molar
ratio to 2-vinylpyridine and in its total molarity in water.

The results obtained in this preliminary study indicate that
there is an effect on the spontaneous polymerization by the ratio
of acid to monomer and by the presence of additional water over
that reported for Table II. This study indicates that at high
ratios of sulfuric acid to 2-vinylpyridine, a leveling effect on
conversion and molecular weight is obtained. At the molar ratio
of 1/1, no polymerization reaction occurred, since only monomeric
salt formed. The lack of significant polymerization at the molar
ratio of 3/1 acid to monomer is unexpected, and this result could
have a bearing on the mechanism of polymerization.

TABLE III.   Spontaneous Polymerization of 2-Vinylpyridine
by Sulfuric Acid in Water

$\dfrac{[H_2SO_4]}{[2VP]}$	M H_2SO_4	Conversion, %	$[\eta]^a$
1/1	0.60	0	----
2/1	1.17	32.7	1.67
3/1	1.70	trace	----
4/1	2.20	58.4	1.51
10/1	4.65	76.5	2.31
15/1	6.18	40.0	2.03
20/1	7.40	40.0	1.89
25/1	8.40	31.1	2.02

[a] obtained for neutralized polymers in benzene at 25°C; dl/g

In contrast to these polymerization reactions in the presence
of added water, 4-vinylpyridine was found not to undergo spontaneous polymerization to protonated poly(4-vinylpyridine) under similar five-day reactions. It is possible that this reaction may occur at a slower rate than that for 2-vinylpyridine. Consequently,
preliminary results suggest that the reactions with "neat" sulfuric
acid appear most favorable for high-polymer formation.

One of the most unusual properties of the poly(2-vinylpyridines)
which were obtained (after neutralization by base) is their rather
narrow molecular-weight distribution. Preliminary gel permeation
chromatographic studies indicate that the ratio $\bar{M}_w/\bar{M}_n$ ranges from
1.1 to 1.4 (14). Further studies are currently in progress.

CONCLUSION

The quaternization and protonation reactions studied in this
work for the preparation of water-soluble, cationic polyelectrolytes from 2- and 4-vinylpyridine are under further investigation.
By adjustment of the reaction conditions with respect to solvent,
concentration, and temperature, it should be possible to optimize
results. Furthermore, it should also be possible to extend the
principles of spontaneous polymerization to the preparation of
other cationic polyelectrolytes.

Acknowledgement is made to the donors of The Petroleum Research Fund, administered by The American Chemical Society, for partial support of this work.

BIBLIOGRAPHY

1.   M. F. Hoover, J. Macromol. Sci. Chem., A4, 1327 (1970).

2.   Terminology used by Rembaum and co-workers for the introduction of ionic amine units in the polymer main chain; see also the chapter by S. P. S. Yen, D. Casson, and A. Rembaum.

3.   V. A. Kabanov, K. V. Aliev, T. I. Patrikeeva, O. V. Kargina, and V. A. Kargin, J. Polym. Sci., Part C, 16, 1079 (1967).

4.   V. A. Kabanov, Pure Appl. Chem., 15, 391 (1967).

5.   V. A. Kabanov, O. V. Kargina, and V. A. Petrovskaya, Vyskomol. Soedin., 13 (2), 348 (1971), and the references cited therein.

6.   J. C. Salamone, B. Snider, and W. L. Fitch, J. Polym. Sci., Part B, 9, 13 (1970); Polym. Preprints, 11, (2), 652 (1970).

7.   J. C. Salamone, B. Snider, and W. L. Fitch, Macromolecules, 3, 707 (1970).

8.   J. C. Salamone, B. Snider, and W. L. Fitch, J. Polym. Sci., Part A-1, 9, 1493 (1971).

9.   I. Mielke and H. Ringsdorf, J. Polym. Sci., Part C, 31, 107 (1970).

10. I. Mielke and H. Ringsdorf, J. Polym. Sci., Part B, 9, 1 (1971).

11. I. Mielke and H. Ringsdorf, Makromol. Chem., 142, 319 (1971).

12. J. C. Salamone, E. J. Ellis, and S. C. Israel, Polym. Preprints, 12, (2), 185 (1971).

13. J. C. Salamone, E. J. Ellis, and S. C. Israel, Polym. Preprints, 13 (1), 276 (1972).

14. J. C. Salamone, E. J. Ellis, and C. R. Wilson, in preparation.

GRAFT COPOLYMERS OF STARCH AND POLY(2-HYDROXY-3-METHACRYLOYLOXY-

PROPYLTRIMETHYLAMMONIUM CHLORIDE). DEPENDENCE OF GRAFT COPOLYMER

STRUCTURE ON METHOD OF INITIATION

George F. Fanta, Robert C. Burr, W. M. Doane, and
C. R. Russell
Northern Regional Research Laboratory
U. S. Department of Agriculture
Peoria, Illinois

Starch is one of the cheapest and most readily available of all natural polymers. Although the more common derivatives of starch and their properties have been extensively studied, grafting high polymers to starch is an area of research that has received much less attention.

Starch is a high polymer composed of repeating $1\rightarrow 4$-α-D-glucopyranosyl units (often called anhydroglucose units, or simply AGU) and is, generally, a mixture of linear (amylose) and branched (amylopectin) components. Starch occurs in living plants in the form of discrete granules, which range from roughly 5 to 40 μ in diameter, depending on the source. These granules contain regions of crystallinity; and if slurried in water at room temperature, their solubility is negligible even though starch is a hydrophilic polymer. When a water slurry of starch is heated, hydrogen bonds are broken, a measurable fraction of the starch is dissolved, and the granules swell until, near the boiling point of water, they largely disintegrate to form a smooth dispersion.

Starch graft copolymers are prepared by generating free radicals on the starch backbone and then allowing these macroradicals to react with monomer. A number of initiating methods have been used to prepare starch graft copolymers, and these may be divided into three broad categories: initiation by chemical methods (generally ceric ammonium nitrate or the ferrous ion - peroxide redox system), initiation by irradiation (generally cobalt-60), and initiation by mastication. In the last method, starch macromolecules are physically ruptured to give free-radical sites where cleavage of the molecule occurred, and the products of polymerization can thus be referred to as block rather than graft copolymers.

Starch graft copolymers are usually characterized by three parameters: the percent synthetic polymer incorporated in the graft copolymer (percent add-on), the average molecular weight of the grafted branches, and the grafting frequency. Grafted branches of synthetic polymer are easily separated for molecular-weight determination by degradation of the starch portion of the copolymer to glucose and other small carbohydrate fragments through treatment of the graft copolymer with either hot mineral acids or enzymes. Grafting frequency, expressed as the average number of anhydroglucose units per grafted branch, is calculated from the percent add-on and the molecular weight of the graft.

The term grafting efficiency, often used in describing graft polymerizations, is defined as the percentage of the total synthetic polymer formed that has been grafted to starch. High grafting efficiencies are, of course, desirable since a polymerization of low grafting efficiency gives mainly a physical mixture of starch and homopolymer.

The graft polymerization of cationic monomers onto starch is currently under active investigation at the Northern Regional Research Laboratory. Previously we reported graft polymerization of the cationic monomer 2-hydroxy-3-methacryloyloxypropyltrimethylammonium chloride (I) onto wheat starch with initiation by ceric ammonium nitrate or ferrous ammonium sulfate - hydrogen peroxide (1).

$$CH_2 = \overset{\overset{\displaystyle CH_3}{|}}{C}CO_2CH_2\overset{\overset{\displaystyle OH}{|}}{C}HCH_2\overset{\oplus}{N}(CH_3)_3 \qquad \overset{\ominus}{Cl}$$

I

Since cationic starch graft copolymers prepared from I are potentially useful as flocculants (2), we studied in detail how the method of free-radical initiation influenced the final structure of the graft copolymer. In particular, we were interested in how graft copolymers prepared by cobalt-60 initiation differ from those prepared with chemical initiators with respect to percent add-on, molecular weight of grafted poly(I), and grafting frequency. When we initiated copolymerization of I with starch by heat and mastication, we found that starch-poly(I) copolymers had structures different from those prepared with either chemical initiators or cobalt-60. Selected graft copolymers were tested as flocculants for silica.

EXPERIMENTAL

Materials

Monomer I, 2-hydroxy-3-methacryloyloxypropyltrimethylammonium chloride (Sipomer Q-1, Alcolac Chemical Corp.), was recrystallized from 2-propanol.

Unmodified wheat starch was Huron Starbake from Hercules, Incorporated. Ferrous ammonium sulfate hexahydrate, 30% hydrogen peroxide, and xylene were Baker Analyzed Reagent Grades. Ceric ammonium nitrate was Fisher Certified ACS Grade.

Graft Polymerizations

Chemical Initiation. Graft polymerizations initiated with ceric ammonium nitrate or ferrous ammonium sulfate hexahydrate - hydrogen peroxide were carried out as described previously (1). Solid products were extracted three times with cold water and dried. Ungrafted poly(I) was isolated from water extracts by dialysis and freeze-drying.

Cobalt-60 Initiation. The cobalt-60 source was a Gammacell 200 unit from Atomic Energy of Canada Ltd. Dose rate at the center of the chamber was 1.34-1.32 Mrad/hr, as calculated from the initial dosimetry provided by the manufacturer and the decay rate of cobalt-60. Starch was irradiated in 2-oz screw-cap bottles; for temperatures below ambient, bottles were irradiated in a Dewar flask packed with ice water. Temperatures during irradiation were determined in separate experiments by means of a thermistor probe imbedded in starch samples. For irradiations under nitrogen, starch was pumped overnight at room temperature under vacuum (0.8-3.0 mmHg). The vacuum was broken with nitrogen, and the sample evacuated and repressured with nitrogen three additional times. The water content of one starch sample was reduced from 13.5% to 4.9% by this procedure. Starch samples irradiated under air contained 13.5% water.

To facilitate addition of dry solids, reactions were run in a 600-ml beaker equipped with paddle stirrer, thermometer, and tube through which nitrogen was passed. The beaker was covered with polyethylene to exclude air. Water (283 ml) was stirred and purged with nitrogen for 1 hr, and 47.5 g (0.2 mole) of monomer I was added. After the mixture had been stirred for 1 min to dissolve I, 32.4 g (dry basis, 0.2 mole) of irradiated starch was added. Addition of starch was made 4 min after removal from the cobalt-60 source. Reactions were run under nitrogen for 2 hr at 25°C and terminated by addition of 1 g of hydroquinone. After mixtures were centrifuged, swollen solids were extracted twice with water and twice with 95% ethanol, and then vacuum-dried. Ungrafted poly($\underline{I}$) was isolated from the combined extracts by dialysis and freeze-drying.

Graft copolymers were further purified by stirring 30 g of solid overnight in 1200 ml of water and precipitating the graft co-polymer by dropwise addition of 2400 ml of absolute ethanol. Un-grafted poly($\underline{I}$) does not precipitate under these conditions. The polymer that remained in solution amounted to only 4-7% of the total product and contained more than 10% carbohydrate, as estimated by infrared analysis.

Initiation by Heat and Mastication. A mixture of 40.5 g (dry basis) of starch (0.25 mole), 40.5 g of water, and 59.4 g (0.25 mole) of monomer $\underline{I}$ was passed through an extruder (15 x 3/4-in. screw; 5:1 compression ratio) attached to a Plasti-Corder (C. W. Brabender Instruments, Inc.). The extruder temperature was 100°C, the screw was rotated at 80 rpm, and the die had 24-1/32 in. holes. The mixture was extruded into petroleum ether and the hydrocarbon-wet extrudate passed again through the extruder. Petroleum ether was selected because its inert vapors displace air from the mixture during extrusion. The mixture was passed a total of eleven times through the extruder, during which time the temperature at the die rose from 89 to 112°C and the material became more difficult to masticate (increased torque measured on the Plasti-Corder), pre-sumably due to gradual buildup of poly($\underline{I}$) within the reaction mass.

The extrudate from the last pass was separable into a rubbery fraction and a brittle fraction, and each was dispersed in 500 ml of water containing 1 g of hydroquinone. An equal volume of 95% ethanol was added to each dispersion; the solid was separated by centrifugation, extracted twice with 95% ethanol, and dried. The soluble polymer (not precipitated by ethanol) was isolated by di-alysis and freeze-drying and contained carbohydrate, as well as poly($\underline{I}$), determined by infrared analysis. The following amounts were isolated: rubbery--27 g total (7% soluble); brittle--37 g total (20% soluble). The insoluble polymers were further fraction-ated by precipitation from a 1% aqueous dispersion by dropwise ad-dition of an equal volume of absolute ethanol. From the rubbery and brittle fractions 85% and 80%, respectively, of the solid was precipitated. Analyses were performed on these final precipitated fractions.

Initiation by Addition to Hot Xylene. Xylene (350 ml) was purged with nitrogen for 15 min and then heated to 105°C. A mix-ture of 32.4 g, dry basis, of starch (0.2 mole), 32.4 g of nitrogen-purged water, and 47.5 g (0.2 mole) of monomer $\underline{I}$ was added in one portion to the xylene as it was being rapidly stirred. The temper-ature dropped to 83°C. The mixture was heated to 98°C over a 15 min

period, then stirred, and allowed to cool to 25°C for 2.5 hr. After
the xylene was decanted, the gummy polymer was dispersed in about
800 ml of water containing 1 g of hydroquinone. An equal volume of
95% ethanol was added, and the precipitated solid was separated by
centrifugation, washed with ethanol, and dried. The combined etha-
nolic supernatant was stripped of ethanol, dialyzed, and freeze-dried
to yield ungrafted poly(I), which contained about 5% carbohydate by
infrared. The insoluble polymer, which contained 42% poly(I), was
further fractionated by precipitation from a 1% aqueous dispersion
by addition of an equal volume of ethanol (91% of the polymer precipitated). The precipitated fraction contained 41% poly(I).

For the reaction initiated by slow heating, an identical mixture of starch, monomer I, and water was added to 350 ml of xylene
at 23°C. The stirred mixture was then heated for 50 min up to 98°C,
held at 98-101°C for 15 min, and allowed to cool for 2.5 hr to 27°C.

Fractional Precipitation of Polymer Prepared by Hot Xylene
Initiation--Proof of Chemical Bonding Between Starch and Poly(I)

A physical mixture was prepared from the following two components: 1. 2.92 g of wheat starch that had been treated with hot
xylene under the same conditions used for polymerization but in
the absence of monomer I and 2. a water dispersion (1070 ml) of
poly(I) obtained by removal of the starch moiety (by acid hydrolysis) from 5.00 g of copolymer prepared by hot xylene initiation
[poly(I) content, 42%]. The pH of the mixture was 6.0. A dispersion was also prepared from 5.00 g of copolymer [poly(I) content,
42%] and 1020 ml of water, and the pH was adjusted from 5.5 to 6.0.
Both dispersions were heated for 10 min on a steam bath at 85°C and
then cooled to 25°C. Respective Brookfield viscosities (25°C,
30 rpm) of the copolymer and the synthetic mixture were 6500 cP and
500 cP. Both dispersions were passed through a Penick & Ford labo-
ratory-model continuous-steam injection cooker (3) at 170°C and a
steam pressure of 100 psi within the cooker. The solutions, which
contained less than 3% insolubles, were freeze-dried.

Freeze-dried solids were stirred in water to form 1% dispersions.
Addition of 300 ml of absolute ethanol to 250 ml of the synthetic
mixture dispersion sharply precipitated starch (1.39 g) and left
poly(I) (1.05 g) in solution. Similar treatment of the copolymer,
however, precipitated only 3% of the solid. When enough ethanol
was added to give a concentration of 90% by volume, only an additional 23% of the polymer was precipitated. Infrared analysis showed a
weak carbonyl absorption for poly(I) in both precipitates. When
the soluble fraction was concentrated to about 3% in water and ethanol added to a concentration of 95% by volume, an additional 28%
of the solid precipitated. Both this precipitate and the final

soluble fraction contained starch and poly($\underline{I}$), by infrared analysis.

Characterization of Graft Copolymers

The percent add-on in graft copolymers was calculated from nitrogen analysis. Starch was then removed from graft copolymers by acid hydrolysis, and number-average molecular weights for grafted poly($\underline{I}$) were determined by membrane osmometry as reported earlier (2).

To determine percent solubility of gelled poly($\underline{I}$) fractions, 0.04-0.06% dispersions in water were allowed to filter slowly through fluted Whatman 54 paper. Solubility was calculated from the weight of polymer (determined by freeze-drying) in a measured amount of filtrate versus the weight of polymer in a portion of the unfiltered dispersion.

Intrinsic viscosities of poly($\underline{I}$) in dl/g at 25°C were determined with Cannon-Fenske viscometers in a 70:30 (by volume) solution of water - dimethyl sulfoxide, which was 0.35 $\underline{M}$ in sodium chloride. Intrinsic viscosities at the 95% confidence interval were calculated with the aid of a computer program developed by Hofreiter, Ernst, and Williams (4).

Flocculation Studies

Graft copolymer slurries (0.5 g in 375 ml of water) were stirred in a boiling water bath for 30 min and then cooled to 25°C. Portions of each dispersion were centrifuged for 20 min at 5000 $\underline{g}$ and a portion of the clear supernatant was freeze-dried. The percent solubility was calculated from the weight of solid in the supernatant versus the weight of solid in a portion of uncentrifuged dispersion. Dispersions were diluted to a final concentration of 0.2 g/l.

Graft copolymers were tested as flocculants for Celite (a diatomaceous silica obtained from Johns-Manville; average particle size 2.1 μ) as described previously (2). The pH of Celite suspensions was 6.3-6.8.

RESULTS AND DISCUSSION

Graft Polymerizations

Graft polymerization of monomer $\underline{I}$ onto wheat starch by chemical initiation (ceric ammonium nitrate or ferrous ammonium sulfate-

TABLE I. Chemical Initiation

| Starting materials[a] | | | Graft copolymer | | | $\overline{M}_n$ of ungrafted poly(I) | Total conversion of monomer, % | Grafting efficiency, f % |
Initiator[b]	I,[c] mole	Starch granule state[d]	Wt-% poly(I)	$\overline{M}_n$ poly(I)	Graft frequency[e]			
Ce^{+4}	0.3	Granular	15	230,000	7900	560,000	94	13
Ce^{+4}	0.3	Swollen	35	220,000	2500	105,000	86	31
Ce^{+4}	0.075	Swollen	15	320,000	11,000	---	100	32
Fe^{+2}/H_2O_2	0.3	Granular	6	85,000	8100	47,000	72	10
Fe^{+2}/H_2O_2	0.3	Swollen	12	52,000	2300	---	78	13
Fe^{+2}/H_2O_2	0.4	Swollen	18	115,000	3300	---	87	12

[a] All reactions used 0.3 mole of starch (unit mole = 162) and 425 ml water.

[b] Ce^{+4}-initiated reactions used 3×10^{-3} mole of ceric ammonium nitrate. Fe^{+2}/H_2O_2-initiated reactions used 3×10^{-4} mole of ferrous ammonium sulfate hexahydrate and 3×10^{-3} mole of hydrogen peroxide.

[c] $CH_2 = C(CH_3)CO_2CH_2CH(OH)CH_2\overset{+}{N}(CH_3)_3 \quad \overset{-}{Cl}$

[d] Granular: starch slurried in water at 25°C. Swollen: starch-water slurry heated for 1 hr at 60°C and then cooled to 25°C.

[e] Expressed as the average number of anhydroglucose units per grafted branch.

[f] Percentage of the total polymer which is grafted to starch.

hydrogen peroxide) is summarized in Table I. A higher percent add-on and more frequent grafting of poly($\underset{\sim}{I}$) were observed when starch granules were swollen by heating in water to 60°C before reaction at room temperature with monomer and initiator. In reactions with granular starch, the starch was allowed to react in a water slurry without prior heating. Initiation with ceric ion gave molecular weights for grafted poly($\underset{\sim}{I}$) in the range 200,000-300,000 whereas ferrous ion - hydrogen peroxide initiation gave lower molecular weights. Grafted branches of poly($\underset{\sim}{I}$) were infrequently spaced along the starch backbone; the most frequently grafted product in Table I had an average of 2300 anhydroglucose units per grafted branch. Total conversion of monomer to poly($\underset{\sim}{I}$) was high in all reactions; however, only 10-32% of the poly($\underset{\sim}{I}$) formed was actually grafted to starch. The remainder was homopolymer.

Graft polymerizations of monomer $\underset{\sim}{I}$ onto granular wheat starch preirradiated by cobalt-60 are shown in Table II. Reaction of monomer $\underset{\sim}{I}$ with preirradiated starch was recently reported by Jones and Jordon (5); however, we have extended their work to include the influence of a number of reaction conditions on graft copolymer structure.

Both grafted and ungrafted poly($\underset{\sim}{I}$) were of higher molecular weight than polymers produced by chemical initiation, possibly because ceric and ferric ions can act as chain terminators (6). Also, poly($\underset{\sim}{I}$) grafts were not so frequent as those formed in chemically initiated reactions. Since the $\bar{M}_n$ of most polymers in Table II was higher than could be accurately determined by our membrane osmometer, values for intrinsic viscosity are given instead. Total conversions of monomer to polymer were somewhat lower than for chemical initiation; however, grafting efficiencies were generally higher.

The temperature of starch during irradiation influenced significantly the course of the reaction. Irradiation of starch to 5.0 Mrad with no external cooling caused the temperature to rise from 25 to 49°C (Run 1). Since starch free radicals have reduced stability at higher temperatures (7), the lower values for percent add-on and grafting efficiency as compared with Run 2, where starch was held at 5°C during irradiation. were not unexpected. The low intrinsic viscosity for grafted poly($\underset{\sim}{I}$) in Run 1 is surprising, however, in view of the low percent add-on compared to Run 2. If a high radiation temperature destroys enough free radicals to reduce the percent add-on to 15%, we might also expect a higher intrinsic viscosity than 1.4 since. as with runs made at lower radiation doses, fewer growing chains would be competing for monomer. One possible explanation is that degradation products from starch irradiation (8) act as chain terminators for growing poly($\underset{\sim}{I}$). At 1.0 Mrad, smaller differences in percent add-on and grafting efficiency were noted between starch samples irradiated at ambient temperature (Run 4) and at 5°C (Run 3), since the shorter irradiation time re-

TABLE II. Cobalt 60 Initiation[a]

Run	Conditions[b]	Radiation temperature, °C	Total dose, Mrad	Graft copolymer Wt-% poly(I)	Graft copolymer [n][c] of poly(I)	[n][c] of Ungrafted poly(I)	Total conversion of monomer,	Grafting efficiency,
1	A	25-49	5.0	15	1.4	2.7	55	18
2	A	5	5.0	25	1.3[d]	2.4	72	32
3	A	5	1.0	25	2.4	2.8	69	41
4	A	25-35	1.0	20	2.8	3.2	66	33
5	A	5	0.1	13	4.8	3.7	22	54
6	B	5	1.0	26	3.0	3.0	68	40
7	C	5	1.0	<3	--	--	<3	--
8	D	5	1.0	15	1.8	2.8	28	54

[a] Reactions used 0.2 mole each of starch and I plus 283 ml of water.

[b] A: Irradiation and reaction performed under nitrogen. Reaction time was 2 hr.
B: Starch irradiated under air. Reaction performed under nitrogen. Reaction time was 2 hr.
C: Irradiation and reaction performed under air. Reaction time was 2 hr.
D: Irradiation and reaction performed under nitrogen. Reaction time was 8 min.

[c] Dl/g at 25°C in 70:30 (vol) water-dimethyl sulfoxide which was 0.35M in sodium chloride.

[d] $\overline{M}_n$ ~600,000; grafting frequency ~11,000 anhydroglucose units per grafted branch.

quired for 1.0 Mrad caused a smaller temperature rise (25-35°C).
The detrimental effect of high irradiation temperatures is most
apparent from the lower values for percent add-on and grafting
efficiency in Run 1 as compared with Run 4, even though the total
irradiation dose in Run 1 was higher by a factor of five.

The effect on graft polymerization of the magnitude of the irradiation dose may be seen by comparing Runs 2, 3, and 5, which were
made with starch preirradiated at 5°C to a total dose of 5.0, 1.0,
and 0.1 Mrad, respectively. Values for percent add-on were similar
at 5.0 and 1.0 Mrad; even a total dose as low as 0.1 Mrad gave a
copolymer that contained a significant amount of grafted poly($\underset{\sim}{I}$).
As expected, there was little or no polymerization of $\underset{\sim}{I}$ in the presence of unirradiated starch. With a lowering of total dose, total
conversion to polymer decreased; however, grafting efficiency increased. The total dose needed to maximize values for both percent
add-on and grafting efficiency apparently lies between 1.0 and 0.1
Mrad. Relative values for percent add-on and intrinsic viscosity
of grafted poly($\underset{\sim}{I}$) in these three reactions were as predicted; i.e.,
a higher total dose produces more free radicals that in turn lead
to more grafts of lower molecular weight.

The effect of a nitrogen atmosphere, as opposed to air, during
both the irradiation of starch and its reaction with monomer is
apparent when Runs 3, 6, and 7 are compared. In Run 6, starch with
its usual moisture content of 13.5% was irradiated under air; whereas
in Run 3, starch with a moisture content of 5% was irradiated under
nitrogen (lower moisture was due to the evacuation technique used
to displace air). In both runs, irradiated starch was reacted with
monomer in water purged with nitrogen. Both reactions gave virtually
the same percent add-on, total conversion of monomer, and grafting
efficiency, although starch irradiation under air led to a somewhat
higher intrinsic viscosity for grafted poly($\underset{\sim}{I}$). In Run 7, starch
was irradiated as in Run 6; however, the reaction with monomer was
carried out in water not purged with nitrogen. Little or no reaction was observed. We can conclude from these three runs that
although the presence of oxygen during starch irradiation is not
detrimental, oxygen severely hinders reaction of irradiated starch
with monomer.

The effect of reaction time of irradiated starch with monomer
is seen by comparing Run 3 (2-hr reaction) with Run 8 (8-min reaction). Irradiation of starch to 1.0 Mrad was performed at 5°C
under nitrogen in both runs. After a reaction time with monomer of
only 8 min, the percent add-on reached 60% of the value found after
2 hr. Also, grafting efficiency increased, and the intrinsic viscosity of grafted poly($\underset{\sim}{I}$) was reduced by the shorter reaction time.

Initiation by Heat and Mastication

The mastication of starch in the presence of polymerizable monomers has been examined by several groups of workers as a route to starch graft (block) copolymers (9-11). As part of our study on methods of initiation, we subjected mixtures of monomer $\underline{I}$ and starch to heat and mastication in order that the structures of reaction products might be compared with those obtained from chemical and cobalt-60 initiation.

An equimolar mixture of starch and monomer $\underline{I}$ was masticated by slurrying the two reactants in an amount of water equal to the dry weight of starch and then extruding the slurry through a die at $100^{\circ}C$. The reaction mass was recycled a total of eleven times through the extruder. During final extrusion, the first portion of the material to come out of the die was rather rubbery, whereas the last portion was brittle. The two fractions were separated and then fractionated by precipitation with ethanol from water dispersions. The rubbery and brittle fractions contained 37% and 51% poly($\underline{I}$), respectively (Table III). In contrast to graft copolymers prepared with either chemical or cobalt-60 initiation, removal of the starch moiety by treatment with hot mineral acid afforded poly($\underline{I}$) which was less than 20% soluble in water and which existed mainly as a viscous and apparently crosslinked gel. This crosslinked structure is probably an indirect result of the small amount of water used, which makes the molar concentration of monomer $\underline{I}$ in the polymerization recipe relatively high.

Accurate figures for percent conversion of monomer to polymer were not obtained, since a significant amount of material was lost due to holdup in the extruder. Also, the soluble polymer that remained after fractional precipitation from water with ethanol contained carbohydrate as well as poly($\underline{I}$).

That starch is indeed mechanically degraded under these conditions was shown by extruding a starch-water slurry in the absence of monomer $\underline{I}$ and determining the intrinsic viscosity and water content of the starch paste after each pass through the extruder. A mixture of 40% water and 60% starch [(η) = 1.6 dl/g in 90:10 by volume dimethyl sulfoxide - water] was changed to a mixture of 22% water and 78% starch [(η) = 0.87 dl/g] after ten extrusions through the die at $100^{\circ}C$.

It seemed probable to us that starch should be degraded to macroradicals not only if masticated, but also if the starch granule matrix were rapidly swollen, so that stresses associated with granule swelling and disruption could not be dissipated in time to prevent bond rupture. A logical method for rapidly swelling starch in the presence of monomer was to add a slurry of starch and monomer $\underline{I}$

TABLE III. Initiation by Heat and Mastication[a]

Method of initiation	Starch-poly(I) copolymer		Total conversion of monomer, %	Grafting efficiency, %
	Wt-% poly(I)	Physical state of poly(I) moiety		
Extrusion, 100°C (rubbery fraction)	37	Gel (19% sol)	---	---
Extrusion, 100°C (brittle fraction)	51	Gel (11% sol)	---	---
Xylene, 105°C	41	Gel (11% sol)	63	78
Xylene, slow heating	22	Solution (79% sol; $[n] = 6.0$)	39	44

a Equimolar amounts of starch and monomer were used. The weight of water was equal to the dry weight of starch.

in water to a hot organic liquid. Xylene was chosen for this purpose due to its high boiling point, inertness, and ready availability.

When equimolar amounts of starch and monomer I in an amount of
water equal to the dry weight of starch were added to xylene rapidly stirred at 105°C, a copolymer was isolated that contained 41%
poly(I) after fractional precipitation with ethanol from a water
dispersion (Table III). Removal of the starch moiety by acid hydrolysis yielded gelled poly(I) which resembled that isolated from ex-
truder-prepared copolymers.

Evidence for chemical bonding between starch and gelled poly(I)
was obtained when the copolymer was first dissolved in water by the
partially degradative process of jet cooking (3,12) and then subjected to fractional precipitation by addition of ethanol(see Experimental). Although a jet-cooked synthetic mixture of starch and gelled poly(I) (isolated from the copolymer by acid hydrolysis) was
sharply separated by this technique, a similar separation of the
copolymer itself could not be achieved. As additional evidence, the
Brookfield viscosity of a water dispersion of copolymer before jet
cooking was significantly higher than that of the synthetic mixture
(6500 cP vs 500 cP).

Treatment of starch with hot xylene under the conditions used
for polymerization, but in the absence of monomer I, reduced intrinsic viscosity (1 N potassium hydroxide) from 1.6 to 1.3 dl/g; this
reduction indicated that rapid starch granule swelling was indeed
causing bond rupture in starch. Moreover, slow heating of a mixture of starch, water, I, and xylene gave a different product from
that produced by rapid granule swelling, as seen from the last entry
in Table III. The copolymer from this reaction contained only 22%
poly(I), and removal of starch by acid hydrolysis did not give a
gel but afforded poly(I) which was 79% water soluble.

Heat-induced polymerization of I in the absence of starch did
not give the same results as polymerization within the starch matrix.
Conversion to poly(I) was more than 50% when the reaction was initiated by addition of a solution of monomer I in water to hot xylene;
however, the polymer was not a gel but was more than 85% soluble in
water. A 16% conversion to poly(I), which was 82% soluble in water,
was realized when a flask containing I and water was placed in a
120°C oil bath for 1 hr.

                              Flocculation

For laboratory testing (2) of graft copolymers as flocculants,
beakers containing aqueous diatomaceous silica (Celite) suspensions
were stirred simultaneously, while graft copolymer solutions of

TABLE IV. Flocculation of Celite

Run	Initiation	Graft copolymer			Per cent Celite remaining at different concentrations of graft copolymer			
		Wt-% poly(I)	$\bar{M}_n$ or $[\eta]$ of poly(I)	Solubility, %	0 ppm	4 ppm	8 ppm	12 ppm
1	Ce^{+4}	35	220,000	87	89	57	35	24
2	5 Mrad at 5°C	25	1.3[a]	100	91	49	32	21
3	1 Mrad at 5°C	25	2.4	94	90	49	35	24
4	1 Mrad at 25–35°C	20	2.8	98	91	46	32	24
5	Fe^{+2}/H$_2$O$_2$	18	115,000	88	89	66	43	32
6	Ce^{+4}	15	320,000	62	89	80	50	39
7	5 Mrad at 25–49°C	15	1.4	94	90	66	45	34
8	1 Mrad at 5°C[b]	15	1.8	88	91	60	43	31
9	0.1 Mrad at 5°C	13	4.8	50	91	72	55	44

a $\bar{M}_n$ ~600,000.
b Reaction time with monomer: 8 min.

varying concentration were added at the same time to each beaker.
After completion of stirring, suspensions were allowed to settle
for 30 min, and the amount of Celite remaining in the supernatant
in each beaker was determined. The percentage of the original Celite still in suspension was then tabulated against the flocculant
concentration in parts per million (ppm) in the beaker. Low percentages are, of course, desirable since they indicate an efficient
precipitation of Celite and thus a high degree of flocculation.
Results are given in Table IV.

The percent solubility of graft copolymers in water at $100^{\circ}C$,
a property of importance in flocculation, depended on the method of
initiation used for their synthesis. Chemical initiation (Runs 1,
5, and 6) afforded products of diminished solubility. Water solubility was also lower with cobalt-60-initiated copolymers when a
low radiation dose (Run 9) was used. This effect was not surprising
since ionizing radiation is known to lower the molecular weight of
polysaccharides (8) and should, therefore, increase water solubility
at the higher doses.

In general, the ability of graft copolymers to flocculate Celite increased with percent add-on; however, low water solubility
(Runs 6 and 9) reduced their effectiveness. The copolymer prepared
by hot xylene initiation was a poor flocculant, apparently owing
to the crosslinked structure of the poly($\underline{I}$) moiety. Flocculation
efficiency was improved, however. when the copolymer was dispersed
by jet cooking. Variations in both percent add-on and percent solubility for the products in Table IV make it difficult to conclude
whether cobalt-60 or chemical initiation is the method of choice
for optimizing flocculating ability. It is apparent, however, that
the two methods do not give large differences in product performance.

The graft copolymers described in this report are first-gener-
ation products and are not equal in flocculating ability to some
commercial polymers now on the market. However, we believe that
starch graft copolymers show considerable promise in this area of
application. We are continuing our study of the graft polymerization
of cationic monomers onto starch using various methods of initiation
and hope to learn more about the influence of the structure of a
graft copolymer on its ability to act as a flocculant.

We thank B. R. Heaton and K. A. Jones for nitrogen analyses
and W. L. Williams for the computer calculations of intrinsic viscosity.
The mention of firm names or trade products does not imply that
they are endorsed or recommended by the U. S. Department of Agriculture over other firms or similar products not mentioned.

BIBLIOGRAPHY

1. G. F. Fanta, R. C. Burr, W. M. Doane, and C. R. Russell, J. Appl.
    Polym. Sci., 15, 2651 (1971).

2. G. F. Fanta, R. C. Burr, C. R. Russell, and C. E. Rist, J. Appl.
    Polym. Sci., 14, 2601 (1970).

3. V. L. Winfrey and W. C. Black (to Penick & Ford, Ltd.), U. S.
    Pat. 3,133,836 (1964).

4. B. T. Hofreiter, J. O. Ernst, and W. L. Williams, J. Appl.
    Polym. Sci., in press.

5. D. A. Jones and W. A. Jordon, J. Appl. Polym. Sci., 15, 2461
    (1971).

6. S. R. Palit, T. Guha, R. Das, and R. S. Konar, in N. M. Bikales,
    Ed., Encyclopedia of Polymer Science and Technology, Interscience
    Publishers, New York, Vol. 2, 1965, pp. 229-266.

7. Z. Reyes, Stanford Research Institute, private communication,
    1965.

8. R. L. Whistler and T. R. Ingle in R. L. Whistler and E. F.
    Paschall, Eds., Starch: Chemistry and Technology, Vol. I,
    Academic Press, New York, 1965, p. 409.

9. B. H. Thewlis, Staerke, 16, 279 (1964).

10. R. J. Ceresa, Polymer, 2, 213 (1961).

11. D. J. Angier, R. J. Ceresa, and W. F. Watson, J. Polym. Sci.,
    34, 699 (1959).

12. G. F. Fanta, R. C. Burr, W. M. Doane, and C. R. Russell,
    J. Appl. Polym. Sci., in press.

THE POLYMERIZATION OF 3-DIMETHYLAMINO-n-PROPYL CHLORIDE AND

THE FORMATION OF STAR-SHAPED AND BRANCHED POLYELECTROLYTES

S. P. S. Yen, D. Casson, and A. Rembaum

Jet Propulsion Laboratory
California Institute of Technology
Pasadena, California

It has been shown recently (1) that, under well-defined conditions, step-growth polymerization is the predominant process in the reaction of a diamine of structure $\underset{\sim}{I}$ with a dihalide of structure $\underset{\sim}{II}$,

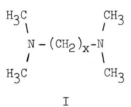

$$Z - (CH_2)_y - Z$$

$$\underset{\sim}{I} \qquad\qquad \underset{\sim}{II}$$

where x and y are the number of CH_2 groups in the diamine and the dihalide, respectively, and Z is Cl, Br, or I.

This polymerization reaction yields cationic polyelectrolytes of unit segment $\underset{\sim}{III}$,

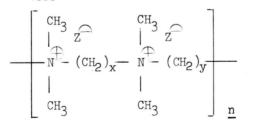

$$\underset{\sim}{III}$$

where $\underline{n}$ is the degree of polymerization. The polymers, referred to
as x,y-ionene halides, are ideally suited for a study of polycations.
Their synthesis is simple and straightforward. The molecular-weight
distribution is fairly narrow (M_w : M_n <2) and the charge density
in the polymer backbone can be varied readily. It should be borne
in mind that with certain specific values of x and y cyclic or linear diammonium salts are formed almost exclusively (1) and the high-
est-charge-density ionene halide of relatively high molecular weight
is the 3,3-ionene bromide. The latter was obtained by Marvel and
co-workers (2) from 3-dimethylamino-n-propyl bromide, and its molecular weight was estimated to be of the order of 2000-4000 on the
basis of determination of ionic and nonionic bromine content in the
isolated polymer.

In view of the present (3) and potential applications of
high-charge-density polyelectrolytes as well as of their academic
interest, we have investigated the mechanism of formation and some
physical and chemical properties of the 3,3-ionene chloride formed
from 3-dimethylamino-n-propyl chloride. As a result of a systematic study of the polymerization process as a function of concentration, solvent, temperature, and time we have established the
conditions for the synthesis of this ionene polymer with a weight-average molecular weight of at least 60,000. A preliminary study of
the reaction of dimethylamino-n-propyl chloride with low-molecular-weight compounds and polymers containing either chloromethyl or
dimethylamino groups was also carried out and it was shown that these
reactions yield star-shaped and branched polyelectrolytes.

In the following description 3-dimethylamino-n-propyl chloride
will be referred to as the AB monomer and, the resulting polymer as
the 3,3-ionene chloride its structure being represented by $\underline{IV}$

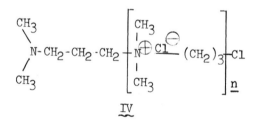

$$\underline{IV}$$

EXPERIMENTAL

Homopolymer

The AB monomer was isolated from its hydrochloride salt by reaction
with sodium hydroxide. 3-Dimethylamino-n-propyl chloride hydro-

chloride (100 g, 0.633 mole) was dissolved in the minimum quantity of water, cooled in an ice bath, and 200 ml of 20% NaOH solution added dropwise with vigorous stirring. The AB monomer was then extracted with several small portions of ether. The ether extracts were combined, washed twice with water, and then dried over anhydrous magnesium sulfate. After a drying period of 12 hours, the ether solution was filtered and evaporated. The AB monomer, together with a small quantity of remaining ether, was finally vacuum distilled. The fraction distilling between 22°C and 25°C, at 5 mmHg pressure, was collected and stored at 0°C until required. Both NMR and IR spectra confirmed the structure and purity of the AB monomer prepared as described above.

The AB monomer was allowed to polymerize under varying conditions. All the products of the reactions were recovered by freeze-drying or evaporation of the final solution to remove some of the solvent and then precipitating the polyelectrolyte by pouring the concentrated solution into excess dry acetone, with vigorous stirring. The products were allowed to stand overnight, filtered, washed several times with acetone, and finally dried in a vacuum oven at 40°C.

Star-Shaped and Branched Polyelectrolytes

Star-shaped polyelectrolytes were prepared by reacting the AB monomer with tri- or tetrafunctional primary halides or with tetrafunctional tertiary amines. The halides used were 2,4,6-tri(chloromethyl)mesitylene, 1,2,4-tri(chloromethyl)benzene, and 1,2,4,5-tetra(chloromethyl)benzene. The latter two compounds were synthesized from p-xylene (4) and their structures confirmed by means of NMR and IR spectra. The tetrafunctional amines were prepared as follows:

(a) 0.1 Mole of hexamethylene-1,6-diisocyanate was added dropwise, with stirring, to 0.2 mole of 1,3-bis(dimethylamino)-2-propanol cooled in an ice bath. Both reactants were freshly distilled under vacuum before use. The reaction mixture was allowed to warm gradually to room temperature and stirring continued for several hours. The viscous liquid was diluted with toluene and then separated from the latter by ether extraction.

(b) 0.1 Mole of toluene-2,4-diisocyanate was added dropwise, with stirring, to 0.2 mole of 1,3-bis(dimethylamino)-2-propanol cooled in an ice bath. Both reactants were freshly distilled under vacuum before use. After being allowed to warm to room temperature, the reaction mixture solidified. The solid was broken up and washed thoroughly with benzene, and then dried in a vacuum oven at 40°C.

$$[(CH_3)_2N-CH_2]_2-CH\text{-}OOCNH(CH_2)_6NHCOO\text{-}CH-[CH_2-N(CH_3)_2]_2$$

<u>V</u>

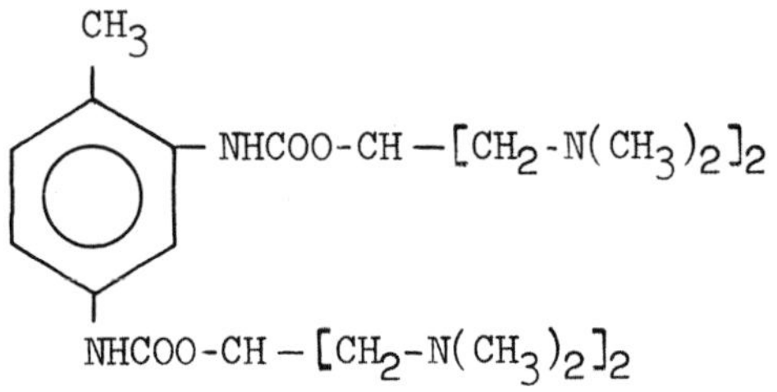

$$\text{CH}_3$$

NHCOO-CH $-[CH_2\text{-}N(CH_3)_2]_2$

NHCOO-CH $-[CH_2\text{-}N(CH_3)_2]_2$

<u>VI</u>

The structures of both tetrafunctional amines (<u>V</u> and <u>VI</u>) were confirmed by means of NMR and IR spectra. Branched polyelectrolytes were prepared by reacting poly(vinylbenzyl chloride), poly(4-vinyl-pyridine), or polyethylenimine with the AB monomer.

Rates of Polymerization

Reaction rates, for the homopolymerization of the AB monomer, were measured by titration of unreacted tertiary amine end groups. An aliquot of the reaction mixture was added to excess dilute hydrochloric acid and the unreacted acid titrated potentiometrically with dilute sodium hydroxide solution.

RESULTS

Polymerization of AB Monomer

Effect of Solvent. The AB monomer was allowed to polymerize at 41°C for five days in various solvent systems. The initial monomer concentration being kept constant at 1.0 molar. The results are summarized in Table I.

TABLE I. Effect of Solvent

Solvent	Volume Ratio	Yield, %	$[\eta]^c$; dl/g
DMF^a/MeOH	1:1	46	0.051
DMF/MeOH	1:2	36	0.047
DMF/MeOH	1:3	30	0.024
DMF/H$_2$O	4:1	100	0.092
DMSO^b/H$_2$O	4:1	100	0.064
DMSO/MeOH	4:1	101	0.014
CHCl$_3$		90	0.050
CH$_3$CN		81	0.053

- - - - - -

^{a}Dimethylformamide
^{b}Dimethyl sulfoxide
^{c}in 0.4 $\underline{M}$ aq KBr.

Effect of Temperature. Table I shows that the highest intrinsic viscosity was achieved at 41°C in the DMF/H$_2$O system. The AB monomer was, therefore, allowed to polymerize at various temperatures using 4:1 DMF/H$_2$O as solvent, an initial monomer concentration of 1.0 mole/l, and a 48 hr reaction time. The results are summarized in Table II.

TABLE II. Effect of Temperature

Temperature, °C	Yield, %[a]	$[\eta]$[b];dl/g
41	114	0.092
54	100	0.100
68	111	0.112
82	101	0.089
96	107	0.073

- - - - - -

[a]The yields over 100% are due to insufficient drying time.
[b]in 0.4 M aq KBr.

Effect of Initial Monomer Concentration. The AB monomer was allowed to polymerize in either 4:1 DMF/H$_2$O or 4:1 DMSO/H$_2$O at 54°C, using various initial monomer concentrations and a 48 hour reaction time. The results are shown in Table III.

TABLE III. Effect of Initial Monomer Concentration

Solvent	Initial Monomer Conc. (mole/l)	Yield, %	$[\eta]$[a];dl/g
DMF/H$_2$O	0.5	98	0.033
DMF/H$_2$O	1.0	99	0.126
DMF/H$_2$O	1.5	100	0.142
DMF/H$_2$O	2.0	86	0.194
DMF/H2O	2.5	91	0.175
DMF/H$_2$O	3.0	100	0.162
DMF/H$_2$O	3.5	100	0.164
DMSO/H$_2$O	1.0	100	0.064
DMSO/H$_2$O	1.5	100	0.102
DMSO/H$_2$O	2.0	100	0.129
None	---	25	0.070

[a]in 0.4 M aq KBr.

Tables I, II, and III indicate that highest molecular weights are obtained in the DMF/H$_2$O (4:1) mixture at temperatures in the range of 50-80°C and at a concentration of 2-3.5 mole/l.   The insolubility of the final polymer in DMF-water mixtures could be a reason for the difficulty in achieving intrinsic viscosities higher than 0.2 dl/g.   The polymerization was, therefore, investigated in pure water in which the polymer is miscible in all proportions.

Polymerization in Water.   The 3,3-ionene chloride (AB polymer) could be obtained by heating a stirred suspension of AB monomer in water.   In Table IV the effects of air, oxygen, and nitrogen on the yield and intrinsic viscosity of the polymer formed in the aqueous system are shown.   The aqueous polymerization system thus offers a convenient technique for the synthesis of 3,3-ionene chloride with viscosities higher than those achieved in other solvents, provided the process is carried out in the absence of air and at high monomer concentration.   Additional studies shown in Table V, of monomer concentration and reaction time confirmed the above conclusions.

TABLE IV.   Polymerization of AB Monomer in Water in the Presence or Absence of Air

Temp, °C	Conditions	Monomer Conc., mole/l	Reaction Time, hr	Conversion %	$[\eta]$[a];dl/g
100	Air	5.96	4	62	0.146
100	Oxygen	5.96	4	66	0.098
100	Nitrogen	5.96	4	70	0.208
100	Vacuum	5.96	50	100	0.224
100	Nitrogen	5.96	50	100	0.223

- - - - -

[a]in 0.4 M aq KBr.

Rates of polymerization were followed by the titration technique.   Figure 1 shows the results.   Curves 1, 2, 3, and 4 reflect the increase in rate at room temperature as the dielectric constant of the solvent increases; the same effect is shown in curves 5 and 6 obtained at 54°C.

The molecular weight of the polymer increases with time of conversion, as expected from a step-growth polymerization system.   This is shown in Table VI.   A similar change of intrinsic viscosity with

TABLE V.   Effect of Reaction Time and Monomer Concentration at
100°C on Polymerization of AB Monomer in Water

Monomer Concentration mole/l	Reaction Time, hr	Conditions	Conversion %	$[\eta]^a$; dl/g
5.96	4	Nitrogen	70	0.208
5.96	50	Nitrogen	100	0.223
4.87	8	Nitrogen	90	0.199
4.46	8	Nitrogen	90	0.189
4.12	8	Nitrogen	---	0.140
6.69	50	Vacuum	100	0.199
5.96	50	Vacuum	100	0.223
5.96	50	Nitrogen	100	0.223
5.96	148	Vacuum	100	0.240
5.96	384	Vacuum	100	0.230
5.54	50	Vacuum	100	0.210
4.87	50	Vacuum	100	0.174
4.46	50	Vacuum	100	0.162
4.12	50	Vacuum	100	0.145

- - - - - - -

[a] in 0.4 $\underline{M}$ aq KBr.

time was observed when the polymerization was carried out in water
alone.   Intrinsic viscosity changes as a function of time in the
DMF-water system at 54°C and in water at 100°C are shown in Figure 2.

Polymer Degradation. Examination of Table VI and Figure 2 indicates that relatively long reaction times lead to a decrease in
molecular weight. In order to ascertain whether this decrease was
due to occurrence of degradation, the polymers were kept at elevated
temperatures in aqueous solutions and the relative viscosity of

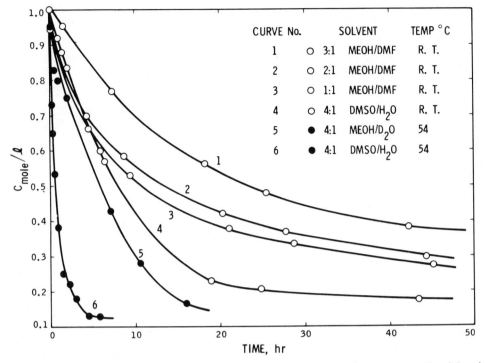

Fig. 1. Rates of polymerization of 3-dimethylamino-n-propyl chloride.

TABLE VI. Intrinsic Viscosities of AB Polymer as a Function of Time[a]

Time, hr	$[\eta]$[b]; dl/g
16.0	0.062
21.5	0.097
47.2	0.122
71.5	0.117
383.5	0.101

- - - - - - - -

[a]Isolated from separate batches of 1 molar AB monomer polymerized at 54°C in DMF/H_2O

[b]in 0.4 M aq KBr.

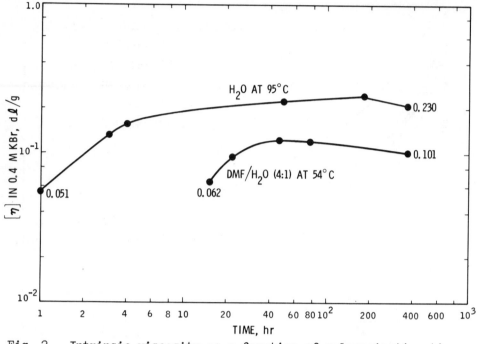

Fig. 2.   Intrinsic viscosity as a function of polymerization time.

isolated samples was determined as a function of heating time.
Figures 3 and 4 show that the polymer degrades slowly in water, much
faster in the presence of sodium hydroxide, but is stabilized in the
presence of hydrochloric acid.

NMR Analysis.   The 220 MHz NMR spectrum of the bulk monomer
shown in Figure 5 indicates the absence of impurities.   The same
applies to the polymer isolated from water solution.   Its 220 MHz
NMR spectrum, also easily interpreted, is shown in Figure 6.
Figure 7 shows the 60 MHz NMR spectra of the AB monomer undergoing
polymerization in $DMSO/D_2O$ 4:1 mixture at 20°C.   As the reaction
proceeds, a new triplet is recorded at τ = 5.85 increasing in intensity with time and gradually disappearing as the reaction approaches
completion.   In order to elucidate its origin, the NMR spectra of
polymer samples isolated at 20, 40, 60, and 80% conversion were determined.   In each case, the NMR absorption corresponded to structure
IV.   In the low-molecular-weight specimens   (e.g., 20% conversion)
a peak due to the proton resonance of the $-N(CH_3)_2$ group was also
noted (τ = 7.7), which was not detectable in high-molecular-weight
samples.

Considering the possibility of cyclic intermeridate formation
in the process of AB polymerization, diethylazetidinium bromide was

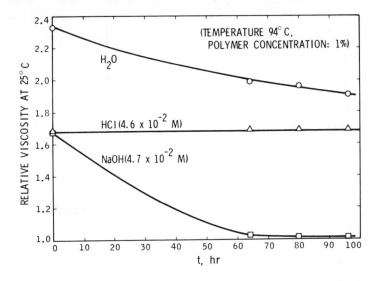

Fig. 3. Stability of 3,3-ionene chloride in solution; original $[\eta]$ = 0.22 in 0.4 M KBr.

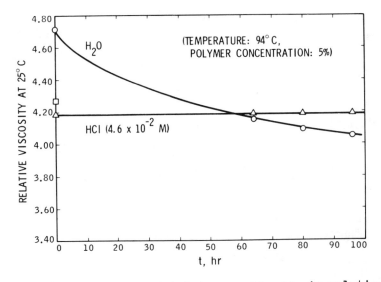

Fig. 4. Stability of 3,3-ionene chloride in solution; original $[\eta]$ = 0.22 in 0.4 M KBr.

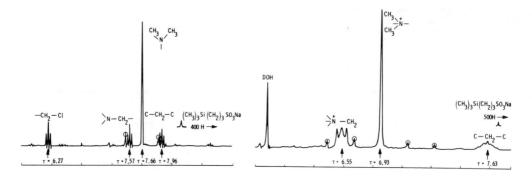

Fig. 5. NMR spectrum of AB monomer, 3-dimethylamino-n-propyl
chloride, in bulk at 20°C.

Fig. 6. NMR spectrum of 3,3-ionene chloride in D₂O at 85°C;
polymerized in water; [η] = 0.24 in 0.4 M KBr.

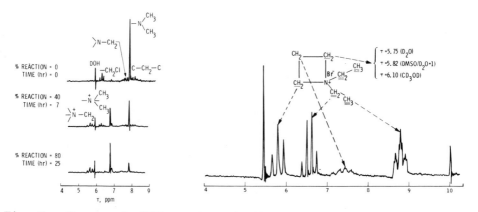

Fig. 7. Changes in NMR spectrum in the polymerization of AB monomer
as a function of time; DMSO/H₂O(4:1) at 20°C.

Fig. 8. NMR spectrum of diethylazetidinium bromide in D₂O.

synthesized according to the previously described procedure (2) and its NMR spectrum recorded in D_2O, $DMSO/D_2O$, and MeOH (Figure 8). The spectrum proves that a cyclic intermediate is formed in the polymerization of AB monomer (See Discussion and Conclusions).

Star-Shaped Polymers

Star-shaped polymers can be obtained in either DMF/H_2O (4:1 by volume) at $54^{\circ}C$ or in H_2O at $95-100^{\circ}C$. The intrinsic viscosity of the product made in water was higher than that made in the water-DMF mixture.

The intrinsic viscosities of star-shaped polyelectrolytes from the reaction of structure V, VI, 2,4,6-tri(chloromethyl)mesitylene, or 1,2,4,5-tetra(chloromethyl)benzene with the AB monomer are listed in Table VII.

TABLE VII   Formation of Star-Shaped Polyelectrolytes from AB
             Monomer and Various Compounds

Compound	Solvent	Temperature, $^{\circ}C$	Time, hr	Moles AB / Moles Compound	$[\eta]^a$; dl/g
b	DMF/H_2O^f	54	168	10	0.150
c	DMF/H_2O^f	54	168	10	0.140
d	DMF/H_2O^f	54	168	10	0.180
e	DMF/H_2O^f	54	168	10	0.150
e	H_2O	95	60	10	0.245

- - - - - - - -

[a] in 0.4 M aq KBr.

[b] Structure V.

[c] Structure VI.

[d] 2,4,6-Tri(chloromethyl)mesitylene.

[e] 1,2,4,5-Tetra(chloromethyl)benzene.

[f] 4:1 ratio.

To confirm the presence of the star-shaped polymer, one gram of the fourth product reported in Table VII was dissolved in methanol. 1,4-Dibromobutene (0.05 g) was added and the mixture was heated at 60 C for ten minutes. A gel was formed which was insoluble in water as well as in common organic solvents. The formation of the gel by crosslinking the tertiary-amine-terminated chains with the reactive bromo groups served as evidence of the presence of AB polymer bonded to 1,2,4,5-tetra(chloromethyl)benzene. Additional evidence for the formation of star-shaped polyelectrolytes is derived from the study of NMR spectra as a function of time. In Figures 9, 10, and 11 the spectral changes occurring as the reaction of 1,2,4,5-tetra(chloromethyl)benzene with the AB monomer proceeds with time are shown. The NMR spectrum of 1,2,4,5-tetra(chloromethyl)benzene consists of two lines (Figure 9), the low-field line being due to the phenyl proton ($\tau = 2.47$) and the $\tau = 5.18$ line to the chloromethyl groups. [A small amount of impurity in the perdeuterated DMSO yielded peaks at $\tau = 6.9$ and 7.7]. The intensity of the proton resonance absorption line at $\tau = 5.18$ is drastically reduced after two minutes of reaction with the AB monomer (compare Figure 9 with Figure 10). With an excess of AB monomer the benzylic chloromethyl group concentration is practically reduced to zero after 120 minutes (Figure 11). The changes in the resonance absorption of the phenyl proton (Figures 9, 10, and 11) may be attributed to the benzylic chloromethyl groups participating in the quaternization reaction and to polyelectrolyte shielding effects. The formation of the cyclic structure (dimethylazetidinium chloride) during the reaction is indicated by the appearance of a triplet at $\tau = 5.75$ which is clearly visible after 60 minutes of reaction time (Figure 10).

The reaction of the AB monomer (Figures 7, 9, 10, and 11) may be followed by the changes at $\tau = 7.93$ due to the $-N(CH_3)_2$ group and at $\tau = 6.5$ due to the chloromethyl group of the AB monomer.

The appearance of a new triplet at $\tau = 5.75$, which is clearly visible after 120 minutes of reaction (Figure 11), was assigned as $\oplus N-CH_2$ from dimethylazetidinium chloride (cyclic structure) on the basis of the NMR spectrum of the model compound (Figure 8).

The NMR spectrum of the star-shaped 3,3-ionene chloride reaction product after long reaction time (Figure 11) is practically identical with a 60 MHz NMR spectrum of the 3,3-ionene chloride except for one additional peak at $\tau = 4.8$ ($\oplus N-CH_2-\Phi$) which results from the quaternization of 1,2,4,5-tetra(chloromethyl)benzene with AB monomer.

Branched Polymers

Branched polymers with a comb-like structure were obtained by reacting AB monomer with poly(4-vinylpyridine), poly(ethylenimine), or poly(vinylbenzyl chloride) (mw 90,000). In the latter case an insoluble crosslinked product was obtained by reaction with the AB monomer at $54°C$ in DMF/H_2O (4:1) for seven days. However, PVBC yielded a water-soluble material when the reaction was carried out

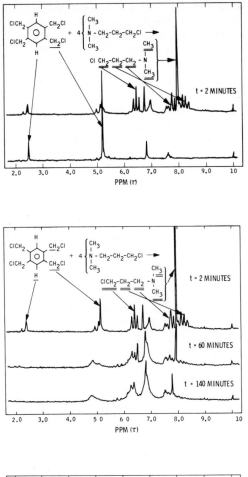

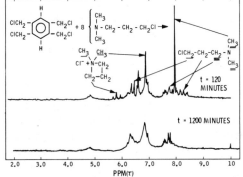

Figs. 9, 10, and 11.   Changes in NMR spectrum during the reaction of 1,2,4,5-tetrachloromethylbenzene with dimethyl-amino-n-propyl chloride in DMSO at 54°C.

as follows. PVBC (31 g) was stirred with the AB monomer (26 g) for 30 minutes and then diluted with water (60 ml). The solution was heated in presence of nitrogen for two hours at 100°C. The isolated product was soluble in water, methanol, and 0.1 M NaNO$_3$, but insoluble in 0.4 M KBr. It was further reacted (1 mole) in water at 95°C with the AB monomer (9 moles) for two hours. The isolated branched polymer was soluble in water, methanol, and 0.4 M KBr. Its intrinsic viscosity in 0.4 M KBr was found to be 0.38 dl/g. When heated at 100°C for two days it became completely insoluble in water or methanol. Elemental analysis of the soluble branched polymer agreed well with a composition of five AB units per benzyl segment; calculated: N, 9.2%; ionic chlorine, 23.4%; observed: N, 9.13%; ionic chlorine, 23.7%.

Properties

The high crystallinity of different ionene bromides, chlorides, iodides, and perchlorates was established by the examination of x-ray diffraction patterns using CuKα radiation (Figures 12 and 13).

The solution behavior of 3,3-ionene chloride and 6,6-ionene chloride is illustrated by a plot of specific reduced viscosity versus ionic strength (Figure 14). The intrinsic viscosity of both polyelectrolytes, determined at different ionic strengths, is shown in Figure 15 as a function of the reciprocal of the square root of ionic strength.

DISCUSSION AND CONCLUSIONS

Synthesis

The systematic investigations of the reaction conditions for the synthesis of 3,3-ionene chloride resulted in a simple and economical procedure for the manufacture of a cationic polyelectrolyte with positively charged nitrogen atoms in the backbone and with the highest charge density known.

The rates of formation of 3,3-ionene chloride are strongly dependent on the dielectric constant of the solvent (Figure 1) and increase with the increase of the amount of DMF present in water (Table I). The increase in rate is even more pronounced with DMSO (Figure 1), in agreement with the results of Tsuchida and co-workers (5). Since the successive steps of the polymerization reaction lead to polar species originating from nonpolar reactants, stabilization of the intermediates formed by increase of dielectric constant is to be expected.

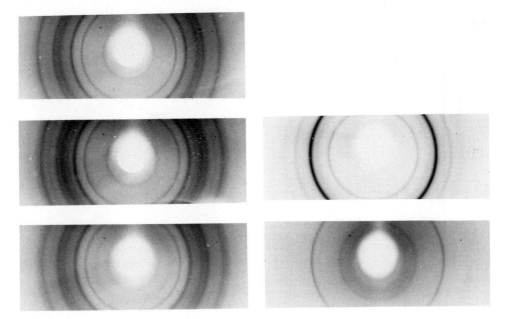

Fig. 12.   X-ray diffraction patterns of 3,3-ionene chloride of
different molecular weights; $[\eta]$ = 0.05, 0.19, 0.24
for top, middle, and lower patterns, respectively.

Fig. 13.   X-ray diffraction patterns of 3,3-ionene perchlorate (top)
and 3,3-ionene triiodide (bottom).

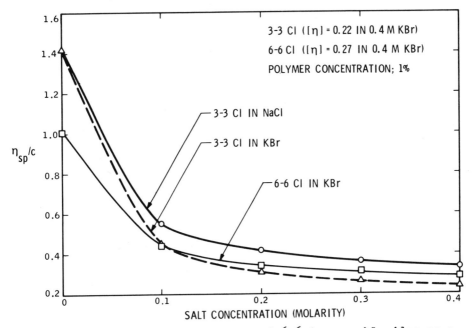

Fig. 14.   Chain coiling of 3,3- and 6,6-ionene chlorides as a
function of ionic strength.

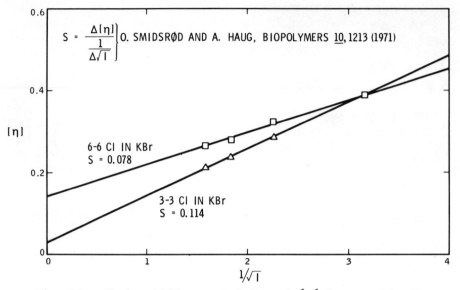

Fig. 15. Chain stiffness of 3,3- and 6,6-ionene chlorides.

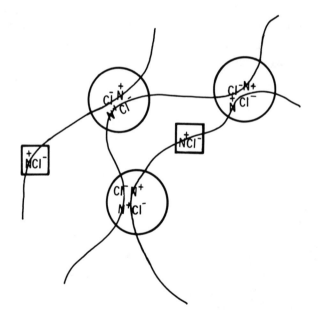

Fig. 16. Pseudocrosslinks of 3,3-ionene chloride (0.5%) in a mixture of acetone/methanol (50:1).

The NMR of the AB monomer undergoing polymerization in a DMSO—water mixed solvent, as a function of time (Figure 7) showed the appearance of a triplet at $\tau = 5.85$ and could not be attributed to the effect of DMSO because the τ value for the $\oplus$NCH$_2$ group in the isolated polymer was equal to 6.55, i.e., radically different and therefore unlikely to be a consequence of shift due to solvent interaction. Furthermore, the formation of a triplet at $\tau = 5.75$ was also observed in DMF/H$_2$O, water, and methanol solvents.

All attempts to isolate an intermediate from the polymerizing solution failed. However, the synthesis of diethylazetidinium bromide and the examination of its NMR spectrum (Figure 8) indicates strongly that the suspected intermediate consists of a cyclic structure, i.e, the dimethylazetidinium chloride. The existence of this type of intermediate was previously postulated (2) but no sufficient evidence for its formation was available.

Dimethylazetidinium chloride is less stable than diethylazetidinium bromide possibly because of either inductive or steric hindrance effects. These observations together with the complete interpretation and the excellent agreement of the assignments in all NMR spectra including those recorded during the formation of star-shaped polyelectrolytes (Figures 9, 10, and 11) gives confidence to the postulate of a cyclic intermediate.

The homopolymerization of the AB monomer may therefore be represented by the following mechanism:

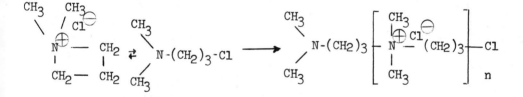

A similar mechanism must also apply to the formation of star-shaped and branched polyelectrolytes, in which case the grafting of AB monomer is accompanied by a homopolymerization reaction since the reactivity of terminal groups on the branches after addition of the first unit is most likely to be identical to that of the homopolymer. The amount of homopolymer present in the isolated branched products remains to be determined.

Properties

The intrinsic viscosity in 0.4 $\underline{M}$ aqueous KBr of the throughly dried 3,3-ionene chloride prepared in highly concentrated water

solution was found to be 0.25 dl/g. A sedimentation equilibrium
study of this material yielded a weight-average molecular weight
value of 75,000 (6). A relationship between molecular weight and
intrinsic viscosity for the 3,3-ionene chloride is being developed
using this ultracentrifugation technique.

Application of the previously determined Equation 1 for 3,4-
ionene bromide (7)

$$[\eta] = 2.94 \times 10^{-4} M^{0.61} \tag{1}$$

yields a viscosity molecular weight average of 63,000 for 3,3-ion-
ene chloride ($[\eta] = 0.25$). Equation 1 can be used, therefore, for
preliminary estimates of the molecular weight of this polyelectrolyte.

A comparison of specific reduced viscosity of 3,3- and 6,6-ion-
ene chloride as a function of ionic strength indicates that an ionene containing a larger number of positive charges in its chain
undergoes more extensive coiling with increasing ionic strength than
the corresponding ionene with comparatively low numbers of positive
charges (Figure 14). This same figure also confirms the previously
noted effect of decreased viscosity in KBr as compared with NaCl
solutions (7).

In order to compare the chain stiffness of 3,3- with 6,6-ionene
chloride, the intrinsic viscosity of both polyelectrolytes was determined at four different ionic strengths. The experimental
points, when plotted as in Figure 15, lie on a straight line in
agreement with previous data (8-13). The slopes S of the lines corresponding to 6,6- and 3,3-ionene chlorides are found to be equal
to 0.078 and 0.114 respectively. Since both ionenes are approximately of the same molecular weight, the S values express the comparative chain stiffness (14). It is concluded, therefore, that
the 3,3-ionene forms stiffer chains, in dilute solutions, than the
6,6-ionene and that chain stiffness increases most probably with
the increase of charge concentration. However at high ionic
strength, the more highly charged 3,3-ionene coils up to a greater
extent than the 6,6. A comparison of molecular weight - viscosity
relationship for the 3,4- and 6,6-ionenes derived from light-scattering results (7) carried out in concentrated salt solutions confirms this conclusion, since for the same molecular weight the intrinsic viscosity of 3,4-ionene was found to be lower than for the
6,6-ionene. This indicates, therefore, that the higher-charge-den-
sity ionene forms more coiled-up chains than its lower-charge-den-
sity homolog (at high ionic strength).

One of the consequences of high charge density of 3,3-ionene
chloride is the formation of thixotropic gels by addition of a
non-solvent, e.g. acetone to a methanolic solution of the AB

polymer. The gels are reversible, yielding a homogeneous solution by addition of water. This phenomenon is probably due to the existence of ionic clusters acting as pseudocrosslinks in the solvent - nonsolvent mixture and is illustrated in Figure 16. Water solvates the $N^{\oplus}$ $Cl^{\ominus}$ links, reverting the gel back to a homogeneous solution. A similar interpretation was proposed for solutions of cationic polyurethanes (15).

The high crystallinity of ionenes as shown by the x-ray diffraction patterns (Figures 12 and 13) reflects the salt-like nature which is enhanced by the charges in the backbone. Rigorous drying in vacuum at 60°C prior to x-ray determination resulted in very diffuse x-ray diffraction patterns. It appears that hydration influences the packing or the orientation of chains. Further studies are necessary to understand this phenomenon.

This paper represents one phase of research performed by the Jet Propulsion Laboratory, California Institute of Technology, sponsored by the National Aeronautics and Space Administration, Contract NAS7-100.

BIBLIOGRAPHY

1. H. Noguchi and A. Rembaum, Macromolecules, 5, 253 (1972).

2. C. F. Gibbs, E. R. Littmann, and C. S. Marvel, J. Am. Chem. Soc., 55, 753 (1933).

3. D. Casson and A. Rembaum, J. Polym. Sci., Part B, 8, 733 (1970).

4. M. Kulka, Can. J. Res., 23, 106 (1945).

5. E. Tsuchida, K. Samada, and K. Moribe, Makromol. Chem., 151, 207 (1972).

6. See the chapter by M. Schmir and A. Rembaum in this book.

7. D. Casson and A. Rembaum, Macromolecules, 5, 75 (1972).

8. R. A. Cox, J. Polymer Sci., 47, 441 (1960).

9. S. A. Rice and M. Nagasawa, Polyelectrolyte Solutions, Academic Press, New York 1961.

10. A. Takahashi and M. Nagasawa, J. Am. Chem. Soc., 86, 543 (1964).

11. P. D. Ross and R. L. Scruggs, Biopolymers, 6, 1005 (1968).

12. D. T. F. Pals and J. J. Hermans, Rec. Trav. Chim., 71, 433 (1952).

13. O. Smidsrod, Carboh. Res., 13, 359 (1970).

14. O. Smidsrod and A. Haug, Biopolymers, 10, 1213 (1971).

15. D. Dieterich, W. Keberle, and H. Witt, Angew. Chem. Internat. Ed., 9, 40 (1970).

III. CHARACTERIZATION

MOLECULAR-WEIGHT DISTRIBUTION OF WATER-SOLUBLE POLYMERS BY

EXCLUSION CHROMATOGRAPHY

G. L. Beyer, A. L. Spatorico, and J. L. Bronson

Eastman Kodak Company

Rochester, New York

The application of gel permeation chromatography (GPC) in organic solvents to polymers having a substantial content of polar groups often produces incorrect molecular-weight-distribution curves (1,2). This is attributed to adsorption of the polar groups to the gel or other components, with consequent delayed elution. Chemical modification of polar groups, for example by esterification of carboxyl groups, has produced satisfactory results by GPC (3). However, the detailed analysis of molecular-weight distributions of many other polar polymers has not been practical by GPC or other methods up to the present time.

Another form of exclusion chromatography differing from GPC in that hydrophilic gels and aqueous eluants are used is also known as gel filtration. This method is often employed for protein separations and in a few cases for molecular-weight-distribution analysis of hydrophilic polymers (4,5,6), but its application to high-molecular-weight polymers has been inhibited by the instability of hydrophilic gels of suitable pore size, and by inadequate calibration means. Porous glass now available in a wide range of pore sizes (7,8) offers a solution to the first problem, while we have overcome the calibration problem by preparation of narrow-distribution poly-(sodium styrenesulfonate) standards of various molecular weights.

EXPERIMENTAL

The apparatus, sketched in Figure 1, consisted of six 120-cm stainless-steel columns of 5-mm internal diameter, each packed with Corning porous glass powder (CPG-10) of different nominal pore

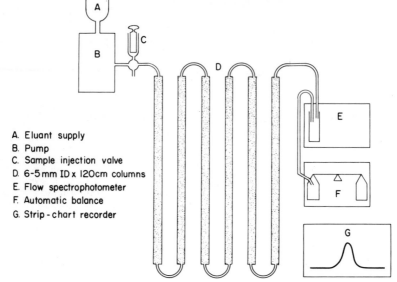

A. Eluant supply
B. Pump
C. Sample injection valve
D. 6-5mm ID x 120cm columns
E. Flow spectrophotometer
F. Automatic balance
G. Strip-chart recorder

Fig. 1.   Schematic of apparatus for exclusion chromatography.

sizes, as follows: 75, 350, 700, 1250, and 2000 angstroms. The aqueous eluant, 0.4 molar NaCl, 0.05 molar borax, pH 8.5, was pumped at 0.5 ml/min through these columns in series. One-ml samples containing 0.1 to 0.5 g/dl of polymer in the salt solution were injected just before the first column, and the composition of the eluate was monitored by a Beckman DB spectrophotometer using a flow cell of 0.5 cm path length and small (0.2 ml) volume. The photometer output (0-100 mv) was fed to a 5-mv strip chart recorder attenuated to give full-scale readings at absorbance values as small as 0.05.

    The poly(sodium styrenesulfonates) were prepared by sulfonation of a series of narrow-distribution polystyrene samples (Pressure Chemicals Co.). The resulting sulfonic acids were then neutralized with sodium hydroxide. The sulfonation by the method of Carroll and Eisenberg (9) used 100% sulfuric acid and silver sulfate catalyst at a temperature of 0°C. Under these conditions it is claimed that complete monosulfonation in the para position occurs without degradation of the polymer chain. We found that some insoluble components were obtained in all products, and the titrations with alkali required about 20 percent less than the stoichiometric quantity. Both results suggest incomplete sulfonation and may explain some of the abnormal results reported below.

RESULTS AND DISCUSSION

Good elution patterns were obtained for these standards by injection of 1-ml samples of concentration 0.1 or 0.2 g/dl. The spectrophotometer was set at 260 nm, near the peak extinction for this polymer. The peak elution volumes observed were plotted versus the log
of molecular weight (calculated for the sodium sulfonate salt), as
shown in Figure 2. The linear plot found over the range of calibration (4×10^4 to 1.3×10^6) was extended linearly to both higher and
lower molecular weights for use as the calibration curve, though it
is uncertain how far this linear region is valid. Based on this
calibration curve, the elution patterns for the standards were converted to molecular-weight distributions which are exemplified in
Figure 3. Here the absissa is the log of molecular weight and the
ordinate is the weight fraction in equal increments of log molecular
weights, with all curves normalized to equal area. Average molecular
weights were calculated from these distributions and are summarized
in Table I, which also includes ratios of the weight- to number-aver-
ages found by this method and by gel permeation chromatography at
Kodak on the starting polystyrenes. No corrections for band broadening have been made for any results.

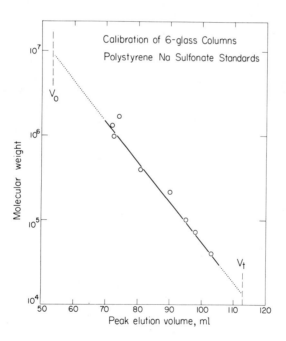

Fig. 2. Calibration of glass-filled columns using poly(sodium
styrenesulfonate) standards.

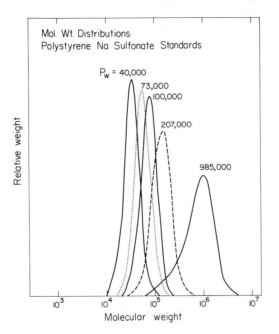

Fig. 3.   Molecular-weight distribution of poly(sodium styrenesulfonate) standards.

TABLE I.   Average Molecular Weights for Polystyrenesulfonates

Polystyrene Sample[a]	By exclusion chromatography			By GPC	
	M_w x 10^{-3}	M_n x 10^{-3}	M_w/M_n	M_w calc.	M_w/M_n
2b	37.5	33.5	1.12	42.7	1.10
7b	57.5	49.5	1.16	74.4	1.10
7a	81	70	1.16	100	----
4b	147	118	1.25	237	1.13
5a	984	564	1.75	1082	1.33
13a	1030	505	2.04	1320	1.41
6a	1540	766	2.01	1700	----

- - - - - -

a. Pressure Chemicals Co. identification

It is evident from the ratios of M_w/M_n that exclusion chromatography provides an estimate of the molecular-weight distribution in fair agreement with that found by GPC for the lower-molecular-weight samples. The results also indicate that the higher-molecular-weight sulfonates have a substantially broader distribution than the original polystyrenes, due to the presence of lower-molecular-weight tailing, which may be caused either by incomplete sulfonation or chain degradation.

Study of Poly(ethyl Acrylate-co-Acrylic Acid)

This chromatographic system has been utilized in a study of copolymers of ethyl acrylate and acrylic acid in the mole ratio 3:1 (see also the chapter by S. P. Gasper and J. S. Tan). These solution polymers, prepared in the Chemistry Division at Kodak, have been shown by several methods to be random copolymers. Fractions of these polymers prepared both by fractional precipitation and by the Baker-Williams double-gradient chromatographic method were anayzed; subsequently a well-characterized polydisperse whole polymer was tested. In all experiments, 1.0 ml injections were made of solutions containing 0.3 to 0.5 g/dl of polymer in the usual eluant composition (0.4 M NaCl, 0.05 M borax). The spectrophotometer was set at 220 nm, near the peak of the strong absorption band present in esters and carboxylic acids. Each elution curve, in absorbance units, was normalized and the elution volumes were converted to the corresponding molecular weights based on the calibration curve (Figure 2). The molecular-weight values for polymers other than the standards are indicated as equivalent values in terms of the polystyrenesulfonate standards and are symbolized as P*. This is necessary because it has been shown (10) that separations in gels are based on differences in hydrodynamic volume, rather than molecular weight. Thus, true molecular weights would be obtained from calibration curves such as Figure 2 only when the hydrodynamic volumes of standard and sample molecules are the same. True molecular weights can be derived from calibration curves in terms of hydrodynamic volume, but this has not as yet been used here because of problems with the polystyrenesulfonate standards.

The distribution curves for four copolymer fractions are plotted in Figure 4. Two samples prepared by column fractionation are seen to have much more narrow distribution curves than the two fractions prepared by fractional precipitation. An alternative method of plotting integral distribution curves on log-probability paper has been found useful for these polymers. Such a representation, shown in Figure 5, has as abscissa the molecular weight on a log scale and as ordinate the weight percent of components below a given molecular weight on a probability scale. It has been shown (11) that a linear plot indicates that the distribution follows the

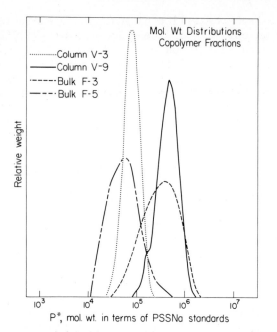

Fig. 4.   Molecular-weight distribution of copolymer fractions expressed as differential plots.

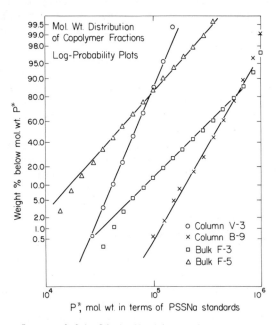

Fig. 5.   Molecular-weight distribution of copolymer fractions expressed as log-probability plots.

log-normal distribution function, in which case various average
molecular weights can be calculated readily. Even if departures
from linearity occur, we have found this plot convenient to display
clearly the content above or below a given molecular weight. The
linear plots for the fractions led to calculated weight averages in
agreement with those by summation of the incremental data from the
exclusion chromatography differential curves. A comparison is given
in Table II of the weight-average equivalent molecular weights, P_W^*,
from exclusion chromatography, with the viscosity averages, M_V,
based on light-scattering data (12) for a series of different fractions. The exclusion chromatography results were found to be within
12% of the viscosity averages for these fractions.

TABLE II. Average Molecular Weights for Copolymer Fractions

Sample	M_V	P_W^*
Column fract V-3	67,000	74,000
Column fract V-9	430,000	404,000
Precip fract 3	343,000	347,000
Precip fract 5	58,000	65,000

Unfractionated Samples. Exclusion chromatography of an unfractionated copolymer which had been more fully characterized (12)
provided a more reliable estimate of the capacity to determine molecular-weight distributions. The copolymer chosen had twice been chro-
matographically fractionated to yield ten fractions in each experiment. One series of fractions had been characterized by intrinsic
viscosity to yield viscosity-average molecular weights, M_V, based
on other characterization data (12). The other series of fractions
had been studied by light scattering and GPC (the last after methylation of each fraction), so that both weight-average molecular weights
and polystyrene-equivalent molecular weights were available on each
fraction. The whole polymer had also been methylated and analyzed
by GPC. Thus, the fullest possible characterization had been carried
out by methods accessible up to now.

The differential distribution curves found by exclusion chromatography (gel filtration) on the unfractionated ionized polymer
and by GPC on the methylated polymer are shown in Figure 6, and the
corresponding log-probability plots are given in Figure 7. Both
results fit the log-normal function fairly well, though they differ
somewhat in slope. The results based on the chromatographic fractions are given on the probability plots in Figure 8, which includes a comparison with the exclusion chromatography results. The
GPC results for methylated fractions appear in closest agreement with
exclusion chromatography, while the molecular weights derived from
viscosity and light scattering indicate somewhat higher values for
each fraction.

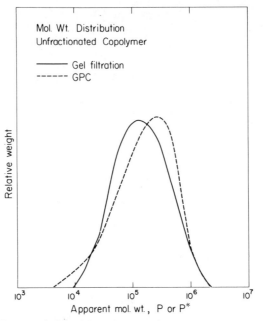

Fig. 6.   Molecular-weight distribution of unfractionated copolymer
determined by two methods.

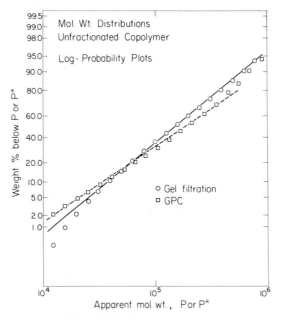

Fig. 7.   Molecular-weight distribution of unfractionated copolymer
determined by two methods and plotted as log probability.

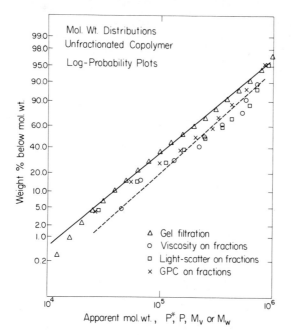

Fig. 8. Molecular-weight distributions for unfractionated copolymer.

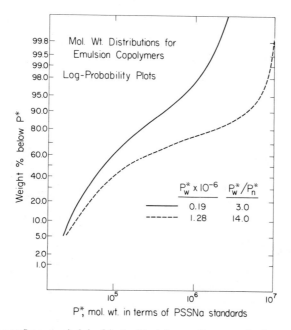

Fig. 9. Molecular-weight distributions for emulsion copolymers.

A summary of the weight-average quantities derived from each type of measurement is provided in Table III. The weight-averages from exclusion chromatography, P_w^*, and from GPC, P_w, on the whole polymer were obtained by summation of about 50 increments on the measured elution curves.

TABLE III.   Average Molecular Weights for Unfractionated Copolymer

Method	Type of Average	Value x 10^{-3}
Viscosity	M_V	297
GPC (after methylation)	P_w	268
Exclusion chromatography	P_w^*	240

The ten percent difference in the values of P_w and P_w^* is partly due to unequal hydrodynamic volumes of the samples and standards in the two systems, and partly to the experimental uncertainties. The larger difference of these two from the viscosity-average on the whole polymer and from both light-scattering and viscosity results for fractions is to be expected, since both GPC and exclusion chromatography are relative measurements until the hydrodynamic volume concept is applied to convert to true molecular weights. Although differences in the average values are shown to be substantial, nevertheless the exclusion chromatography method is seen in Figure 8 to provide distributions approaching those by other methods for this copolymer.

Other Polymers

Further tests have been made on polymers for which no other molecular-weight-distribution data were available; these are given below to indicate the possible utility and limitations of the method.

The exclusion chromatography distribution curves in Figure 9 are for ethyl acrylate - acrylic acid copolymers of the same compositions as those above, but prepared by an emulsion polymerization method. The apprent molecular weights extend over a much wider range (thus larger ratios of R_w^*/P_n^* are found), and marked departures from the log-normal distribution function are noted. Although these samples had no components in the interstitial volume limit of the present column set, we have tested related polymers for which a large fraction of the components are in the interstitial volume, and indicate apparent molecular weights approaching 10^7 (assuming a linear calibration curve).

In addition to the two polymer types discussed above, we have examined uncharged polymers (dextran and polyacrylamide), and several acrylic polymers with sulfonate substituents. In all these cases, elution has been normal and reasonable molecular-weight-distribution results were obtained. However, we have found that two different polymer types which contained cationic groups eluted abnormally from the porous glass columns. Gelatin which contains both cationic and anionic groups, but bears a substantial net negative charge at the pH of the borax buffer (8.5), produced a long elution curve extending beyond the volume at which monomers normally elute, indicating some adsorption. An acrylic copolymer with quaternary ammonium side groups was apparently strongly adsorbed since no visible elution pattern was obtained. Further studies are under way to extend the range of these columns to higher molecular weights, and means of minimizing adsorption problems are under consideration.

FIBLIOGRAPHY

1. K. H. Altgelt and J. C. Moore, in M. J. R. Cantow, Ed., Polymer Fractionation, Academic Press, New York, 1967.

2. R. M. Screaton and R. W. Seamon, J. Polymer Sci., C21, 297 (1968).

3. A. L. Spatorico, et al, Eastman Kodak Co., unpublished results.

4. J. G. McNaughton, W. Q. Yean, and D. A. I. Goring, Tappi, 50, 548 (1967).

5. K. A. Gramath and B. E. Kvist, J. Chromatogr., 28, 69 (1967).

6. W. Brown, S. I. Falkenhag, and E. B. Cowling, Nature, 214, 410 (1967).

7. Controlled-Pore Glass, Corning Glass Co., Corning, New York.

8. Porasil, Waters Associates, Framingham, Mass.

9. W. R. Carroll and H. Eisenberg, J. Polymer Sci., Part A-2, 4, 599 (1966).

10. Z. Grubisic, P. Rempp, and H. Benoit, J. Polymer Sci., Part B, 5, 753 (1967).

11. H. Wesslau, Makromol. Chemie, 20, 111 (1956).

12. S. P. Gasper, Eastman Kodak Co., private communication, 1971.

SEDIMENTATION EQUILIBRIUM OF HIGH CHARGE DENSITY CATIONIC POLYELECTROLYTES

Maurice Schmir and Alan Rembaum

Jet Propulsion Laboratory, California Institute of

Technology, Pasadena, California

Although a number of theoretical treatments and applications of sedimentation equilibrium have been reported in recent years (1-9), as yet no systematic theory or method is available for the study of nonideal, polydisperse polyelectrolyte systems. In view of the increasing importance of polycations in polymer chemistry, microbiology, and industry, the development of convenient, reliable methods of physical characterization of polycationic polymer preparations has become an immediate problem. The analytical ultracentrifuge, by virtue of the wide range of operating conditions accessible (from 800 to 68,000 rpm) and the sensitivity of the available optical systems is an excellent candidate for the nucleus of a routine, high-precision system of polymer molecular-weight determination. The molecular weights accessible by current centrifuge methods range over nearly four orders of magnitude from several thousand to the tens of millions. The ability of the analytical ultracentrifuge to handle concentrations of samples as low as 0.05% (in some cases, lower) allow concentration-dependent studies to be extrapolated with reliability into the zero concentration range.

Using the water-soluble polyammonium compound 3,3-ionene chloride, I,

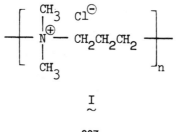

I

as a test case, we have attempted to assess the usefulness of the
analytical ultracentrifuge as a tool in the determination of the
average molecular weight and molecular-weight range of a synthetic
cationic polyelectrolyte preparation. Sedimentation equilibrium
experiments were carried out over a variety of conditions in order
to evaluate the applicability of current sedimentation equilibrium
theory to nonideal, polydisperse, polyelectrolyte systems. As a
necessary part of this work, computer software specifically
designed for these experiments was developed not only to handle
the large amount of data generated by sedimentation equilibrium
studies, but also to program a computer-controlled scanning system
capable of far greater speed and precision than current manual
methods of data accumulation.

It should be noted that the work described here represents an
initial evaluation of the entire working system, from the theoretical framework to the scanning hardware. Certain theoretical
points, discussed below, are still under investigation. However,
they do not affect the nature of the methods and results described
in this report.

EXPERIMENTAL

The 3,3-ionene chloride was prepared according to the method
described by Noguchi and Rembaum (10). See also the chapter by
S.P.S. Yen, D. Casson, and A. Rembaum. The intrinsic viscosity of
this material was found to be 0.22 dl/g before rigorous drying.
The material used in actual runs was dried under high vacuum in a
methanol drying pistol for forty-eight hours before preparation of
solutions. This drying procedure reduced the sample weight by 12%
due to loss of water.

A stock solution was prepared at 6.0 mg/ml of 3,3-ionene
chloride in 0.5 M KBr. This solution was used for all subsequent
dilutions. Dilution was always performed gravimetrically using an
analytical balance so as to avoid pipetting errors.

The partial specific volume of the 3,3-ionene chloride, 0.75
cm^3/g, was determined at $20^{\circ}C$ in a 25 ml pyncnometer. The index of
refraction increment was taken as 0.160 ml/g, giving a fringe shift
in the Rayleigh interference optical system of 0.9986 mm/(mg/ml).

All centrifugation was carried out on a Beckman Spinco Model E
analytical ultracentrifuge, serial number 442, using an An-D
aluminum rotor. The centrifuge was equipped with temperature
control which was adjusted so that all runs were maintained at
$21.5^{\circ}C$.

Rayleigh interference patterns were recorded on Kodak Spectroscopic II-G plates and developed in Kodak D-19 developer for 5 minutes. Typical exposure times were between 30 and 60 seconds.

The attainment of equilibrium was checked by photographs taken at six-hour intervals. Within one series, forty-eight hours provided a safe margin for the initial speed and twenty-four hours for all subsequent speeds.

Photographs were measured on a Nikon comparator equipped with a rotating stage for plate alignment. The 20X lens was used throughout the measurements.

All computations were performed on an IBM 370/155 and plots made on an associated Calcomp plotter. The pilot automated plate scans were controlled by a time-shared PDP-10 computer.

RESULTS AND DISCUSSION

Background and Application

The theory and application of sedimentation equilibrium in the analytical ultracentrifuge has been described extensively in the literature (1-9) and dealt with in depth by Schachman (11).

Under the influence of the gravitational field experienced by the sample in the ultracentrifuge, a concentration gradient along the radial direction is established. If the conditions are chosen properly, this gradient will stabilize after a suitable period of time. The concentration gradient in the ultracentrifuge sample cell at sedimentation equilibrium is a function of the molecular weight or molecular-weight distribution of the sample. This gradient is visualized by a Rayleigh interference optical system built into the centrifuge. The experiment is performed with a double-sector cell, one sector of which contains the sample and the other the solvent in which the sample is dissolved. The concentration gradient within the sample sector causes a continuously increasing difference in the indices of refraction between the two sectors. This in turn causes curvature in the emerging interference pattern. Figure 1 shows such a pattern as it is seen in the ultracentrifuge. Region I is an image of the sample cell. The sample is between r_m, the radial position of the solution meniscus with respect to the axis of rotation, and r_b, the position of the cell bottom. Region II is an image of the reference counterbalance. The radial distance of the wire from the axis of rotation is a constant, so that all measurements of radial position

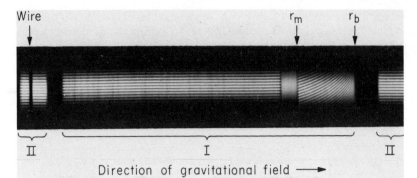

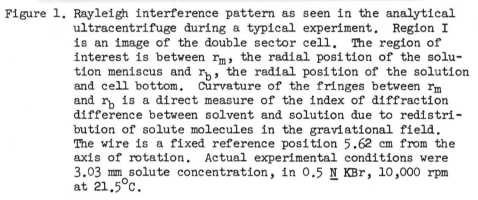

Figure 1. Rayleigh interference pattern as seen in the analytical
        ultracentrifuge during a typical experiment. Region I
        is an image of the double sector cell. The region of
        interest is between r_m, the radial position of the solu-
        tion meniscus and r_b, the radial position of the solution
        and cell bottom. Curvature of the fringes between r_m
        and r_b is a direct measure of the index of diffraction
        difference between solvent and solution due to redistri-
        bution of solute molecules in the graviational field.
        The wire is a fixed reference position 5.62 cm from the
        axis of rotation. Actual experimental conditions were
        3.03 mm solute concentration, in 0.5 $\underline{N}$ KBr, 10,000 rpm
        at 21.5°C.

within the sample cell can be made with respect to the wire.
Subsequent calculations convert this relative distance to an
absolute distance coordinate.

    For an ideal, homogeneous, two-component system the concen-
tration gradient at any point in the centrifuge cell is given by,

$$\frac{1}{c_r} \frac{dc_r}{dr} = \frac{M(1 - \bar{\upsilon}\rho)\omega^2 r}{RT} \tag{1}$$

where c_r is the concentration at distance r from the axis of
rotation, M is the molecular weight at that point, $\bar{\upsilon}$ is the
partial specific volume of the solute, ρ is the solvent density,
and ω is the rotational speed in radians per second. This equation can be rewritten so as to express the molecular weight at a
given point as a function of the gradient and radial position,

$$M = \frac{2RT}{(1 - \bar{\upsilon}\rho)\omega^2} \frac{d \ln c}{dr^2} . \tag{2}$$

The slope of the plot of ln c vs. r^2 corresponds to the molecular
weight at each point r.

Equation 1 can be solved to give an explicit expression for the concentration at any point r in the cell,

$$\ln \frac{c_r}{c_m} = \frac{M(1 - \bar{v}\rho)\omega^2}{2RT} (r^2 - r_m^2) \quad , \tag{3}$$

where c_m and r_m are the concentration at the meniscus and the radial position of the meniscus, respectively. An alternative solution to Equation 1 is given by the expression

$$M = (c_b - c_m)/c_o\lambda \tag{4}$$

where c_b is the solute concentration at the bottom of the cell, located at r_b. We have summarized the experimental parameters in the parameter λ, where

$$\lambda = \frac{(1 - \bar{v}\rho)\omega^2(r_b^2 - r_m^2)}{2RT} \quad . \tag{5}$$

Equation 4 allows us to determine the molecular weight of the sample taken across the entire cell.

Synthetic polymer preparations do not generally meet the requirements of ideality and monodispersity so that Equations 3 and 4 are not adequate descriptions of the situations encountered in polymer chemistry. Fujita has extensively treated the problem of nonideality and polydispersity in neutral polymers (2) and developed an expression for the observed weight-average molecular weight, M_{app}, as a function of the true weight-average molecular weight, M_w, and higher-order terms in concentration and the parameter λ,

$$\frac{1}{M_{app}} = \frac{1}{M_w} + B_{LS}c_o(1 + \lambda^2 M_z^2 /12 + \ldots) \tag{6}$$

$$+ \text{ higher terms in } c_o \quad ,$$

where c_o is the initial concentration of polymer in the centrifuge cell and B_{LS} is the second virial light-scattering coefficient. The expansion in powers of λ reflects the speed dependence of M_{app} arising from the polydispersity of a nonideal polymer solute. The effect of changing speed is to alter not only the distribution of macromolecule in the cell but also the effects of nonideality. Equation 6 is based on the model of a neutral coil polymer treated as a pseudo-two-component system and does not deal with the problem of charge on the macromolecule.

Utiyama and co-workers (8) have systematically examined the applicability of Equation 6 to a range of polydisperse polystyrene samples prepared by mixing two different monodisperse samples in varying amounts. They demonstrated that it was possible to obtain weight-average molecular weights to within a few percent of values calculated from the known mixture composition by using an extrapolation procedure based on Equation 6.

Using Equation 4, the apparent molecular weight is determined for one initial concentration, c_o, over a range of speeds. A plot of $(M_{app})^{-1}$ vs. λ^2 is then extrapolated to zero λ^2. This gives an extrapolated value of $(M_{app})^{-1}$ which is a speed-independent value for the particular c_o chosen. This procedure is repeated for a range of different c_o's. For each c_o it will be possible to obtain a speed-independent value of $(M_{app})^{-1}$. A plot of these values of $(M_{app})^{-1}$ vs. c_o will yield B_{LS} as a slope of the true $(M_w)^{-1}$ as the intercept.

The theory embodied in Equation 6 does not specifically deal with charged systems, so that the experimental test of Equation 6 using polystyrene does not tell us about the validity of the Fujita formulation for polyelectrolytes such as ionenes. At this time, no comprehensive theoretical treatment for nonideal, polydisperse polyelectrolytes exists. We now describe the results obtained in this initial empirical evaluation of the usefulness of the Fujita equation for the determination of the molecular weight of ionene polymers at high ionic strength.

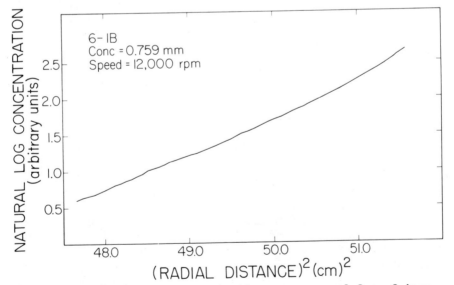

Figure 2. Natural log of concentration as measured from fringe shift plotted against the square of the radial position.

Molecular Weights at One Speed

The results of a typical single-concentration, single-speed run are shown in Figure 2. The data have been plotted according to Equation 2. The slope at any distance r from the axis of rotation gives the weight-average molecular for the distribution of macromolecule at that radial position and solute concentration. Figure 3 summarizes the values for M_w obtained across the cell using Equation 2. The single most striking feature of this plot is the curvature. This clearly reveals the polydispersity only suggested by the slight curvature of the ln c vs. r^2 plot, Figure 2. In certain systems the effects of nonideality are more pronounced than those of polydispersity and the plot in Figure 2 would show a downward curvature. Occasionally these effects are found to cancel each other to within experimental error producing a straight line plot of ln c vs. r^2 and a corresponding constant value of M_w as a function of radial position. If such a result is obtained, it is necessary to carry out successive runs as a function of concentration and speed in order to test the apparent ideality and monodispersity.

The weight-average molecular weight determined for the entire sample according to Equation 4 is indicated on the vertical axis in Figure 3. The scatter in the values of M_w at lower radial

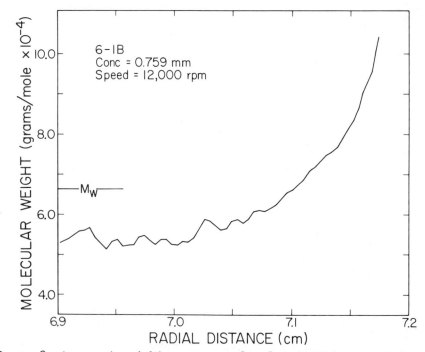

Figure 3. Apparent weight-average molecular weights computed according to Equation 2 from point-by-point slopes of data in Figure 2, plotted as a function of radial position.

positions is typical of results obtained in sedimentation equilibrium experiments and reflects the error associated with reading
Rayleigh interference fringes. When the absolute fringe shift is
substantially larger than the error in reading fringes ($\pm 5\ \mu$) the
results obtained are considerably smoother, as can be seen in
Figure 3 for radial positions greater than 7.08 cm.

Since the concentration at all points in the cell is known,
it is also possible to display the apparent molecular weight as a
function of concentration, Figure 4. Here the region of relatively
high scatter is compressed into a smaller region of the plot than
in Figure 3, indicating that the data most subject to large
apparent scatter corresponds to only a small part of the overall
concentration range achieved in a given run. It is important to
note that the values for M_w plotted in Figures 3 and 4 correspond
to the weight-average molecular weights of the distribution of
polymer at any given r or c.

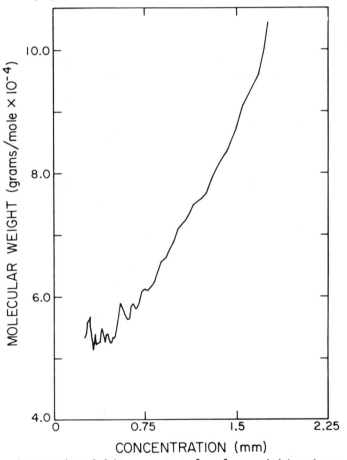

Figure 4. Apparent weight-average molecular weights shown in
Figure 3 plotted as a function of concentration.

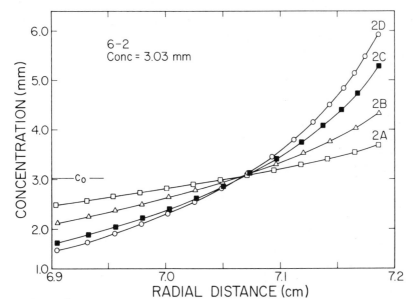

Figure 5. Redistribution of solute as a function of speed. The distribution designated 2A is for 5,200 rpm, 2B, 7,200 rpm, 2C, 9,000 rpm and 2D, 10,000 rpm. Initial solute concentration is 3.03 mm in 0.5 $\underline{N}$ KBr, 21.5°C.

Speed-Dependent Studies

In order to obtain values of M_w that reflect only the non-ideality of the ionene polymer and not its polydispersity it is necessary to obtain a set of apparent M_w's as a function of speed for any one given initial concentration. This was discussed earlier in connection with the Fujita expression Equation 6. The effect of varying the speed of rotation on the distribution of solute at sedimentation equilibrium is shown in Figure 5. The range of speeds is from 5,200 rpm (2A) to 10,000 rpm (2D). The initial loading concentration is indicated on the vertical axis. Note that the change in concentration gradient along the cell as a function of speed occurs about a pivotal point slightly to the right of the center of the solution column and at approximately the initial concentration.

The data represented by the four plots in Figure 5 can now be treated according to Equation 2 to determine the dependence of molecular weight on radial position and concnetration as was done for one speed in Figures 3 and 4. The results of such computations are shown in Figures 6 and 7. The increase in speed from 5,200 rpm to 10,000 rpm depletes the region of the meniscus of higher-molecular-weight components. This is reflected in the monotonically decreasing value of M_w at the meniscus in Figure 6. The effect of

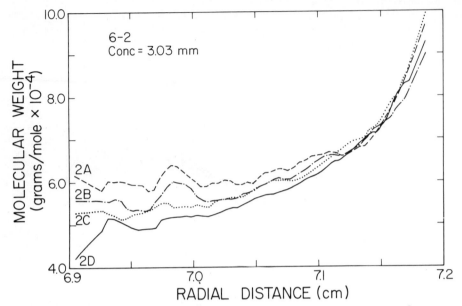

Figure 6. The dependence of apparent weight-average molecular
          weight on radial position for the four distributions of
          Figure 5.

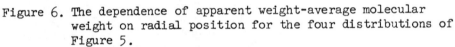

altering speed is not as clearly evident at the cell bottom where
higher molecular weight components predominate. This is because
the addition of a small amount of lower molecular component to the
mass at the cell bottom is a proportionally lower perturbation

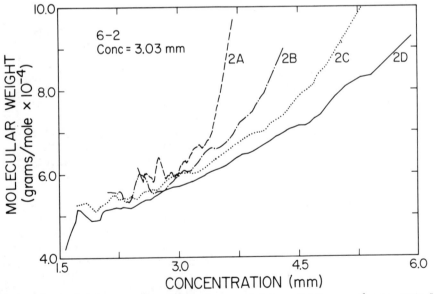

Figure 7. Weight-average molecular weights of Figure 6 converted
          to show dependence on concentration.

than the removal of higher-molecular-weight components from the mass at the meniscus. Figure 7 shows the effect of the distribution of molecular weights along the cell superimposed on the successive increase of the concentration gradient with increasing speed.

Concentration Dependence

For each set of data in Figures 6 and 7 it is possible to compute a corresponding value for the apparent weight-average molecular weight using Equation 4. This gives an apparent M_w for each of the speeds used. This procedure is now used for three additional initial concentrations. Figure 8 summarizes the results of these computations. The reciprocal of the apparent weight-average molecular weight has been plotted against λ^2 according to Equation 6 for the four concentrations used.

The dependence of M_w on initial concentration is apparent from Figure 8, the lowest concentration of 0.759 mm of fringe (open circles) giving the lowest average $1/M_w$ and hence highest M_w. On the basis of the data shown in this figure it is not possible to say, however, that there is any appreciable speed dependence of M_w above experimental error. We have therefore not attempted to extrapolate the plot to $\lambda^2 = 0$. Table I summarizes the results

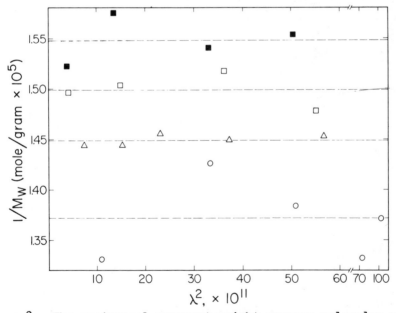

Figure 8.   The reciprocal apparent weight-average molecular weight $1/M_w$, as a function of the experimental parameter, λ^2. Dashed lines correspond to mean values of $1/M_w$ for each set.

TABLE I. Weight-Average Molecular Weight, M_w, for Each of
Four Different Initial Concentrations, c_0

Initial Conc, c_0 (mm)	Average M_w (g/mole x 10^{-4})	Speed Range (rpm x 10^{-3})
0.759	7.290 ± 0.192	6.8 - 12.0
1.503	6.900 ± 0.025	6.0 - 10.0
2.276	6.670 ± 0.072	5.2 - 10.0
3.030	6.460 ± 0.095	5.2 - 10.0

obtained in terms of average M_w's.

The dependence of M_w on concentration is shown explicitly in
Figure 9, where we have plotted $1/M_w$ as a function of $(c_b + c_m)/2$.
This concentration variable, which is the average concentration
across the cell, has been proposed by Williams and co-workers (1)
as a more valid variable than c_0. The dependence of $(c_b + c_m)/2$
on speed has been considered by Doenier and Williams (12) who have
written the expression

$$(c_b + c_m)/2 \ = \ c_0(1 + \lambda^2 M_z M_w/12 + \ldots).\tag{7}$$

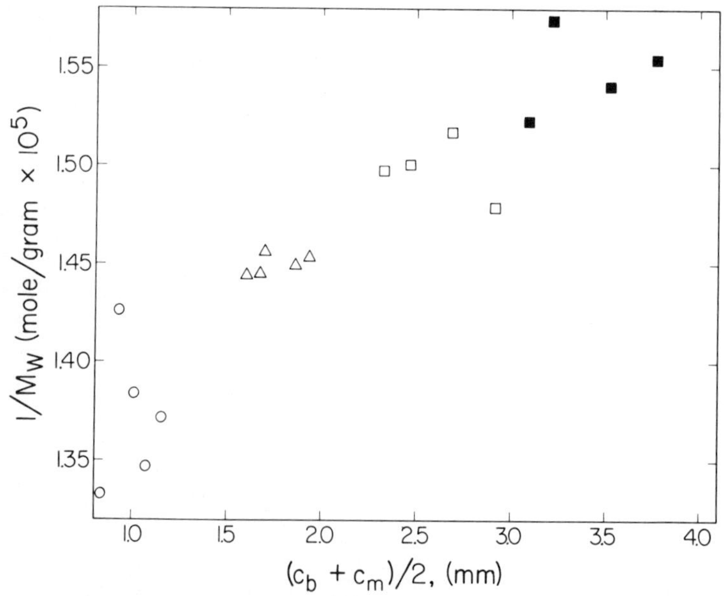

Figure 9. $1/M_w$ as a function the concentration parameter
$(c_b + c_m)/2$ of Ref. 12.

Figure 9 therefore summarizes both the speed and concentration dependence of the apparent M_w's. Although trends are suggested with each group of data, it is not possible to say that there is a speed dependence greater than the experimental error associated with the runs.

The primary difficulty in observing the speed dependence is that we are measuring a perturbation (speed dependence) on a perturbation (nonideality). If we wish to test for speed dependence we can examine the effect of speed on redistribution of the solute. This can be done by making use of Equation 7. Dividing both sides by c_0 normalizes all the data to the same scale. Figure 10 is a plot of $(c_b + c_m)/2c_0$ vs. λ^2. The high degree of colinearity exhibited and the intercept near 1.0 indicate that the Fujita-Williams formulation does apply to the ionene system and that a measurable speed dependence exists.

An additional cause for the difficulty in detecting a λ-dependence in the M_w data is the magnitude of the expansion in λ^2. The scatter in the data plotted in Figure 8 suggests that the effects due to speed dependence are of the order of the experimental error associated with the determination of M_w. To test this possibility, an estimate was made of the magnitude of the higher order terms of the expansion in λ^2 in Equation 6. The expansion in

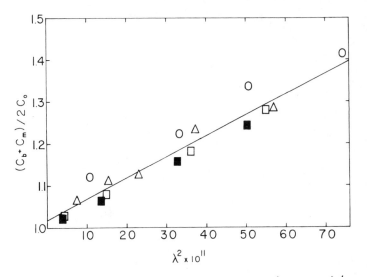

Figure 10. Normalized concentration parameter $(c_b + c_m)/2c_0$ as a function of λ^2. The speed dependence of the concentration variable is apparent from the slope. The colinearity of the data supports the applicability of current theory to charged systems.

λ^2 as derived by Fujita is

$$f(\lambda) = 1 + \lambda^2 M_z^2 /12$$
$$+ \left[(M_{z+1}/M_z)^2 - 2(M_{z+1}M_{z+2})^2/M_z^4 \right]$$
$$\times \lambda^4 M_z^4 /720 + \ldots \tag{8}$$

Using the average value of M_z obtained by the method described
below, it is possible to calculate the higher order molecular
weight moments M_{z+1} and M_{z+2} if some assumption is made about the
nature of molecular weight distribution for the polymer preparation.
In the present case we have assumed a Schulz distribution (13) of
the form

$$f(M) = \frac{p^{h+1}}{\Gamma(h+1)} M^h \exp(-pM) \tag{9}$$

where h and p are adjustable parameters and Γ denotes the Gamma
function. Successively increasing molecular weight moments can be
expressed in terms of the parameters h and p according to the
expression

$$M_i = (h + i)/p \qquad i = 1, \ldots n \tag{10}$$

Typically $M_w = (h + 1)/p$, $M_z = (h + 2)/p$, and so on. The values
obtained for M_w and M_z are used to solve for h and p so that higher
order moments can then be evaluated. The results of such calculations are shown in Table II.

TABLE II. Estimation of High Order Terms of Expansion in λ

Run[a]	Speed (rpm x 10^{-3})	$\lambda(\times 10^5)$	2nd	3rd	Resultant
A	5.2	0.61	0.037	-0.0021	1.035
B	7.2	1.16	0.135	-0.027	1.108
C	9.0	1.81	0.332	-0.168	1.164
D	10.0	2.24	0.510	-0.392	1.118

$M_w = 0.75 \times 10^5$ g/mole
$M_z = 1.10$ " "
$M_{z+1} = 1.44$ " " b
$M_{z+2} = 1.79$ " " b

[a] All runs for an initial solute concentration of 3.03 mm.
[b] Estimated assuming a Schulz distribution (Ref. 13).

TABLE III.   Speed Dependence of Reciprocal Molecular Weights

Speed (rpm x 10^{-3})	λ(x 10^5)	$1/M_w$ (mole/g x 10^5)
5.2	0.61	1.522
7.2	1.16	1.576
9.0	1.81	1.540
10.0	2.24	1.554

The values designated as the second term refer to the λ^2 term of Equation 8 and the third term to the λ^4 term. The increasing, negative value of the third term as a function of speed result in net values of $f(\lambda)$ that vary over a range of only 3% for a 4-fold change in λ in going from 5,200 to 10,000 rpm, as shown in the third column. Higher order terms than those estimated here may further alter the net value of the expansion.

Although the values obtained for this estimation procedure are not exact, they do provide an explanation for the observed difficulty in isolating speed dependence from the molecular weight values alone.

Albright and Williams (14) have pointed out that if the condition $M_w\lambda < 1$ is not met, contributions of higher order terms of the Fujita expressions will become significant. This is the case in these series of experiments where the criterion is satisfied for only the lower speeds used. This can be seen in Table III for the set of runs with c_0 = 3.03 mm. Practically, it is difficult to have $\lambda < 1/M_w$ for all runs since at too low a speed the total fringe shift across the cell will be too small for reliable readings to be made. (With a reading error of ± 5 μ it is necessary to have at least two fringes of concentration gradient across the cell. This corresponds to about 570 μ).

Final Results

Since the speed dependence of the molecular weight is obscured by higher-order terms, it is appropriate to perform the final extrapolation to infinite dilution using average molecular weights for each concentration. The average values of M_w are plotted as $1/M_w$ vs. c_0 in Figure 11. The intercept gives the concentration- and speed-independent weight-average molecular weight and the slope corresponds to the second virial light-scattering coefficient, B_{LS}.

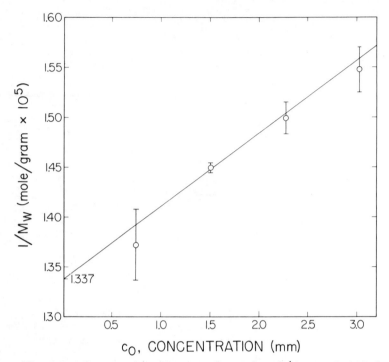

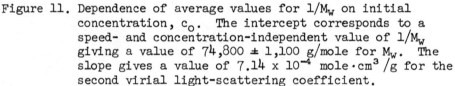

Figure 11. Dependence of average values for $1/M_W$ on initial
concentration, c_O. The intercept corresponds to a
speed- and concentration-independent value of $1/M_W$
giving a value of 74,800 ± 1,100 g/mole for M_W. The
slope gives a value of 7.14×10^{-4} mole·cm^3/g for the
second virial light-scattering coefficient.

TABLE IV. Molecular Weight and Second Virial

Light-Scattering Coefficient of Ionenes

Method	Material	M_W (g/mole)	M_z (g/mole)	B_{LS} (mole·cm^3/g)
Low-speed sed. equil.	3,3-ionene	74,800	110,000	7.41×10^{-4}
High-speed sed. equil.	3,3-ionene	67,000		
Light-scattering[a]	3,4-ionene	57,800		4.17×10^{-4}

[a]Ref. (10).

The value obtained for the weight-average molecular weight of
the 3,3-ionene chloride is 74,800 ± 1,100 g/mole. For the second
virial lightscattering coefficient we obtain a value of 7.41 ± 0.11
x 10^{-1} mole·cm^3/g. These values are tabulated in Table IV, along
with the results of Casson and Rembaum (15) and results we have
obtained using high-speed sedimentation equilibrium in the ultracentrifuge.

Casson and Rembaum have used light-scattering methods to obtain intrinsic viscosity - molecular weight relationships for the
3,4- and 6,6-ionenes in 0.4 $\underline{M}$ KBr. For a sample of 3,4-ionene
bromide having an intrinsic viscosity of 0.231 dl/g, they obtained
an apparent molecular weight of 57,800 g/mole. The corrected intrinsic viscosity of the 3,3-ionene chloride used in the present
study is 0.24 dl/g in 0.4 $\underline{M}$ KBr. The solvent used for the centrifuge study was 0.5 $\underline{M}$ KBr. In view of the difference in structure
and experimental conditions, the comparison between the light-
scattering results and the sedimentation equilibrium results can be
given only moderate weight. However, it is indicative of a better
than order-of-magnitude agreement between the methods.

Our earlier experiments using high-speed sedimentation equilibrium methods have given a value of M_w of 67,000 g/mole for the
3,3-ionene chloride in 0.5 $\underline{M}$ KBr. We have not used the high-speed
or meniscus-depletion method of sedimentation equilibrium because
of restrictions imposed by the solubility and molecular weight of
the ionenes on the range of speeds and equilibrium times accessible. The high-speed method has been developed extensively by
Yphantis (16,17). Future work will explore the possibility of
broader application of this method to the problem of determining
the molecular weights of the ionenes.

The value of M_z shown in Table IV is an average of the value
obtained from each of the runs performed for this study. In
general, the apparent z-average molecular can be obtained from the
expression (3)

$$\frac{\overline{M}_z(1 - \overline{v}\rho)\omega^2}{RT} = \frac{\frac{1}{r_b}\left(\frac{dc}{dr}\right)_{r_b} - \frac{1}{r_m}\left(\frac{dc}{dr}\right)_{r_m}}{c_b - c_m} . \tag{11}$$

The values of M_z obtained in this manner are subject to high-order
error resulting from the use of the derivatives in Equation 11. As
can be seen in Figure 7, at the meniscus the derivative of concentration with respect to radial position is subject to a larger
relative error than at the cell bottom. This error, arising from
the relatively small fringe shifts at the meniscus, appears both in

the numerator and denominator of Equation 11. The error associated with the mean value for M_z is, however, only of the order of 10%, so that we can regard the result as meaningful, if not exact.

The ratio of M_z to M_w is found to be 1.47, so that we would expect the ratio of M_w to M_n to be approximately 1.2. For a condensation polymerization following Flory-type statistics (18), we expect a ratio of 2.0. On the basis of this argument, the preparation of 3,3-ionene chloride used for this study appears to have a considerably sharper molecular-weight distribution than predicted for a theoretical condensation polymerization. This is consistent with the difficulty in observing any strong dependence of M_w on λ, since the expansion of λ^2 expresses the polydispersity of the polymer preparation. (See also the Chapter by S.P.S. Yen, D. Casson and A. Rembaum.)

EXPERIMENTAL LIMITATIONS AND FUTURE PROSPECTS

Although the results discussed thus far indicate that sedimentation equilibrium is a promising method for the determination of nonideality and molecular-weight parameters, we have not exhaustively explored the problems arising from the high positive charge associated with the ionene polymers. It is important to note that the results obtained here apply to a specific macromolecular component defined by the conditions of the experiment, namely 3,3-ionene chloride in 0.5 $\underline{M}$ KBr. Insofar as we understand this there are no difficulties arising from the charged nature of the polymer. However, under different solvent conditions, e.g. 0.5 $\underline{M}$ KCl, the molecular weight may appear to be different. For this reason it will be necessary to explore more thoroughly the effect of counterions, both individually and competitively, on the molecular weights and virial coefficients.

To facilitate the continuation of these experiments, a major effort has been made to improve the method of data gathering. Current methods require the use of a human operator recording fringe positions from a projected image of the Rayleigh interference photograph on a microcomparator. Since approximately four hours is required to read one plate and a study such as the one described here requires of the order of twenty-five plates, the reading process becomes a major task. Furthermore, the reliability of a human observer in reproducing a given reading is ± 5 μ. So that for runs in which a fringe shift of only two fringes is obtained, the error is ± 1%. Since errors accumulate very rapidly in sedimentation equilibrium work, it would be beneficial to reduce this error as much as possible.

With these goals of speed and accuracy in mind, we have developed computer programs to control the automated cathode-ray-

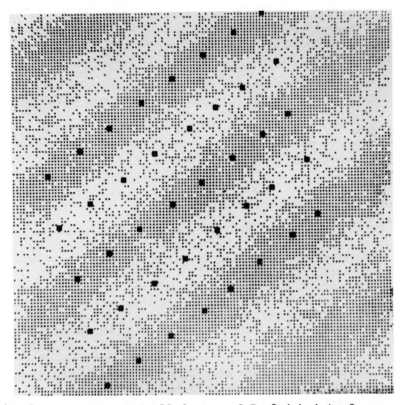

Figure 12.   Computer-controlled scan of Rayleigh interference
pattern. The centroid of each scanned area is the
lower left corner of the squares. An arbitrary number
of fringes may be scanned simultaneously and the cen-
troids averaged. Scan areas are of the order of 100 μ
by 200 μ. The cathode-ray-tube spot size is 15 μ, but
is moved in 1.3 μ increments. The dimensions of the
scan are adjusted for fringe thickness and curvature
by the computer. Compensation is also made for varying
emulsion density. The fringe pattern as shown is from
a cathode ray display of the interference photograph
using a ten-fold reduction in the number of actual
intensity points. A complete image is significantly
smoother in appearance.

tube scanning device of the California Institute of Technology.
The scanning procedure, computer-controlled but accessible to a
human operator, is capable of scanning a plate in about two minutes
with a ± 1 μ reproducibility. Data smoothing is accomplished as an
integral part of the scanning procedure. Figure 12 shows the
results of a scan over a portion of the fringe pattern. The scanning procedure is intelligent and does not merely digitize the
entire photograph. The fringe is followed from meniscus to bottom

with continuous compensation for curvature, change of fringe width, and varying emulsion density. The smoothed data are then introduced into the final computer programs for analysis and reduction to interpretable results.

We are currently engaged in testing procedures to determine optimal system parameters for the control algorithms developed.

By making use of this computerized plate reading it will be possible to expand and refine the types of experiments accessible to us on the analytical ultracentrifuge.

CONCLUSION

We have shown that sedimentation equilibrium in the analytical ultracentrifuge is a precise and exhaustive method for studying the molecular weight of a class of high-charged positive polyelectrolytes, the ionenes. The behavior of 3,3-ionene chloride in 0.5 $\underline{M}$ KBr can be adequately described by current sedimentation equilibrium theory which takes account of both nonideality and polydispersity. However, more work must be done in the area of counterion effects to elucidate the functional effects of the high charge content of the ionenes. To facilitate the expansion of future sedimentation experiments into this area, and to extend the general usefulness of the analytical ultracentrifuge as a physical chemical tool, computer-controlled scanning methods have been developed to improve plate-reading techniques.

We wish to thank Richard Doenier for extensive discussions regarding the theory and technique of sedimentation equilibrium, Laura Schwartz for her unceasing help in the development of the computer software required by this work, and Alan Adler for the development of the scanning computer programs.

This paper represents one phase of research performed by the Jet Propulsion Laboratory, California Institute of Technology, sponsored by the National Aeronautics and Space Administration, Contract NAS7-100.

BIBLIOGRAPHY

1. J. W. Williams, K. E. Van Holde, R. L. Baldwin, and H. Fujita, Chem. Rev., <u>58</u>, 715 (1958).

2. H. Fujita, <u>Mathematical Theory of Sedimentation Analysis</u>, Academic Press, Inc., New York, N.Y., 1962.

3. K. E. Van Holde and R. L. Baldwin, J. Phys. Chem., 62, 734 (1958).

4. D. A. Yphantis, Ann. N.Y. Acad. Sci., 88, 586 (1960).

5. P. E. Hexner, L. E. Radford and J. W. Beams, Proc. Nat. Acad. Sci. U.S.A., 47, 1848 (1961).

6. F. E. LeBar, Proc. Nat. Acad. Sci. U.S.A., 54, 31 (1965).

7. Q. M. Griffith, Anal. Biochem., 19, 243 (1967).

8. H. Utiyama, N. Tagata, and M. Kurata, J. Phys. Chem., 73, 1448 (1969).

9. P. Munk and D. J. Cox, Biochemistry, 11, 687 (1972.

10. H. Noguchi and A. Rembaum, J. Polymer Sci., Part B, 7, 383 (1969).

11. H. K. Schachman, Ultracentrifugation in Biochemistry, Academic Press, Inc., New York, N. Y., 1959.

12. R. C. Doenier and J. Williams, Proc. Nat. Acad. Sci. U.S.A., 64, 828 (1969).

13. G. V. Schulz, Z. Physik. Chem. (Leipzig), A193, 168 (1944).

14. D. A. Albright and J. W. Williams, J. Phys. Chem., 71, 2780 (1967).

15. D. Casson and A. Rembaum, Macromolecules, 5, 75 (1972).

16. D. A. Yphantis, Biochemistry, 3, 297 (1964).

17. D. A. Yphantis and D. E. Roark, Biochemistry, 11, 2925 (1972).

18. P. J. Flory, J. Am. Chem. Soc., 58, 1877 (1936).

VISCOELASTICITY OF HIGHLY CONCENTRATED SOLUTIONS OF POLY(ACRYLIC ACID) SALTS

A. Eisenberg, M. King, and T. Yokoyama

McGill University

Montreal, Canada

The incorporation of ions into a polymeric material can produce structural changes of considerable magnitude, the effect of which can generally be observed in the form of greatly modified viscoelastic behavior. In recent years, the pace of research into the structure and viscoelasticity of ion-containing polymers has increased sharply. Synthetic ion-containing polymers can be classified into three main groups. The first of these includes the inorganic ion-containing polymers; extensive studies have been made on the phosphate (1) and the silicate (2-4) systems. The second group, which has been the subject of a recent review (5), comprises the organic copolymers in which the ionizable fraction is small; several systems of this type have been studied recently, including the ethylene ionomers (6-9), the styrene ionomers (10,11), and the carboxylic rubbers (12).

The third major classification of ion-containing polymers comprises those materials which are generally termed <u>polyelectrolytes</u> - water-soluble organic homopolymers or copolymers containing a high percentage of ionizable material. The solution properties of polyelectrolytes have received considerable attention (13,14). However, few of the studies made on polyelectrolytes have been in the area of their viscoelastic properties. Viscoelastic studies on undiluted polyelectrolytes have shown that ionization produces dramatic changes in their thermomechanical behavior (10,15). As might have been expected, the magnitudes of these changes are much greater than in the ionic polymers mentioned previously. However, the difficulties involved in the preparation of unplasticized polyelectrolytes have hindered a more complete examination of the effect of the solid-state polyelectrolyte

structure on viscoelasticity. Investigations of the viscoelastic
behavior of polyelectrolytes in moderately concentrated solutions
(up to 5% by weight) have proved interesting (16,18), but except
for effects at very low ion concentrations (19), the vast changes
in physical properties that are apparent in the solid state are not
seen in moderately concentrated solutions. In the region of polymer
concentration between the point of gelation and the undiluted polymer, however, the dramatic effects of ionization on the viscoelastic
properties of polyelectrolytes can still be observed, and because
many of the problems which have inhibited the study of unplasticized
polyelectrolytes can be avoided, this area was chosen for the present study.

The polyelectrolytes used in this study were plasticized sodium salts of poly(acrylic acid), the plasticizers being water,
formamide, glycerine, and ethylene glycol. Stress-relaxation measurements taken over 3.5 decades of time demonstrate that time-
temperature superposition of viscoelastic data is not valid in this
polymer system. X-ray diffraction measurements, which indicate
that microphase separation occurs in these polymers, suggest that
stress-relaxation data can be analyzed in terms of two concurrent
relaxation mechanisms. The results of this analysis provide an
insight into the viscoelastic processes in solid-state polyelectrolytes.

EXPERIMENTAL

The poly(acrylic acid) - PAA - used in this study was prepared
by the polymerization of acrylic acid in toluene with benzoyl peroxide initiator. Polymerizations were carried out at ca. 50°C with
conversions up to 80%. The polymer was dried under vacuum at 80°C
to constant weight. Several initiator-monomer ratios in the range
0.01% to 0.1% (by weight) were employed to produce a range of molecular weights, which were calculated from intrinsic viscosities
measured in dioxane solution at 30°C, using published correlations
(20). In subsequent thermomechanical studies, PAA samples with molecular weights ranging from 2.9×10^5 to 9.9×10^5 were used; no
significant differences in either T_g or in the viscoelastic properties in the region investigated were observed to result from the
variation in molecular weight.

Films of plasticized poly(sodium acrylate) - PNaA - suitable
for stress-relaxation measurements were prepared as follows: aqueous solutions of PAA were neutralized to degrees varying from 0%
to 100% by aqueous NaOH. The solutions were concentrated by evaporation and cast into flat-bottomed containers. Excess plasticizer
was added to the aqueous solutions (except where the plasticizer
was water itself) and evaporation was continued until the desired

level of plasticizer content was reached. In the case of form-
amide-plasticized polymer, the addition of excess plasticizer was
repeated, because of the high volatility of formamide, in order to
ensure the elimination of the water. It is possible that a small
percentage of water remained in these samples, but it was assumed
to be inconsequential. For comparative purposes, the plasticizer
contents were expressed as volume percentages, assuming a negligible volume of mixing. Density measurements on similar ion-containing polymers (21) indicate that the error involved in this conversion is small.

Stress-relaxation measurements were made under nitrogen atmosphere using apparatus similar to that described previously (22,
23). For each sample, the stress at constant strain was measured
for periods of time up to 2 x 10^4 sec (although for the majority
of samples, only up to 10^3 sec) for a series of temperatures in the
vicinity of the glass transition. The temperature variation during
any one run was held below $\pm 0.1°$. High modulus measurements were
made in the bending mode, while those at lower moduli (below ca. 5
x 10^8 dyn/cm^2) were done in stretching. Since volatile plasticizers
were used, the variation in sample weight was monitored, and measurements were discontinued after a weight loss of 1% was observed.
It is reasonable to assume that the weight loss of plasticizer is
essentially diffusion controlled, since measurements could be taken
at relatively high temperatures for the low-plasticizer-content
materials before significant plasticizer loss occurred. Occasional
measurements were taken in random order of temperatures to ensure
that the loss of plasticizer to the extent of 1% did not produce a
loss of reproducibility.

A brief dynamic mechanical study of PNaA was also undertaken
using a torsion pendulum, described elsewhere (23). Measurements
of the loss tangent at ca. 1 Hz were made on three samples of PNaA,
containing formamide (48%), glycerine (55%), and water (50%), respectively. [Unless otherwise stated, all plasticizer contents are
expressed in volume percentages.]

A number of cesium salts of PAA were prepared in a manner
similar to that described above. X-ray diffraction measurements
were made on these salts, as well as on some of the sodium salts
which had been used for mechanical measurements. The x-ray scattering patterns were obtained at room temperature using nickel-fil-
tered copper radiation generated by a Rich Seifert x-ray source
operating at 40 kV and 20 mA. The patterns were recorded photo-
graphically in a conventional wide-angle camera with a 0.05-cm pin
hole and 6.6-cm sample-to-film distance. The photographs were
microdensitometered using a Joyce, Loebl double-beam recording
microdensitometer. Exposure times varied with the type of material,
but were generally 4-6 hr.

RESULTS

Stress Relaxation

In Figure 1, an example of the type of stress-relaxation data observed in this polymer system is presented. The data shown are for the material PNaA-glycerine (66%). In addition to the individual curves of modulus versus time at various temperatures, Figure 1 also shows an attempt to construct a master curve in which the overlap between successive curves is maximized in their short-time regions. It is apparent that the result of this procedure is not a true master curve, but what may be termed a "pseudo-master curve," which describes only the short-time relaxation behavior at any given temperature. It is worth stressing that the same pattern is obtained whether measurements are made in order of increasing temperature or not. Also, the lack of superposition cannot be attributed to the loss of plasticizer, because this would result in deviations of the opposite direction and would certainly not lead to reproducible results.

The failure of time-temperature superposition was found to occur with every material studied, although the magnitude of the deviations of the long-time moduli varied with the amount and type of

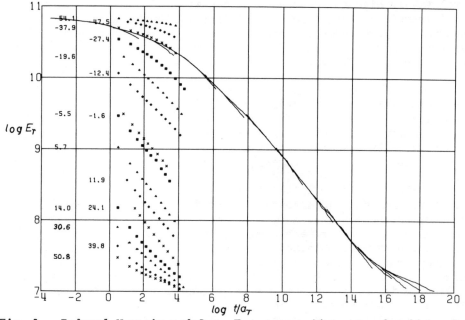

Fig. 1.   Reduced Young's modulus, E_r, versus time as a function of temperature ($^\circ$C) for PNaA-glycerine (66%). Solid lines represent pseudo-master curve.

plasticizer and with the degree of neutralization of the PAA.   In
every case, the deviations were of the same nature as in the example shown in Figure 1; that is, the long-time deviations were toward lower values of moduli.   The relative deviations were most pronounced for water-plasticized PNaA, which also showed the most rapid
rate of relaxation, and the least pronounced for formamide-plasti-
cized PNaA, which had a much slower rate of relaxation.   The magnitude of the deviations reached a maximum at intermediate levels of
formamide content for fully neutralized PAA and at intermediate
degrees of neutralization for constant plasticizer content.   The
treatment of the stress-relaxation data from this polymer system
will be dealt with more thoroughly in subsequent sections of this
paper.

Modulus - Temperature Curves

Because the pseudo-master curves determine short-time behavior
only, it is not surprising that the variations in their shapes are
closely paralleled by the variations in the shapes of curves of
10-sec modulus versus temperature.   Because of their greater ease
of presentation and simplicity in interpretation, modulus - temperature curves are shown here to illustrate the variation in mechanical
behavior of these polymers with the percentage and type of plasticizer and with the degree of neutralization.   In Figure 2, the
10-sec Young's modulus is given as a function of absolute temperature for PAA samples with degrees of neutralization between 0% and
100% and with a common plasticizer content (ca. 27% formamide).
The stars on each curve indicate the position of T_g, as measured
by thermal expansion (24).   It is worthwhile to note the rise in
glassy modulus - by about a factor of 3 - and the great increase
in the breadth of the transition as the ion content increases.   It
is clear that the high glassy moduli must be due to additional intermolecular bonding which results from the presence of ions.

Figure 3 illustrates the variation in the modulus of the fully
neutralized polymer as a function of the percentage of formamide.
One can see that the breadth of the transition and the glassy modulus decrease as the plasticizer content increases, as one might expect.   In Figures 2 and 3, an inflection point below T_g appears in
the modulus - temperature curves for low percentages of formamide
and high degrees of neutralization.   This inflection point seems
to correlate with the occurrence of a broad shoulder in the loss
tangent below T_g, as indicated by the dynamic mechanical measurements to be discussed in more detail below.

The shapes of some of the modulus - temperature curves presented in Figures 2 and 3 are reminiscent of those obtained for
partially crystalline polymers such as low-density polyethylene
(25).   However, x-ray diffraction patterns obtained from a variety
of PAA salts showed no evidence of any crystallinity.

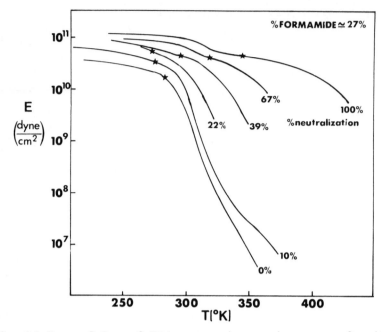

Fig. 2. 10-Sec modulus of PAA versus temperature as a function of
the degree of neutralization. The stars on each curve in
Figures 2-6 indicate the position of T_g.

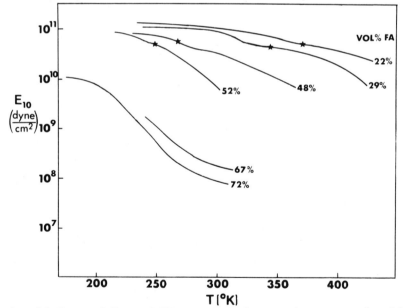

Fig. 3. 10-Sec modulus of PNaA versus temperature as a function
of the formamide content.

Figures 4 and 5 show the modulus - temperature behavior of PNaA plasticized with glycerine and water, respectively. In these two cases, the range of plasticizer content is considerably less than for formamide, so the great variation in the shapes of the curves is not as apparent. However, it is readily seen that the rate of relaxation, which parallels the decrease in modulus with temperature, is considerably greater in the latter two cases.

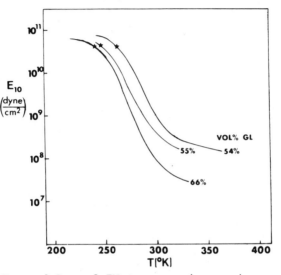

Fig. 4.   10-Sec modulus of PNaA versus temperature as a function of the glycerine content.

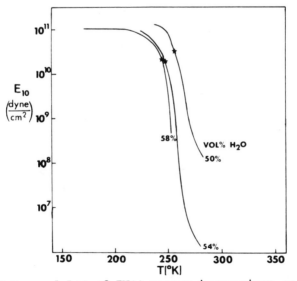

Fig. 5.   10-Sec modulus of PNAA versus temperature as a function of the water content.

A direct comparison of the effect of plasticizer on the modulus - temperature behavior of PNaA is shown in Figure 6. The curves drawn in this figure were all obtained for approximately the same volume fraction of plasticizer. The glassy moduli and the glass-transition temperatures of these polymers are all very similar. However, above T_g, the great variation in the efficiency of these plasticizers is apparent. This variation is a good indication that the plasticizer plays a very important role in determining the lifetimes of the ionic bonds.

Dynamic Mechanical Measurements

Measurements of the loss tangent, tan δ, at ca. 1 Hz as a function of temperature were made on three samples of PNaA, plasticized with water (50%), formamide (48%), and glycerine (55%), respectively. The data are shown in Figure 7, normalized with respect to the primary transition, for the sake of clarity. The dotted lines in Figure 7 indicate the regions which were inaccessible to measurement of tan δ. Although the relative intensities are different, each curve shows a primary maximum and a broad shoulder just below it in temperature. It should be recalled that inflection points below T_g were observed in the modulus - temperature curves of PNaA at low formamide contents and high degrees of ionization. This shoulder in the loss tangent thus seems to correlate with the inflection points in the modulus - temperature curves of the PNaA-formamide series. The fact that no inflection point is seen in the PNaA-water or PNaA-glycerine series is likely due to the fact that, since the relative intensities of the primary maximum and the shoulder are quite different in these two cases, the effect of the low-temperature shoulder on the modulus would be relatively small.

X-Ray Diffraction

X-ray diffraction patterns were obtained for a number of plasticized salts of PAA. Since the x-ray study was not intended to be comprehensive in itself but was undertaken in support of the thermomechanical study, diffraction patterns were obtained only for selected samples chosen mainly to correspond to materials on which mechanical measurements had been made. All of the x-ray patterns recorded showed either one or two diffuse halos in the range $4° < 2θ < 40°$. In each case, the amorphous spacing (26), $d_g = 1.22\ d_{Bragg}$, was calculated. The results are tabulated below.

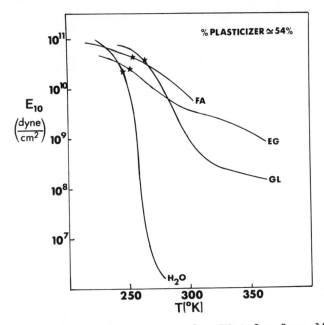

Fig. 6. Modulus-temperature curves for PNaA for four different plasticizers.

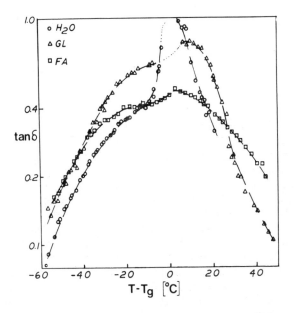

Fig. 7. Tan δ versus $(T-T_g)$ for PNaA for three different plasticizers.

TABLE I. X-Ray Studies of Plasticized PAA Salts

Plasticizer Content	% Ion Content	Counterion	d_g (Å) Inner Halo	Outer Halo
Formamide (23%)	0	-	12.1	6.7
Water (~5%)	0	-	12.1	6.7
Water (~5%)	1	Cs	12.3	6.7
Water (~5%)	5	Cs	13.4	6.2
Water (~10%)	25	Cs	14.3	-
Water (~20%)	100	Cs	15.0	-
Formamide (38%)	100	Cs	15.4	-
Formamide (29%)	100	Na	15.0	5.0
Formamide (48%)	100	Na	14.4	4.8
Water (low)	100	Na	15.[a]	

- - - - - - -

[a]Calculated from Ref. 27.

DISCUSSION

In this section, correlations between the various types of experimental findings will be presented. First of all, it will be shown that the x-ray diffraction measurements on the acrylate salts support a structure in which ion aggregation occurs. Then the rheological data will be analyzed according to the two-mechanism response expected of some microphase-separated materials. It will be shown by direct and indirect evidence that the mechanism which dominates the short-time behavior above T_g is adequately described by the W.L.F. equation and thus can likely be attributed to a normal diffusional process. The mechanism which corresponds to the long-time deviations in the pseudo-master curves is of low activation energy (17-32 kcal) and is essentially a pure viscosity. The possible relationship of this mechanism to the ionic phase is discussed. Finally, a correlation between dynamic and static mechanical data suggests the existence of a third relaxation mechanism in this polymer system.

The x-ray diffraction patterns obtained in unionized PAA are similar to those observed in other amorphous polymers, such as polystyrene and some of its derivatives (26). With increasing cesium content, the longer spacing increases, while the shorter spacing disappears. For the fully neutralized polymers, similar values of the longer spacing are seen in both the cesium and sodium salts, and for both water and formamide as plasticizers; however, in the sodium salts, an outer halo is still observable. The fact that an amorphous scattering pattern is also obtained for the unionized material indicates that some sort of order exists even in this case. This is not surprising, since the extent of hydrogen bonding in PAA should be considerable, and associations between the carboxyl groups could lead to a partially ordered structure. This amorphous order would manifest itself in an x-ray scattering pattern because of the higher electron density of oxygen.

It is assumed that the cesium and sodium salts of PAA have essentially the same structure. This was not tested specifically in this system, but from rheological considerations, it was found to be true for styrene - methacrylic acid ionomers (28). Non-crystalline x-ray maxima are due to an average separation between recurrent centers of relatively high electron density. Because the electron density of cesium is much higher than that of the other atoms involved, the single halo observed in PCsA samples can likely be attributed to an average spacing between regions of high cesium content. The fact that a second halo is observed in PNaA is reasonable since sodium, with an electron density not much greater than carbon, would not dominate the diffraction pattern as would cesium. This evidence is, therefore, consistent with a microphase-separated structure (regions of high and low ionic content) for this polymer system. This conclusion is supported by a number of other factors, including predictions from thermodynamic arguments (29), evidence of phase separation in other organic ion-containing polymers (11, 27,28), and rheological evidence in the present system, to be discussed below.

The longer spacings in the fully neutralized PAA samples provide a measure of the size of the ionic aggregates. Based on simple stoichiometry and assuming the complete incorporation of carboxyl groups, the number of ion pairs per aggregate is estimated to be about 25. The fact that the longer spacings do not vary much with the type of plasticizer indicates that the structure in glassy PNaA is not very dependent on the type of plasticizer. However, x-ray diffraction is not sensitive to dynamic phenomena. The time dependence of the structure, on which the rheological properties depend, must be investigated by other means.

The type of stress-relaxation data presented in Figure 1 is characteristic of the viscoelastic response of a thermorheologically complex material; it suggests the existence of two concurrent re-

laxation mechanisms, each with a different temperature dependence.
A viscoelastic response of this nature would be in line with a
two-phase structure in which each phase could be expected to yield
with the applied stress. In such a system the contributions of the
two mechanisms to the total compliance should be additive (30). A
general method for the analysis of this type of data from ion-containing polymers is presented in another publication (31); a brief
summary of the method is given here.

The moduli from the original stress-relaxation curves at each
temperature, as well as the moduli from the upper envelope of
modulus versus reduced time, are converted to compliances (32).
This procedure results in a pseudo-master curve of compliance versus reduced time which is similar to but not the same as the mirror
image of Figure 1. It is assumed that, at any given temperature,
the short-time compliance is mainly due to the primary relaxation
mechanism; the upward deviations from the lower envelope of the
compliance - time curves are, therefore, due to the contribution of
the second mechanism at that temperature. The separation of the
contributions of the two mechanisms is accomplished by point-by-point
subtraction of the compliance values from the upper envelope and
values corresponding to the deviations at long times.

The results of this subtractive procedure for PNaA-glycerine
(66%) are shown in Figure 8. In addition to the curves of secondary compliance versus time, Figure 8 shows an attempt to prepare a
second master curve by shifting the individual compliance curves
horizontally relative to the curve at the highest temperature. In
view of the errors involved in the subtractive procedure, the degree
of overlap obtained seems quite reasonable. The significance of
the dashed line in Figure 8, representing a theoretical slope of
unity, will be discussed later.

The natures of the two relaxation processes and their relationships with the polymer structure are not perfectly clear; however,
an analysis of their temperature dependences will be useful in relating them to physical processes. The mechanism whose contribution
dominates the short-time behavior above T_g can be investigated by
an analysis of the primary shift factors, a_T, which contribute to
the primary master curves.

It is of interest to determine whether the shift factors corresponding to the primary mechanism follow an Arrhenius-type temperature dependence, or whether they obey the W.L.F. equation (33)

$$\log a_T = - c_1{}^g (T - T_g)/(c_2{}^g + T - T_g) \qquad (1)$$

where the W.L.F. parameters $c_1{}^g$ and $c_2{}^g$ are referred to T_g.

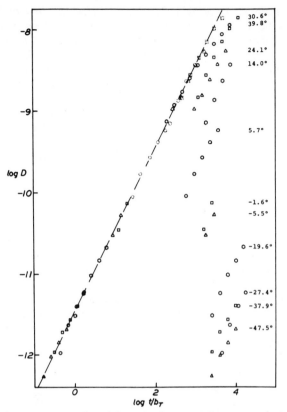

Fig. 8. Master curve of subtracted compliance, D, for
         PNaA-glycerine (66%). Individual curves and corresponding
         temperatures (°C) are shown. The dashed line represents
         a theoretical slope of unity.

The W.L.F.-type temperature dependence of the short-time shift
factors is clearly observed in the system PNaA-glycerine, where a
wide range of temperature above T_g is accessible for stress-relaxation measurements (34). As examples, W.L.F. parameters established from measurements in the temperature range $(T_g + 30°) < T$
$< (T_g + 90°)$ are summarized as follows: for PNaA-glycerine (55%),
$c_1{}^g = 35$, $c_2{}^g = 127$, referred to $T_g = -28°C$; and for PNaA-glycerine
(66%), $c_1{}^g = 38$, $c_2{}^g = 125$, referred to $T_g = -34°C$.

With volatile plasticizers, notably formamide and water, the
accessible temperature range above T_g is small because of plasticizer loss. Thus, the establishment of W.L.F. parameters from
stress-relaxation measurements is impossible in these plasticized
systems. However, there is other evidence, to be presented below,
to support the applicability of the W.L.F. equation to the primary
relaxation process even in these cases.

In all of the materials studied, the short-time shift factors, a_T, could be drawn to fit an Arrhenius-type temperature dependence in the immediate region of T_g. Examples of this can be seen in Figure 9, in which log a_T versus 1/T is plotted for the series PNaA-formamide. The fact that the shift factors fit well with an Arrhenius-type temperature dependence in the vicinity of T_g cannot be considered significant in establishing the true temperature dependence, however, since the two forms of temperature dependence are generally indistinguishable in this range.

The W.L.F. equation can be used to predict the apparent activation energy at T_g, $(\Delta H_a)_{T_g}$; the relationship, derived from Eq. 1, is given by (33)

$$(\Delta H_a)_{T_g} = 2.303 \ R \ (c_1{}^g/c_2{}^g) \ T_g{}^2 \tag{2}$$

Although the constants $c_1{}^g$ and $c_2{}^g$ could not be established for the formamide series, the proportionality between $(\Delta H_a)_{T_g}$ and $T_g{}^2$ can be demonstrated. From linear relationships of the type obtained in Figure 9, values of $(\Delta H_a)_T$ were computed. For the PNaA-formamide series, these values were found to increase monotonically with decreasing plasticizer content and, as can be seen from Figure 10, a linear relationship exists between $(\Delta H_a)_T$ and $T_g{}^2$.

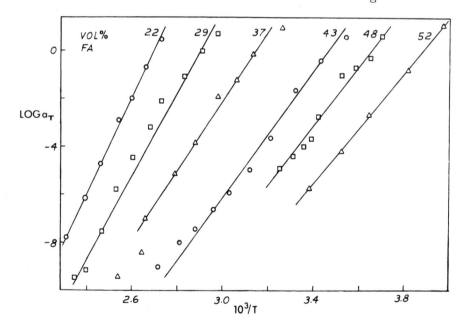

Fig. 9. Log a_T versus 1/T for PNaA-formamide series.

This linear relationship can be predicted from Equation 2 if it is assumed that the ratio c_1^g/c_2^g is constant in the range of composition studied. The W.L.F. parameters are given formally by (33)

$$c_1^g = B/2.303f_g$$

$$c_2^g = f_g/\alpha_f \tag{3}$$

where f_g is the fractional free volume at T_g, α_f is the free volume expansion coefficient, and $B \sim 1$. Within a single plasticizer-polymer system, f_g is likely to be relatively constant, and experimentally, the variation in α_f is small. Thus, it is reasonable to assume that c_1^g/c_2^g is either constant or, at worst, a slowly varying function of the plasticizer content. Therefore, it can be concluded that the W.L.F. equation applies to the formamide series as well.

In view of the evidence to support the applicability of the W.L.F. equation to the primary relaxation process in PNaA, it is reasonable to attribute its mechanism to the same type of diffusional process that characterizes the glass transition in most nonionic polymers. This assignment is also reasonable in view of the magnitude of the activation energy associated with the primary mechanism. By extrapolating the linear relationship obtained in Figure 10 to the T_g of undiluted PNaA [$T_g = 250°C$, as reported previously (35)], $(\Delta H_a)_{T_g} = 175$ kcal is obtained. This value of $(\Delta H_a)_{T_g}$ is in general agreement with values which can be obtained for a wide variety of thermorheologically simple polymers by the application of Equation 2.

The activation energy associated with the secondary process can be calculated from the shift factors used in obtaining master curves of the type shown in Figure 8. The activation energies obtained varied from 32 kcal for PNaA-H_2O (50%) to 17 kcal for both PNaA-glycerine (66%) and PNaA-formamide (48%). The variations in these values with the percent plasticizer were not tested because of the insufficient number of long-time runs and the tediousness of the method. It is difficult to specifically define the relaxation process involved in this polymer, but it is quite possible that it could correspond to the decomposition of an ion aggregate by the removal and subsequent migration of one ion pair. The amount of energy involved in this process would depend on the size and geometry of the aggregate and on a number of other factors, including the solvating effect of the plasticizer. These values of activation energy, which lie between 17 and 32 kcal, seem reasonable for such a mechanism.

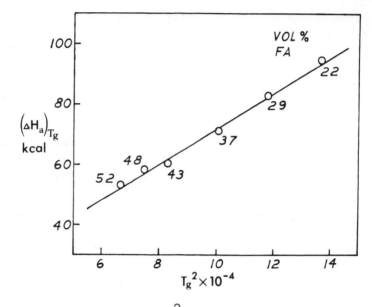

Fig. 10. $(\Delta H_a)_{T_g}$ versus T_g^2 for PNaA-formamide series.

Whatever its form, if the secondary relaxation process were due to the yielding of an ionic phase, its contribution to the compliance would correspond to a pure viscosity in the same fashion as bond interchange (36). In this case, the relationship

$$D(t) \sim t/\eta \qquad\qquad (4)$$

should hold. It can be seen in Figure 8, by comparison with the doubly logarithmic slope of unity, that this is a reasonable approximation. Similar slopes were found in the other plasticized systems studied.

The correlation of the sub-T_g shoulder observed in the dynamic mechanical study with one of the relaxation processes observed in the stress relaxation experiments should be possible through an analysis of the activation energies of the two dynamic mechanical peaks. This would involve dynamic mechanical measurements in a frequency range in which the peaks would be separated. Assuming that the sub-T_g transition is of lower activation energy than the primary transition, as might be expected, the separation of the peaks would require lower frequencies than those available on a torsion pendulum.

One method of investigating dynamic mechanical behavior at low frequencies is by the conversion of stress relaxation data to a dynamic mechanical function. This was done for some members of the PNaA-formamide series. The static creep compliance, $D(t)$, which had been obtained from the stress relaxation data in the manner described above, was converted to dynamic compliance by the method of Schwarzl and Struik (37), assuming a semilogarithmic approximation of third order for $D(t)$. Although the conversion to dynamic compliance is highly sensitive to experimental errors and is subject to large uncertainties because of truncation errors, it was nevertheless possible to qualitatively reproduce the pattern in tan δ that can be obtained from direct dynamic mechanical measurements using a torsion pendulum (34).

Preliminary calculations place the value of the activation energy for the sub-T_g process in the PNaA-formamide series in the order of 50 kcal. Since this value is close to values observed for the primary process in the same series, this mechanism could be expected to be competitive in stress-relaxation behavior. Thus, it is possible that the pseudo-Arrhenius region observed in the vicinity of T_g is the result of two competing relaxation mechanisms. It should be stressed that this sub-T_g process is not the same as that seen at long times above T_g , the latter being of much lower activation energy. A more extensive study correlating the static and dynamic mechanical behavior in salts of PAA is currently in progress.

The authors would like to acknowledge the assistance of Karl Taylor in the preparation and characterization of some of the poly(acrylic acid), and R. St. J. Manley of the P.P.R.I.C. for the use of his x-ray equipment. This work was supported in part by the Petroleum Research Fund.

BIBLIOGRAPHY

1. A. Eisenberg, Advan. Polymer Sci., 5, 59 (1967).

2. V. O. Frechette, Ed., Non-crystalline Solids, Wiley, New York (1960).

3. J. A. Prins, Ed., Physics of Non-crystalline Solids, North-Holland, Amsterdam (1965).

4. A. Eisenberg and K. Takahashi, J. Non-Cryst. Solids, 3, 279 (1970).

5. E. P. Otocka, J. Macromol. Sci., C5, 275 (1971).

6. T. C. Ward and A. V. Tobolsky, J. Appl. Polymer Sci., 11, 2403 (1967).

7. W. J. MacKnight, L. W. McKenna, and B. E. Read, J. Appl. Phys., 38, 4208 (1967).

8. E. P. Otocka and T. K. Kwei, Macromolecules, 1, 401 (1968).

9. K. Sakamoto, W. J. MacKnight, and R. S. Porter, J. Polymer Sci., Part A-2, 8, 277 (1970).

10. W. E. Fitzgerald and L. E. Nielsen, Proc. Roy. Soc. (London), A282, 137 (1964).

11. A. Eisenberg and M. Navratil, J. Polymer Sci., Part B, 10, 537 (1972).

12. A. V. Tobolsky, P. F. Lyons, and N. Hata, Macromolecules, 1, 515 (1968).

13. S. A. Rice and M. Nagasawa, Polyelectrolyte Solutions, Academic Press, New York, 1961.

14. F. Oosawa, Polyelectrolytes, Dekker, New York, 1971.

15. L. E. Nielsen, Polym. Eng. Sci., 9, 356 (1969).

16. A. Silberberg and P. F. Mijnlieff, J. Polymer Sci., Part A-2, 8, 1089 (1970).

17. A. Konno and M. Kaneko, Makromol. Chem., 138, 189 (1970).

18. M. Sakai, I. Noda, and M. Nagasawa, J. Polymer Sci., Part A-2, 10, 1047 (1972).

19. M. King and A. Eisenberg, J. Phys. Chem., 57, 482 (1972).

20. S. Newman, W. R. Krigbaum, C. Laugier, and P. J. Flory, J. Polymer Sci., 14, 451 (1954).

21. A. Eisenberg and M. King, Macromolecules, 4, 204 (1971).

22. L. Teter, Ph.D. Thesis, U.C.L.A., 1966.

23. B. Cayrol, Ph.D. Thesis, McGill Univ., 1972.

24. A. Eisenberg and T. Sasada, in J. A. Prins, Ed., Physics of Non-Crystalline Solids, North-Holland, Amsterdam, 1965, p. 99.

25. M. Takayanagi, Mem. Fac. Eng. Kyushu Univ., 23, 41 (1963).

26. R. F. Boyer and H. Keskkula, in N. M. Bikales, Ed., Encyclopedia of Polymer Science and Technology, Interscience Publishers, New York, Vol. 13, 1970, pp. 224-229.

27. F. C. Wilson, R. Longworth, and D. J. Vaughn, A.C.S. Polymer Preprints, 9, 505 (1968).

28. M. Navratil, Ph.D. Thesis, McGill Univ., 1972.

29. A. Eisenberg, Macromolecules, 3, 147 (1970).

30. J. D. Ferry, W. C. Child, Jr., R. Zend, D. M. Stern, M. L. Williams, and R. F. Landel, J. Colloid Sci., 12, 53 (1957).

31. A. Eisenberg, M. King, and M. Navratil, to be published.

32. I. L. Hopkins and R. W. Hamming, J. Appl. Phys., 28, 906 (1957).

33. J. D. Ferry, Viscoelastic Properties of Polymers, 2nd ed., Wiley, New York (1970), Chap. XI.

34. M. King, Ph.D. Thesis, McGill Univ., 1972.

35. A. Eisenberg, H. Matsuura, and T. Yokoyama, J. Polymer Sci., Part A-2, 9, 2131 (1971).

36. A. V. Tobolsky, Properties and Structure of Polymers, Wiley, New York (1960).

37. F. R. Schwarzl and L. C. E. Struik, Advan. Mol. Relaxation Processes, 1, 201 (1968).

SOME GEL PERMEATION CHROMATOGRAPHIC STUDIES ON CYCLOCOPOLYMERS OF

MALEIC ANHYDRIDE

George B. Butler and Chester Wu

University of Florida

Gainesville, Florida

The regularly alternating 1:2 cyclocopolymer of divinyl ether and maleic anhydride was first synthesized in 1951. Evidence for the proposed structure and for a bimolecular alternating inter-intramolecular chain propagation mechanism, now commonly referred to as the cyclocopolymerization mechanism, for its formation was published in 1958 (1-5). The following scheme (Scheme I) illustrates the mechanism by which the cyclocopolymer is produced:

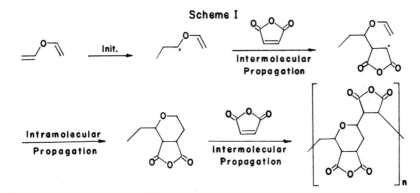

Scheme I

An interesting and significant property which this copolymer has been found to possess is its physiological action. It was observed as the result of tests conducted at the Cancer Chemotherapy National Service Center, National Institutes of Health, that the copolymer of $\overline{M}_n$=15-20,000 exhibited interesting antitumor properties (6). For example, weights of tumors developed in test animals were as low as 11% of those developed by control animals. In more recent results (7-9), clinical studies have shown that the copolymer of

certain molecular-weight ranges possesses the ability to induce
interferon formation. Interferon, first discovered in 1957 (10),
is a substance of protein-like structure that is produced in the
cells of vertebrates in response to viral infection, and possesses
antiviral action. Today, it is generally accepted that interferons
play an essential role in the formation of a host's nonspecific
resistance to superinfection with a second virus (11). Thus, the
role of interferon in combating viral infections may be similar to
that of antibodies toward bacterial infections. It is immediately
apparent that a successful and readily available interferon inducer,
such as the divinyl ether - maleic anhydride cyclocopolymer (DVE-MA)
could perhaps play an extremely significant role not only in aiding
in recovery from a viral infection, but in its use in prior generation of interferon to prevent a viral attack. (See also the chapter by W. Regelson).

It was the purpose of this investigation to develop a gel
permeation chromatography (GPC) method which would eventually permit
fractionation of DVE-MA into fractions of narrow molecular-weight
distribution for further evaluation as an interferon inducer, particularly since the present evidence indicates that molecular weight
is an important factor in controlling this property.

A previous attempt to fractionate this copolymer by use of a
stepwise continuous solvent-gradient column method was unsuccessful
(12). However, Allen and Turner successfully applied a fractional
precipitation method to a copolymer having $[\eta]=0.282$ dl/g, using
hexane as precipitant from acetone solution, to obtain nine fractions varying from $[\eta]=0.493$ to 0.019 dl/g. However, it was necessary to add sodium tetraphenyl boron to the system in order to at-
tain adequate separation. Subsequent application of this technique
in preparation of samples for laboratory evaluation as an interferon inducer led to serious toxicity in the samples, apparently
due to residual boron impurities.

RESULTS AND DISCUSSION

GPC Fractionation of DVE-MA Copolymer as a Polyelectrolyte

In 0.1 M Na_2SO_4 Aqueous Solution. Two copolymer samples having
inherent viscosities of 1.56 and 0.51 dl/g, respectively (0.2 g/100
ml DMSO, 30.0°C), were converted to their sodium carboxylate forms
and injected into Deactivated Porasil columns (3 ft each of the
type AX, BX, CX, DX, EX arranged in series) in 0.1 M Na_2SO_4 aqueous
solution. Fractionation was realized as indicated by the observation
that the former sample was eluted within 32-37 counts with a peak
maximum at 34.5 counts while the latter came out within 37 and 42
counts having a peak maximum at 39.6 counts. This work was not

further pursued because GPC analysis of polyelectrolytes presents
a rather complicated problem of neutralizing the charge on the polymer chain and dealing with the effect of charge on molecular size
under these conditions. Besides, there are no narrowly distributed
water-soluble polymer standards available to provide a calibration
curve. Most importantly, it was found that the copolymers were adsorbed by the packing gel as indicated by the gradual increase of
pressure in the columns as more samples were injected. Unfortunately,
the Deactivated Porasil columns were thoroughly flushed with acetone
several times in the process of changing carrier from water to organic solvents and vice versa. Therefore, the special treatment on
the Porasil surface was apparently washed out. Successful GPC fractionation of these copolymers as polyelectrolytes will require more
study. (See also the chapter by G. L. Beyer, A. L. Spatorico, and
J. L. Bronson).

In Dimethylformamide as carrier. Allen and Turner (12) characterized DVE-MA copolymers and observed that intrinsic viscosities
and osmotic pressures measured in DMF showed unusual slopes. No
suggestions to explain this behavior were made. These authors found
$[\eta]$ to increase very sharply at higher dilution, suggestive of polyelectrolyte behavior. We reasoned that the copolymers may behave
as a polyelectrolyte in DMF as the result of reaction of the anhydride
with the solvent. This was confirmed by IR spectroscopic analysis.

Fujimori (13) of this laboratory attempted to fractionate
DVE-MA copolymers in DMF using Styragel columns. The copolymers
showed GPC peaks at the void volume count number regardless of varying known number-average molecular weights. The copolymers were
excluded from the gel pores in the column.

IR spectra (see Figures 1 and 2) of maleic anhydride and DVE-MA
copolymer in DMF both showed the presence of strong carboxylate
anion at 1400 cm^{-1} and around 1600 cm^{-1}. Hence it is concluded that
DMF reacts with the anhydride rings in both maleic anhydride and the
copolymer backbone according to Scheme II. Both Structures A and
B could exist in DMF. When the copolymer concentration in DMF is
decreased the charges on the polymer may be separated from each
other to some extent, resulting in the mutual repulsion of the same
charge in the main chain to expand rapidly (14). Due to this chain
expansion at high dilution in DMF, $[\eta]$ of the copolymer increases
immensely. In GPC fractionation, the sample size is usually small,
and the greatly extended chains of the copolymers in DMF would not
permit them to penetrate through the various pore sizes of the gel
which happen to be smaller than the copolymer molecules. As a
result, the copolymers were eluted at the void volume count number.

Scheme II

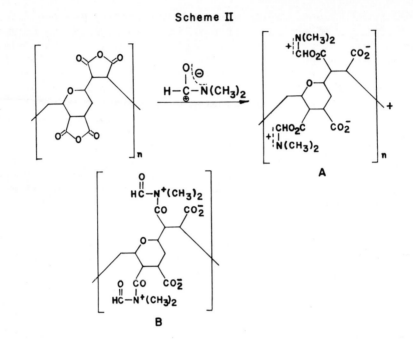

A

B

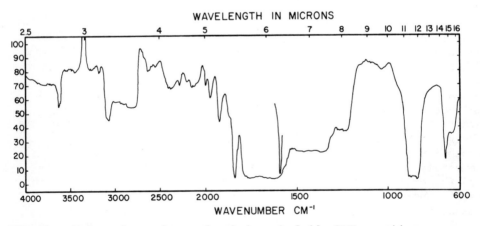

WAVELENGTH IN MICRONS

WAVENUMBER CM⁻¹

Fig. 1.   Infrared spectrum of maleic anhydride-DMF reaction
          product in DMF.

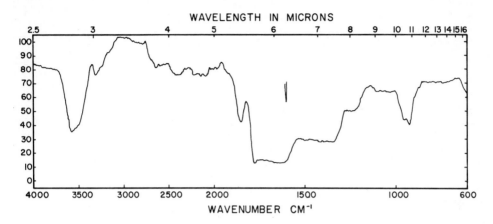

Fig. 2.   Infrared spectrum of the copolymer of maleic anhydride
and divinyl ether in DMF.

GPC Fractionation of DVE-MA Copolymer
and Its Dimethyl Ester in Organic Media

DVE-MA Copolymers in Mixed Solvent (80% Acetone and 20% Tetrahydrofuran.   The GPC analyses of the copolymers are shown in Table I.   Mixed solvents of 80% acetone and 20% tetrahydrofuran were
used, since this mixture dissolves both the polystyrene standards
and the copolymers.   Polystyrene is only partially soluble in acetone which appears to be a good solvent for DVE-MA copolymers.   On
the other hand, THF dissolves polystyrene but not the copolymers.
The mixed solvents were therefore employed in this study.   It can
be seen that copolymers A-AH, B-AH, C-AH, D-AH, E-AH, and F-AH all
had about the same $\bar{M}_w$ and $\bar{M}_n$.   They all showed similar patterns in
the chromatograms, ranging from 28 to 37 counts with peak maxima
centered at about 30 counts.

The above copolymers were expected to possess different mole-
ular weights as they were prepared under the same conditions except the amount of initiator was varied. $\bar{M}_n$ obtained from vapor-
phase "osmometry" measurements indicated that copolymer A-AH had
higher $\bar{M}_n$ than copolymers B-AH and C-AH and that copolymer E-AH had
higher $\bar{M}_n$ than F-AH.   This result is contrary to what one would
predict from free-radical polymerization theory and cannot be

TABLE I.   GPC Analyses of the Copolymers AH Series[a]

Copolymer	A-AH	B-AH	C-AH	D-AH	E-AH	F-AH	M-AH	N-AH
Polymerization solvent	b	b	b	b	b	b	c	c
Mole %, AIBN	5.0	0.5	0.1	0.2	1.0	5.0	5.0	0.2
Polymerization time, hr	14	16	16	0.83	0.83	1.3	0.50	1.5
$\bar{M}_w \times 10^{-4}$	3.55	3.39	3.11	4.08	3.61	3.93	4.74	4.38
$\bar{M}_n \times 10^{-4}$	2.27	2.02	2.03	2.47	2.08	2.15	3.52	3.07
$\bar{M}_w/\bar{M}_n$	1.59	1.68	1.53	1.65	1.74	1.83	1.35	1.43
Peak maximum (elution count)	30.1	30.2	30.6	29.7	29.9	29.6	29.6	29.6
Peak base width (elution count)	28.0-38.0	28.0-38.0	28.5-37.5	27.5-38.0	28.0-38.0	27.5-38.5	28.0-38.0	27.5-37.5
$\bar{M}_n$(VPO) $\times 10^{-4}$	3.73	1.33	0.927	0.408	0.253	0.273	0.184	0.255

[a] AH = copolymer of divinyl ether and maleic anhydride
[b] Acetone
[c] 1,4-Dioxane

adequately explained at present. Possibly different modes of termination may be occurring. Also, $\bar{M}_n$ determined by GPC is much higher than that by the VPO method, the ratios ranging from 1.5 to 20. The VPO method gives absolute values of $\bar{M}_n$ and these values therefore must be accepted as the more accurate and reliable data. The reasons for the higher values of $\bar{M}_n$ by GPC may be due to the fact that the copolymers used in this investigation are of relatively low molecular weight. The specific column set-up does not give good resolution for low-mw copolymers. It would be necessary to use more columns packed with low-pore-size Styragel in order to detect and separate rather low-mw copolymers.

These results are consistent with the fact that there were also difficulties in the GPC calculations as the tail-end of the GPC chromatogram overlapped with two smaller solvent peaks at 37-39 count which necessitated the arbitrary cut-off point at 37.5 count for GPC calculation. Elution count at 37.5 corresponds to mw of 1,600 from polystyrene calibration. (See Figure 3). Those species of mw lower than 1,600, even though they may be present, were not counted. If those copolymers having mw lower than 1,600 had been taken into consideration for the calculation, they would affect $\bar{M}_w$ only slightly but would lower $\bar{M}_n$ substantially thus bringing $\bar{M}_n$ by GPC close to that obtained by VPO. Also, the copolymers are subject to hydrolysis, and the mixed solvent might contain small amounts of water which may be enough to react with the copolymers so as to change their structure and refractive index, and even to expand the chain and alter the conformation.

$\bar{M}_w/\bar{M}_n$ ratios of the AH copolymers are in the range of 1.35-1.83 which are somewhat smaller than would be expected. The $\bar{M}_w(GPC)/\bar{M}_n(VPO)$ ratio perhaps gives a better approximation of the true polydispersity of the copolymers. $\bar{M}_w/\bar{M}_n$ values of 4-6 were reported by Allen and Turner (12).

The GPC curves for copolymers E-AH and F-AH prepared in 1,4-dioxane seemed to be sharper than those prepared in acetone and seemed to have less overlapping in the tail of the GPC curve. However, there are greater discrepancies between $\bar{M}_n(GPC)$ and $\bar{M}_n(VPO)$.

Dimethyl Ester of the Copolymer in Mixed Solvents (80% Acetone and 20% THF). The cyclocopolymer of divinyl ether [M_1] and dimethyl fumarate [M_2] (DMFMT) having the composition of $M_1:M_2 = 1:2$ is expected to have a structure similar to those copolymers prepared by esterification of DVE-MA copolymers. When the DVE-DMFMT copolymer was found to be fractionated in the above mixed solvents using Styragel, it was decided to convert the DVE-MA copolymers to their dimethyl ester forms to determined molecular weights under the same conditions. The results are presented in Table II.

TABLE II. GPC Analyses of Copolymer DME Series[a]

	A-DME	B-DME	C-DME	D-DME	E-DME	F-DME	M-DME	N-DME	X-DME	DVE+DMFMT
$\bar{M}_w \times 10^{-5}$	2.22	2.26	2.07	2.15	2.76	2.91	1.38	1.47	3.22	0.993
$\bar{M}_n \times 10^{-5}$	1.47	1.54	1.48	1.59	2.21	2.26	1.08	1.15	1.99	0.759
$\bar{M}_w/\bar{M}_n$	1.51	1.47	1.40	1.35	1.25	1.29	1.28	1.27	1.61	1.31
Peak maximum (elution count)	25.2	25.0	25.0	25.0	24.1	23.9	27.2	25.8	24.8	27.2
Peak base width (elution count)	20.0-32.0	20.5-32.0	20.5-31.0	21.0-31.0	20.5-30.0	20.5-30.0	21.0-32.0	23.0-32.5	19.5-31.0	24.0-34.0
$\bar{M}_n$(VPO) $\times 10^{-5}$	0.210	0.435	3.05	1.13	0.278	0.954	0.509	0.244	0.555	0.142
$\bar{M}_w$(GPC) $\times 10^{-5}$	6.25	6.67	6.65	5.27	7.65	7.39	2.91	3.36		
$\bar{M}_n$(GPC) $\times 10^{-5}$	6.65	7.62	7.29	6.44	10.63	10.51	3.07	3.75		

[a] DME = Dimethyl ester of copolymer of divinyl ether and maleic anhydride.

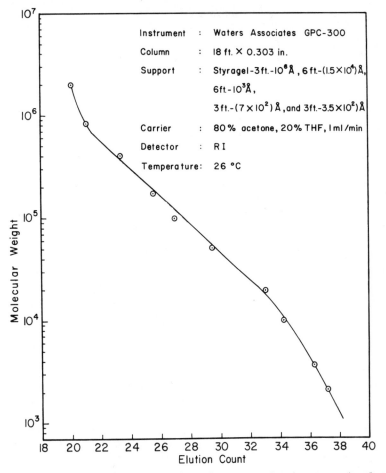

Instrument : Waters Associates GPC-300

Column : 18 ft. × 0.303 in.

Support : Styragel-3ft.-10^6Å, 6ft.-(1.5×10^5)Å, 6ft.-10^3Å, 3ft.-(7×10^2)Å, and 3ft.-3.5×10^2)Å

Carrier : 80% acetone, 20% THF, 1 ml/min

Detector : R I

Temperature: 26 °C

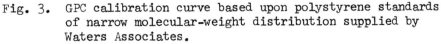

Fig. 3.   GPC calibration curve based upon polystyrene standards of narrow molecular-weight distribution supplied by Waters Associates.

The copolymers were predicted to have different molecular weights as controlled by the amount of initiator used in the previous cyclocopolymerizations. However, they were eluted in about the same count range, and have similar peak shapes with peak maxima at about the same position. The copolymers showed a rather high mw based on the polystyrene calibration as they all eluted in the range from 20 to 31 counts. Two possible explanations may be offered to account for these observations: (a) in the methylation reaction to prepare the dimethyl esters of the copolymer (as shown in Scheme III), diazomethane was used to introduce the second methyl group. It is conceivable that diazomethane might combine two polymer chains, thus increasing the molecular weights. From consideration of the known chemistry of this system, it is difficult to

Scheme III

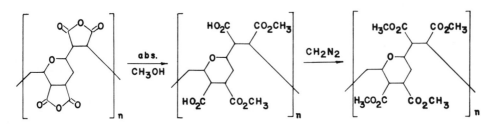

postulate a reaction between CH_2N_2 and the copolymer main chain
that could link two or more chains: (b) steric effects of adjacent
carbomethoxy groups might prevent the polymer chain from assuming
a random coil and thus extending the chain to a certain degree.

The experiments were repeated with particular caution exercised
in the methylation procedure by adding the diazomethane solution
dropwise to a very dilute solution of the acid esters of the copolymers in absolute methanol. However, the same GPC results were
obtained. The steric effect seems to come into play in this case.
The molecular weight is calculated to increase by only 19.22% by
conversion of the anhydride units to the dimethyl ester structure
in the copolymers. However, the GPC data indicated an increase of
five to seven fold in both $\bar{M}_w$ and $\bar{M}_n$ for copolymers A, B, C, and D;
seven to ten fold for copolymers E and F, and three to four fold
for copolymers M and N, as shown in Table II. From the independent
values of $\bar{M}_n$ obtained by the VPO method, $\bar{M}_n$'s for the dimethyl esters
are still in the range from three to thirty-five times higher than
those for the corresponding original copolymers in their acid anhydride forms, as can be calculated from Tables I and II. Further investigation into this unusual phenomenon will be necessary before
a satisfactory explanation can be made.

Figure 4 shows the integral molecular-weight distributions of
copolymers B, F, and N in their anhydride and dimethyl ester forms.
Copolymers B-AH, F-AH, N-AH contained a fairly large amount of
lower-mw species which disappared when they were converted to their
dimethyl ester. The polymer chains might expand more in the low-mw
species than in the higher-mw range. There was always a sharp increase of slope at the curve center and there was not much high-mw
species found from the curve.

GPC chromatograms for one copolymer of DVE-DMFMT and for a
DVE-MA copolymer having η_{inh}=1.56 which was prepared in benzene by
the procedure previously published (15), clearly pointed out that
fractionation is possible for the dimethyl ester copolymer under
the present experimental conditions provided that a new calibration

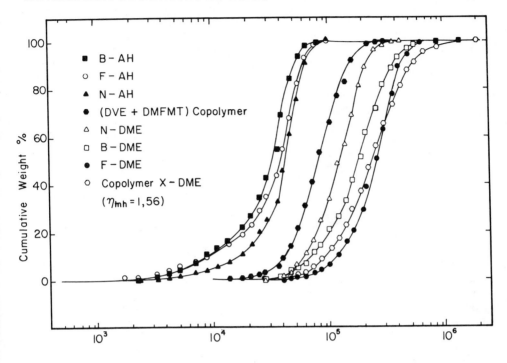

Fig. 4. Integral molecular-weight distribution of copolymers of maleic anhydride and divinyl ether and their methyl esters.

curve suitable for this particular copolymer is made. Since polystyrene is a vinyl polymer, it is quite different from the dimethyl ester of the DVE-MA copolymer, a structure which tends to expand the polymer chain because of two pendant carbomethoxy groups adjacent to each other. Therefore, in order to get more accurate molecular-weight information on this copolymer, narrowly fractionated samples of the dimethyl ester copolymers should be prepared to use in preparation of the calibration curve. On the basis of this calibration curve, accuracy in calculating the molecular weights of the copolymers can be expected to improve. Moreover, GPC characterization of the MA-DMFMT copolymers can be performed according to the suggested procedure described above. Instead of mixed solvents, THF can be used as the GPC carrier. Another fruitful approach would be to use the "universal calibration" approach (16) which would permit use of the present polystyrene calibration. However, the values of the constants K and a of the Mark-Houwink equation are not presently known for this copolymer system.

EXPERIMENTAL

Materials and Equipment

Maleic anhydride (Fisher Scientific Co.) was sublimed under
vacuum twice prior to use. Divinyl ether (Merck, Sharpe and Dohme)
was purified by washing several times with dilute aqueous KOH solution at $0^{\circ}C$, with ice cold water, followed by drying over CaH_2,
then distilling under N_2 atmosphere. 1,4-Dioxane was refluxed over
Na and freshly distilled before use. Acetone for polymerization
was refluxed over P_2O_5 and distilled in N_2 just before use. Thick-
walled polymerization tubes were soaked in dichromate cleaning solution, cleaned by steam washing, rinsed with deionized water and
dried in an oven. Polymerizations were done in a Sargent constant
temperature oil bath. Infrared spectra were recorded using a Beckman IR-8 Infrared Spectrophotometer. NMR Spectra were recorded
using a Varian A-60 Spectrometer. Number-average molecular weights
were measured in acetone at $37^{\circ}C$ by use of a Mechrolab Vapor Pressure Osmometer Model 302. Waters Associates Gel Permeation Chroma-
tograph, Model GPC-300, was used to determine molecular weights and
molecular-weight distribution of the copolymers. A calibration
curve (Figure 3) was prepared using narrowly-distributed polystyrene
standards to aid in interpretation of the results of this study.

Cyclocopolymerizations

Solutions of maleic anhydride (MA) (1.0 g, 0.01 M) and various
concentrations of azobisisobutyronitrile (AIBN) in 20 ml freshly
prepared acetone or 1,4-dioxane were charged by means of a pipet
into thick-walled polymerization tubes maintained at $0^{\circ}C$ in ice
water bath which had been evacuated and filled with purified N_2
while N_2 was passing into the tube. 0.5-1.0 ml (0.005-0.01 M) of
pure divinyl ether (DVE) was charged into the tubes immersed in a
Dewar Flask containing liquid nitrogen. Following conventional
freeze-thaw treatment, the tubes were sealed and placed in a stirred
oil bath at $60^{\circ}C$ for a fixed period of time. The solutions became
cloudy within 15 minutes. The cloudiness of the copolymer solutions
in acetone disappeared when the polymerization tubes were removed
from the oil bath and cooled to room temperature. In a Dri-train
Dry Box filled with N_2, the copolymer solutions were precipitated
by adding dropwise into 200 ml of stirred hexane which had been
previously dried over 4A molecular sieves, followed by filtering
the copolymers which precipitated. The white fluffy copolymers
were then dried in vacuo at room temperature. The copolymers were
transferred to vials capped with rubber septums in slightly positive
N_2 pressure and stored in a desiccator over P_2O_5. The NMR spectrum
of the copolymer in its anhydride form is shown in Figure 5.

Chemical Reactions on the Cyclocopolymers

Hydrolysis and Sodium Carboxylate Formation. A 30 mg sample
of each of the copolymers was weighed out and dissolved in 10 ml
deionized water with stirring on a hot plate. The solution turned
strongly acidic in a few minutes. Then 0.1 $\underline{N}$ NaOH solution was added
dropwise until the copolymer solutions became slightly basic. The
solutions were dried to solids which were retained for use in the
GPC experiments.

Esterification. A 0.2 g sample of each of the copolymers and
15 ml absolute methanol were placed in a 50 ml one-neck round-bot-
tom flask and refluxed for 2 hr. The copolymer solutions were then
cooled in an ice-water bath and diazomethane solution in ether maintained at -15°C was transferred by pipet and propipet and added
dropwise to the copolymer solutions with magnetic stirring. The
diazomethane solution was prepared from p-tolylsulfonylmethyl-nitro-
samide (17) which was in turn made from toluene sulfonyl chloride
(18). The white dimethyl ester copolymer precipitated out as soon
as the diazomethane solution was added. The methylation was stopped when a pale yellow color appeared and persisted for a few minutes.
The resulting yellow solution was filtered and the solid copolymers
were dried in vacuo at 60°C. The yields were well above 90%. Typical analysis; calculated for $C_{16}H_{22}O_9$: C, 53.63; H, 6.19; O, 40.18.
Found: C, 53.72; H, 6.25. The IR spectrum of the dimethyl ester
copolymer is shown in Figure 6 and the NMR spectrum is shown in
Figure 7.

Gel Permeation Chromatographic Measurements

Calibration. Narrowly-distributed polystyrene standards supplied by Waters Associates were used to plot the calibration curve.
The calibration curve is shown in Figure 3. The GPC column arrangement was a series of 6 (3'x3/8" O.D.) stainless steel columns packed
with Styragel having pore sizes of 3 ft.-10^6 Å, 6 ft.-(1.5×10^4)
Å, 6 ft.-10^3 Å, 3 ft.-(7×10^2) Å and 3 ft.-(3.5×10^2) Å respectively.
Mixed solvent of 80% acetone and 20% THF was used as carrier; tech-
nical grade acetone was dried overnight and 4A molecular sieves and
mixed thoroughly with reagent grade THF. The operating temperature
was at room temperature of about 26°C. The flow rate was adjusted
at 1 ml/min and polymer concentration used was 0.2%. Typical GPC
chromatograms are shown in Figure 8a-d.

GPC Calculations. A Fortran IV computer program was used to
calculate weight-average molecular weights ($\bar{M}_w$), number-average
molecular weights ($\bar{M}_n$), the ratios of $\bar{M}_w/\bar{M}_n$, and integral molecular-weight distribution from each GPC chromatogram. Integral
molecular-weight-distribution curves of several copolymers are

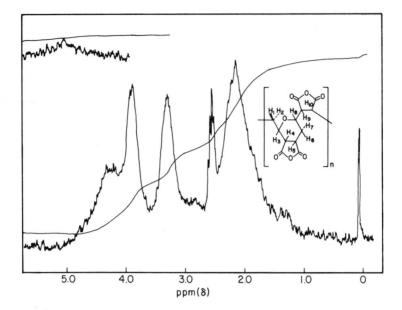

Fig. 5.   60 MHz NMR spectrum of maleic anhydride - divinyl ether
copolymer (DMSO-d6).

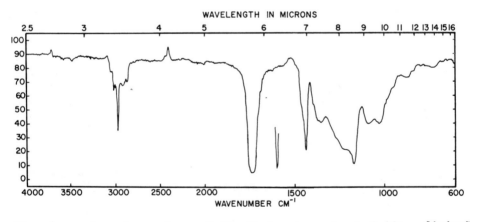

Fig. 6.   Infrared spectrum of dimethyl ester of anhydride - divinyl
ether copolymer.

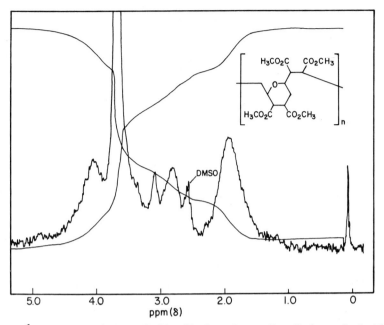

Fig. 7. 60 MHz spectrum of dimethyl ester of maleic anhydride -
divinyl ether copolymer (DMSO-d$_6$).

shown in Figure 4. $\bar{M}_w$ and $\bar{M}_n$ were computed in accordance with the
following equations:

$$\bar{M}_w = \frac{\Sigma H_i M_i}{\Sigma H_i} \qquad\qquad \bar{M}_n = \frac{\Sigma H_i}{\Sigma M_i / H_i}$$

where H_i is the relative height of a GPC curve measured at each
half-elution count, and M_i is the corresponding molecular weight
which was found from the calibration curve. No correction on the
effect of zone broadening was applied.

CONCLUSION

Procedures for fractionating DVE-MA copolymers and their dimethyl esters have been developed and studied. The results indicate that further investigations are necessary before the procedures
can be considered satisfactory. For DVE-MA copolymers, the stringent
requirement that the mixed solvents to be used be free of water,
which causes hydrolysis of the copolymers, should be met to insure
no complication of structural changes during the GPC runs. Moreover,

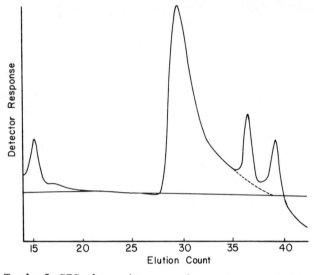

Fig. 8a.   Typical GPC chromatogram of copolymer of divinyl ether
and maleic anhydride.

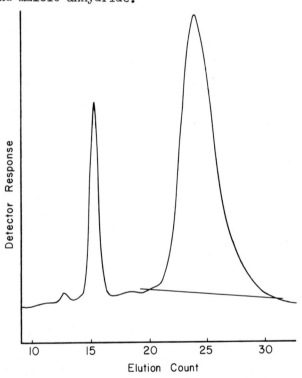

Fig. 8b.   Typical GPC chromatogram of dimethyl ester of copolymer
of divinyl ether and maleic anhydride.

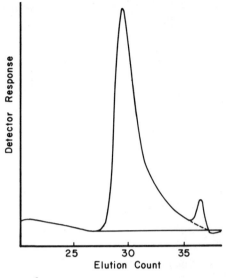

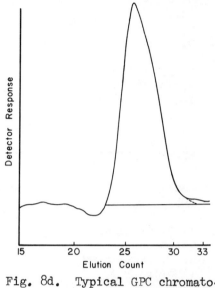

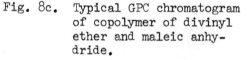

Fig. 8c. Typical GPC chromatogram   Fig. 8d. Typical GPC chromato-
of copolymer of divinyl   gram of dimethyl ester
ether and maleic anhy-   of copolymer of divinyl
dride.   ether and maleic anhy-
                                                             dride.

additional columns packed with Styragel having low pore sizes should
be used to improve resolution and permit more accurate calculation
for the copolymers having low-mw species. As for the dimethyl ester
of the copolymer, the polymer chain tends to expand because of steric
crowding of adjacent carbomethoxy groups, resulting in unusually
high molecular weights calculated from the polystyrene calibration.
In order to obtain accurate molecular weights, narrowly distributed
fractions of the dimethyl ester of the copolymer should be prepared
by some fractionation method, and be used to provide a true calibration curve for the system. As pointed out above, use of the
"universal calibration" (16) approach may also be used to advantage
in this system. Future work will be directed along these lines.

We are grateful to the National Institutes of Health for financial
support of this work under Grant No. CA-06838. C. Wu was a post-
doctoral fellow during the period 1971-1972.

BIBLIOGRAPHY

1. G. B. Butler, Abstracts, 133rd American Chemical Society Meeting, San Francisco, Calif., April 13-18, 1958, p. 6R.

2. G. B. Butler, Abstracts, 134th American Chemical Society Meeting, Chicago, Ill., Sept. 7-12, 1958, p. 32T.

3. G. B. Butler, J. Polymer Sci., 48, 279 (1960).

4. G. B. Butler (to Peninsular ChemResearch, Inc.), U. S. Pat. 3,320,216 (May 16, 1967); U. S. Pat. Reissue 26,407 (June 11, 1968).

5. G. B. Butler, "Cyclopolymerization," in N. M. Bikales, Ed., Encyclopedia of Polymer Science and Technology, Interscience Publishers, New York, Vol. 4, 1966, pp. 568-599.

6. Based upon structural information submitted to NIH on Oct. 12, 1959, sample on June 14, 1960 and test results of July, 1961.

7. T. C. Merigan, Nature, 214, 416 (1967).

8. T. C. Merigan and W. Regelson, New Engl. J. Med., 277, 1283 (1967).

9. A. E. Munson, Ph.D. Dissertation, Dept. of Pharmacology, Medical College of Virginia, Richmond, Va., June 1970.

10. A. Isaccs and J. Lindenmann, Proc. Roy. Soc. (London), Ser. B, 147, 258 (1957).

11. Y. K. S. Murthy and H. P. Anders, Angew. Chem. Internat. Edit., 9, 480 (1970).

12. V. R. Allen and S. R. Turner, J. Macromol. Sci.-Chem., A5, 229 (1971).

13. K. Fujimori, Progress Report, 2/26/70, NIH Grant No. CA-06838.

14. H. Morawetz, Macromolecules in Solution, Chapter VII, Interscience, New York, 1965.

15. G. B. Butler, J. Macromol. Sci.-Chem., A5, 219 (1971).

16. H. Coll and D. K. Gilding, J. Polymer Sci., Part A-2, 8, 89 (1970).

17. Org. Syntheses, Collective Vol. IV, p. 250.

18. Org. Syntheses, Collective Vol. IV, p. 943.

CONFORMATIONAL STUDIES OF POLY(ETHYL ACRYLATE-co-ACRYLIC ACID)

Susan P. Gasper and Julia S. Tan

Eastman Kodak Company

Rochester, New York

In recent years, extensive complementary light-scattering and viscosity studies have been carried out by a number of workers for nonionic polymers (1-7) but very few for polyelectrolytes (8-10). Our present work concerns a systematic study of a poly(ethyl acrylate-co-acrylic acid), an interesting polyelectrolyte with both hydrophilic and hydrophobic groups. The copolymer contains ethyl acrylate and acrylic acid residues in a mole ratio of three to one, respectively (Figure 1). The weight fractions of each comonomer in the polymer were found to be constant over the entire molecular-weight range studied. The different monomer groups are also randomly distributed along each chain.

The copolymer displays normal chain characteristics identical to those of nonionic polymers when studied in dilute organic solutions in the nonionized form. In aqueous solutions, however, it displays both typical and atypical polyelectrolyte behavior. In the fully ionized form in aqueous solution, the polyions increase in chain dimension as do other polyelectrolytes owing to the electrostatic repulsion exerted by the charges on one another along the chain. This copolymer, in addition, displays some unusual properties in aqueous solutions owing to the presence of hydrophobic

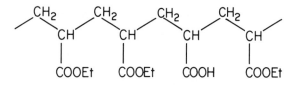

Fig. 1. The structure of poly(ethyl acrylate-co-acrylic acid)(3:1).

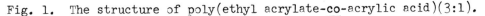

387

ester groups along the chain. The objectives of the first studies
of this copolymer were then to examine the unusual solution behavior of a polyelectrolyte with a high content of hydrophobic side
chains and also to supplement the rather limited literature data
for a systematic study of polyelectrolytes in terms of various theories of the excluded volume effect and viscosity.

The initial characterization work involved <u>light-scattering
measurements</u> on a series of fractions of the copolymer in tetrahydrofuran, 4-methyl-2-pentanone, and 2-heptanone. and in aqueous
sodium chloride media as a function of ionic strength at pH 7 (11).
The molecular weights, second virial coefficients, and radii of
gyration were determined at 25°C. The data obtained were interpreted in terms of current theories of the excluded volume effect on
polyelectrolyte solutions.

Changes in molecular parameters as a function of <u>ionic strength</u>
revealed unusual polyelectrolyte behavior. Figure 2 shows a plot
of radius of gyration ($\langle s^2 \rangle^{1/2}$, Å) against ionic strength (C_s) for
the series of fractions. The plot indicates that the polymer is
an extended coil at low ionic strength and becomes a very compact
coil at high ionic strength. In the ionic strength region between
these two extremes, the data indicate that there may be a conformational change taking place as the ionic strength is progressively
increased. Figure 3 shows a double logarithmic plot of the second
virial coefficient against ionic strength. The breaks in the curves
suggest the existence of a structural transition region in this
system.

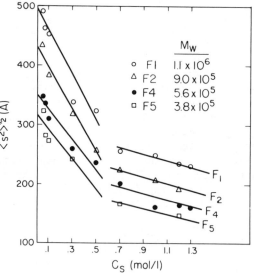

Fig. 2. Radius of gyration vs. ionic strength for fractions of
        the sodium salt of poly(ethyl acrylate-co-acrylic acid).

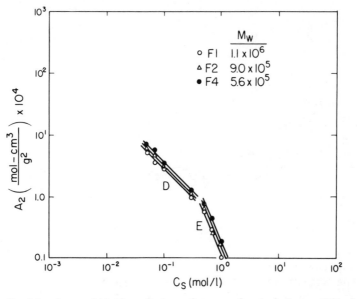

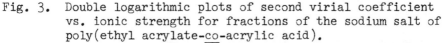

Fig. 3.   Double logarithmic plots of second virial coefficient vs. ionic strength for fractions of the sodium salt of poly(ethyl acrylate-co-acrylic acid).

Intrinsic viscosities and hydrodynamic expansion factors were also measured for several of the same fractions in the set of solvents identical to that of the light-scattering studies (11). Unusual behavior was observed in the hydrodynamic properties of the copolymer. Figure 4 shows a plot of intrinsic viscosity against ionic strength for several of the fractions. The transition from an extended to a compact coil is also evident here.

The occurrence of a conformational transition region, therefore, has been indicated both thermodynamically and hydrodynamically. The transition can be interpreted as a change from an open, extended coil at low C_s to a tightly coiled molecule at high C_s. In the low C_s region, the coil is extended by the mutual repulsion of the charged groups, whereas the insoluble hydrophobic ester groups are dispersed in the aqueous medium. In the presence of a moderate concentration of sodium chloride, a screening effect by the counterions is present. The coil begins to collapse owing to the reduction in the repulsion of the charged groups as well as to the hydrophobic attraction of the ester groups. The transition is somewhat broad and the molecule does not reach the rodlike state even in extremely low C_s media, presumably because this copolymer lacks a high charge density along the chain.

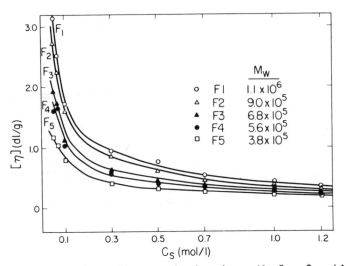

Fig. 4.   Intrinsic viscosity vs. ionic strength for fractions of
          the sodium salt of poly(ethyl acrylate-co-acrylic acid).

In the present work, conformational studies of this copolymer
were extended with fluorescence intensity measurements and potentiometric titrations. The dye, 2-p-toluidinylnaphthalene-6-sul-
fonate (TNS), which has been used extensively in protein studies
to detect hydrophobic binding sites (12), was chosen as the fluo-
rescent probe for the copolymer. The emission of this probe is
completely quenched in water or aqueous sodium chloride medium,
but gives a very intense emission band peaking at 440 nm in alcohols
or other organic solvents when excited with ultraviolet radiation.
Thus the probe would be responsive to the amount of hydrophobic
region available to it within the coil when the solvent is changed
from low C_S to high C_S.

Potentiometric titrations were performed in a 2-propanol-
water (4:1) mixture owing to the limited solubility of the nonionized or partially ionized copolymer in aqueous solution. Titration
curves characteristic of a polyelectrolyte undergoing a conformational transition during the ionization were obtained for this system. Titration studies were made on polymers with the same degree
of polymerization but different acid contents in media of different
ionic strengths. In addition, the inherent viscosities of the same
polymer solutions were measured in media having ionic strengths and
degrees of ionization. These potentiometric and viscosity studies
give additional information concerning the roles of ester and acid
groups in determining the conformations of the chain.

to a polymer solution, and the reaction was allowed to continue for several days. After the addition of a known amount of hydrochloric acid, sufficient to neutralize both the excess base and the polymer, the samples were titrated in a manner similar to that already described. No attempt was made to isolate the products. The ionic strength of each solution was calculated, knowing the amount of acid added to each sample; additional sodium chloride was added when necessary to obtain the desired ionic strength. The acid content of the new copolymers was determined from the resulting titration curves, and the curves were analyzed in a manner similar to that used for the original polymer. The mole ratios of ester to acid for the copolymers studied were 3:1 (F4), 2:1 (F4A), and 1:1 (F4B).

Measurements of the inherent viscosities, ln η_{rel}/c, where η_{rel} is the relative viscosity and c is the polymer concentration, of the copolymer F4 in the mixed solvent at various degrees of ionization and in media of different ionic strengths were also performed in conjunction with potentiometric studies. Viscosities were determined at 0.2 g/dl polymer concentration in Cannon-Fenske viscometers at $25.00 \pm 0.02°C$.

RESULTS AND DISCUSSION

Relative Fluorescence Intensity

The relative fluorescence intensity of aqueous solutions of polymer, TNS, and sodium chloride measured at 440 nm has been plotted as a function of C_s in Figure 5, along with the intrinsic-viscosity data obtained previously for the same fraction (11). Both curves indicate that the transition region is near 0.3 - 0.5 $\underline{N}$. The viscosity curve never reaches a constant value, suggesting that the coil is expanded but never reaches a rodlike state at low C_s. It also shows that the coil is compact at high C_s and reaches a constant size beyond $C_s \sim 0.7$ $\underline{N}$. The intensity curve at low C_s reflects that the dye has considerable affinity for the polymer, even in the expanded form of the latter, possibly owing to small hydrophobic clusters of the ester groups. As the size of the molecule decreases and the hydrophobic regions available to the probe become more pronounced, the fluorescence intensity of the dye increases. In the high ionic strength region, the intensity curve does not reach a plateau as quickly as the viscosity curve. Whereas the overall size of the coil changes very little at increasing high ionic strength, it is possible that van der Waals attraction continues to increase as C_s becomes larger, and hence the fluorescence intensity of the dye continues to increase.

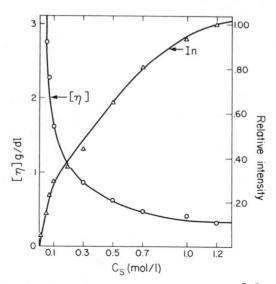

Fig. 5.   Intrinsic viscosity of the copolymer, [η], and relative
fluorescence intensity, In, of TNS vs. ionic strength
for F3.

Potentiometric Titration

The ionization process of a polyelectrolyte can be described
as (13,14) in equation 1:

$$pH = pK_0 - \log \left(\frac{1-\alpha}{\alpha} \right) + \frac{0.434}{RT} \left(\frac{\delta G_{ion}}{\delta \alpha} \right)$$

or $$pK_a = pH + \log \left(\frac{1-\alpha}{\alpha} \right) = pK_a + \frac{0.434}{RT} \left(\frac{\delta G_{ion}}{\delta \alpha} \right) , \qquad (1)$$

where K_0 is the equilibrium constant for an isolated carboxylic
acid, K_a is the apparent equilibrium constant of the chain, α is
the degree of ionization of the polymer, and ($\delta G_{ion}/\delta \alpha$) is the
additional work required to remove H^+ against the strong electrostatic forces of the polyion. For a polyelectrolyte such as poly-
(acrylic acid), which does not undergo a conformational transition
during ionization, pK_a is a monotonic increasing function of α
(15,16). In the case of poly(methacrylic acid) (PMAA)(15,16) or
poly(glutamic acid) (PGA)(14,15,17,18), the titration curve has a
sharp rise at low α, followed by a maximum or plateau region, and
rises again at high α. The abnormal shape of this curve is attributed to a conformational transition within the molecule during ionization, i.e., the helix-coil transition in PGA or the compact-ex-
tended coil transition in PMAA. Figure 6 shows the titration curves for the copolymer of this study (F4) in 2-propanol-water (4:1)
with ionic strengths of 0.0 N, 0.001 N, 0.005 N, and 0.01 N NaCl.

EXPERIMENTAL METHODS

Fluorescence intensity measurements were carried out on a series of polymer-dye solutions. The polymer selected for the measurements was a fractionated sample of the copolymer with M_w of 680,000 (F3). The polymer concentration was kept constant in the series at 1.1×10^{-2} M in the acid moiety and 3.17×10^{-2} M in the ester moiety. The fluorescence probe employed was 2-p-toluidinylnapthalene-6-sulfonate (TNS). A dye concentration of 3.0×10^{-6} M was used in all of the experiments. Solution and dye concentrations were held constant, and a series of eleven solutions were prepared in media of different ionic strengths ranging from 0.0 to 1.2 N, by the appropriate addition of sodium chloride.

Fluorescence measurements were performed with a spectrofluorimeter built in the Eastman Kodak Research Laboratories. The excitation source was a zenon arc lamp of 150 W coupled with a Bausch and Lomb monochromator. The samples were positioned for $45°$ excitation. The detection unit consisted of a Beckman DK-2A spectroreflectometer. Relative fluorescence intensities were determined for the eleven samples. The solutions were excited with radiation of 325-nm wavelength. The intensity of fluorescence of each sample was taken as the intensity at the emission maximum at 440 nm. No shift in the wavelength of this maximum was observed for the various samples. In addition, the wavelength of the absorption maximum and the absorbance at this wavelength were constant for the entire series. In none of the cases did the absorption and emission bands overlap.

An additional fraction of the copolymer of $M_w = 560,000$ (F4) was selected for studying the conformational transition using potentiometric titration. Since the copolymer itself is not soluble in water, it was necessary to find a titration solvent in which the polymer in both the nonionized and completely ionized form would be soluble. A 2-propanol-water mixture (4:1) was selected. Solutions for titration were prepared by dissolving the copolymer in 2-propanol and adding water and sodium chloride to give the desired composition and ionic strength. An initial polymer concentration of 0.4 g/dl was used in each titration. Titrations were carried out by the addition of 0.01 N NaOH in 2-propanol-water (4:1) mixture at an ionic strength identical to that of the polymer solution, in order to maintain a constant chemical environment. A Corning model 12 research pH meter was used in conjunction with a Sargent combination pH electrode. Titrations were performed in the solvent described with no added salt and with sodium chloride concentrations of 0.001 N, and 0.01 N.

In addition, a portion of the same copolymer sample was subjected to basic hydrolysis to convert some of the ester groups into carboxylic acid groups. An excess of sodium hydroxide was added

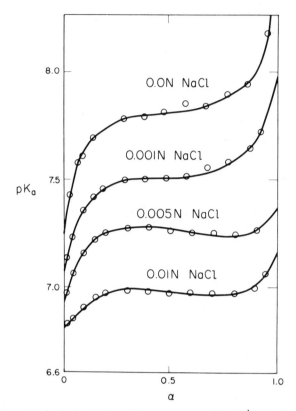

Fig. 6. Potentiometric titration curves for F4 in 2-propanol
         (4:1) at the indicated NaCl concentrations.

The titration curves in 0.01 $\underline{N}$ NaCl and 2-propanol-water (4:1) for
the three copolymers of different acid contents (F4, F4A, and F4B)
with the same chain length are shown in Figure 4. A conformational
transition is clearly indicated from these curves. This transition
becomes more pronounced as the acid content of the copolymer is increased (F4B in Figure 4). These curves can also be represented
by the Henderson-Hasselbach equation (19), i.e.,

$$pH = pK_a - n \log \left(\frac{1-\alpha}{\alpha}\right) \qquad (2)$$

and are shown in Figure 8, where n is an empirical constant depending on the polymer-solvent system. The linear portions of these
plots represent two conformations and the middle part represents
the transition region. The linear parts of the plots in Figure 8
represent the regions labeled a and b in Figure 7. The conformation
at low degree of ionization (see Figures 5 and 6) near 0.2 may be
described as a compact coil with some -COOH groups clustering together, presumably by hydrogen bonding. The coil starts to open
and extends on further ionization and goes through a transition

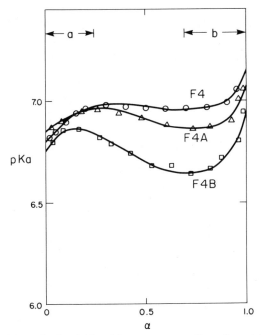

Fig. 7.   Potentiometric titration curves for fractions F4, F4A,
and F4B in 2-propanol-water (4:1) at 0.01 N̲ NaCl.

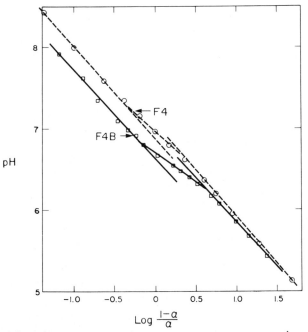

Fig. 8.   Henderson-Hasselbach plots for fractions F4 and F4B in
2-propanol-water (4:1) at 0.01 N̲ NaCl.

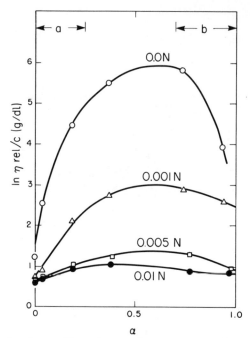

Fig. 9.   Inherent viscosity at a polymer concentration of 0.2 g/dl
          vs. degree of ionization for F4 in 2-propanol-water (4:1)
          at the indicated NaCl concentrations.

region between approximately 0.2 and 0.6.   Ionization beyond this
region will not alter the structure but merely requires more work
for ionization, and the potentiometric titration curve continues to
rise.   This interpretation was confirmed by independent viscosity
measurements for F4 samples at the corresponding degree of ionization and ionic strength.   Results of the inherent viscosity work are
shown in Figure 9.   Viscosity increases as α increases throughout
the region a, which corresponds to the conformation state a as described by the potentiometric titration curve (Figure 7).   The
curves continue to rise but less steeply through the transition region.   Beyond α = 0.6, the curves for 0.001 N, 0.005 N, and 0.01 N
show plateau regions indicative of a constant size and conformation
of the molecule.   The downward trend in the curves is probably due
to the increase in effective ionic strength of the solution as the
titration proceeds and to the insolubility of the fully ionized
polymer in this mixed medium.

Extra free energy is required to ionize a polyion that undergoes a conformational transition during ionization.   The change in
free energy during the conformational transition in the unionized

chain is given by equation 3 (20):

$$\Delta G_o = 2.3 \; RT \int_o^1 (pK - pK_b) d\alpha$$

$$= 2.3 \; RT \int_o^1 (pH - pH_b) d\alpha \tag{3}$$

The actual pH of the chain at α is pH, and pH_b is the hypothetical pH for pure state b at the same α. The variation of pH_b as a function of α can be obtained from the linear extrapolation of part b in Figure 8. Curves of pH and pH_b vs. α for F4, for example, were constructed in Figure 10. Values of ΔG_o for F4, F4A, and F4B in various media were calculated from the areas of the shaded portions, and are listed in Table I. Although ΔG_o appears to decrease with ionic strength from 0.0 N-0.005 N or increase with polymer acid content at a given ionic strength, the differences are small and an average value of 60-90 cal/mole was obtained. This value is comparable to the value of PMAA ($\sim$ 110 cal/mole in 0.1 N NaCl) (19,21).

The fraction of state b (extended coil) present in the transition region can be estimated as follows:

$$\frac{C_b}{C_t} = \frac{\alpha - \alpha_a}{\alpha_b - \alpha_a} \tag{4}$$

where

$$\alpha C_t = \alpha_a C_a + \alpha_b C_b \tag{5}$$

TABLE I. Calculated Values of ΔG_o

Polymer	Ester/Acid, Mole Ratio	ΔG_o,* 0.0 N	ΔG_o,* 0.001 N	ΔG_o,* 0.005 N	ΔG_o,* 0.01 N
F4	3:1	91.3	53.7	37.2	56.6
F4A	2:1				92.6
F4B	1:1			59.6	98.5

- - - - - - -

* cal/mole

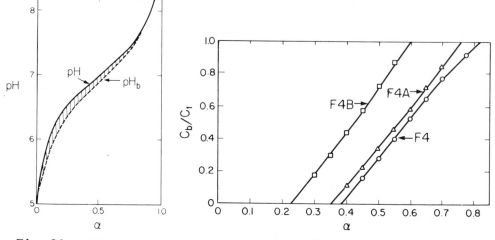

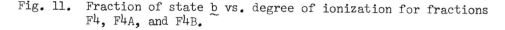

Fig. 10. pH vs. degree of ionization for F4 in 2-propanol-water
(4:1) at 0.01 N NaCl. Curve marked pH_b is a curvilinear
extrapolation obtained from Figure 8.

Fig. 11. Fraction of state b vs. degree of ionization for fractions
F4, F4A, and F4B.

Here C_a, C_b, and C_t are the concentrations of the states a, b, and
total polymer, respectively, and α_a and α_b are the hypothetical
values of α for the two states a and b during the transition at α.
By extrapolating the linear portion of the curve described in Figure 8 into the transition region, values of α_a and α_b can be obtained at a given pH. The fractions of state b (C_b/C_t) plotted
against α for the three polymers studied are shown in Figure 11.
These curves show that the transition begins at lowest α for F4B
and begins at later α values with decreasing acid content of the
copolymers. Using the α values representing the start of the transition for the three polymers (α_0), it is possible to calculate the
fraction of charge per comonomer residue existing at the beginning
of the transition, $f_{\alpha = \alpha_0}$ from the following relationship:

$$f_{\alpha = \alpha_0} = \alpha_0 \left(f_{\alpha = 1.0} \right) \tag{6}$$

where $f_{\alpha = 1.0}$ is the fraction of charge per comonomer residue at
full ionization. The observed values of α_0 , and the calculated
values of $f_{\alpha = 1.0}$ and $f_{\alpha = \alpha_0}$ are listed in Table II. The tabu-
lation shows that although the transition begins at different de-
grees of ionization (column under α_0), the fraction of charge per
residue at the start of the transition (column under $f_{\alpha = \alpha_0}$) is
nearly the same for the three copolymers.

TABLE II.   Transition Parameters for F4, F4A, and F4B

Copolymer	Ester/Acid Mole Ratio	α_o	$f_{\alpha = 1.0}$	$f_{\alpha = \alpha_o}$
F4	3:1	0.38	0.25	0.09_5
F4A	2:1	0.35	0.33	0.11_5
F4B	1:1	0.22	0.50	0.11_0

CONCLUSIONS

The hydrophobic character of poly(ethyl acrylate-co-acrylic acid) in aqueous salt solutions has been investigated by observing the fluorescence behavior of a probe molecule as the ionic strength of the system was progressively increased. The study indicated that the hydrophobic character of the polymer does become increasingly more pronounced as the ionic strength is increased.

The potentiometric studies revealed that a conformational change does take place during ionization. The free-energy change for the transition was determined and was found to be comparable to that of other polymeric systems. Finally, an analysis of the titration data showed that the number of charges per comonomer residue at the onset of the transition was nearly identical for the three copolymers of differing acid contents.

BIBLIOGRAPHY

1. G. C. Berry, J. Chem. Phys., 44, 4550 (1966); ibid., 46, 1338 (1967).

2. T. Norisuye, K. Kawahara, A. Teramoto, and H. Fujita, J. Chem. Phys., 49, 4330 (1968).

3. K. Kawahara, T. Norisuye, and H. Fujita, J. Chem. Phys., 49, 4339 (1968).

4. G. Tanaka, S. Imai, and H. Yamakawa, J. Chem. Phys., 52, 2639 (1970).

5. K. Takashima, G. Tanaka, and H. Yamakawa, Polym. J., 2, 245 (1971).

6. T. Kato, K. Miyaso, I. Noda, T. Fujimoto, and M. Nagasawa, Macromolecules, 3, 777 (1970).

7. I. Noda, K. Mizutani, T. Kato, T. Fujimoto, and M. Nagasawa, Macromolecules, 3, 787 (1970).

8. A. Takahashi and M. Nagasawa, J. Amer. Chem. Soc., 86, 543 (1964).

9. A. Takahashi, T. Kato, and M. Nagasawa, J. Phys. Chem., 71, 2001 (1967).

10. I. Noda, T. Tsuge, and M. Nagasawa, J. Phys. Chem., 74, 710 (1970).

11. J. S. Tan and S. P. Gasper, Bull. Amer. Phys. Soc., 17, 373 (1972); also submitted as Parts I and II to Macromolecules, (1972).

12. G. M. Edelman and W. O. McClure, Acc. Chem. Res., 1, 65 (1968).

13. A. Arnold and J. Overbeek, Rec. Trav. Chim., 69, 192 (1950).

14. M. Nagasawa and A. Holtzer, J. Amer. Chem. Soc., 86, 538 (1964).

15. T. N. Nebrasova, Ye. V. Anufriyeva, A. Yel'yashevich, and O. B. Ptitsyn, Vysokomol. Soyed., 7, 913 (1965).

16. M. Nagasawa, T. Murase, and K. Kondo, J. Phys. Chem., 69, 4005 (1965).

17. A. Wada, J. Mol. Phys., 3, 409 (1960).

18. D. S. Olander and A. H. Holtzer, J. Amer. Chem. Soc., 90, 4549 (1968).

19. J. C. Leyte and M. Mandel, J. Polymer Sci., Part-A, 2, 1879 (1964); Makromol. Chem., 80, 141 (1964).

20. B. H. Zimm and S. A. Rice, J. Mol. Phys., 3, 391 (1960).

21. T. N. Nebrasova, A. G. Gabrielyan, and O. B. Ptitsyn, Vysokomol. Soyed., A10, 297 (1968).

STUDIES ON THE DISTRIBUTION OF SUBSTITUENTS IN HYDROXYETHYLCELLULOSE

E. D. Klug, D. P. Winquist, and C. A. Lewis

Hercules Incorporated

Wilmington, Delaware

Cellulose is a naturally-occurring linear polymer of β-D-glu-copyranose residues (anhydroglucose units) joined through 1→4-glu-cosidic linkages. Each anhydroglucose unit (AGU) in every molecular chain (except those at chain ends) contains three hydroxyl groups, one primary and two secondary. Cellulose derivatives, usually ethers or esters, are obtained by substituting some of these hydroxyls with the desired substituent groups under appropriate conditions. Since most substitution reactions take place under heterogeneous conditions, the substitution pattern which develops is nonuniform.

It is customary to express the substitution level of a cellulose derivative in terms of its degree of substitution (DS), i.e., the average number of hydroxyl groups substituted per anhydroglucose unit. Obviously, the range for DS is zero to three, but not every AGU is necessaryily substituted to the same extent at any DS level except zero or three. For example, at DS 1.0 most AGU's will have a single substituent, but some may have two (or even three), while others may remain unsubstituted.

The substitution pattern is even more complex in the case of certain cellulose ethers such as hydroxyethylcellulose (HEC). Since the substituent group itself contains a hydroxyl group, there is no theoretical limit to the amount of reactant (ehtylene oxide) which may be incorporated. For this reason the substitution of this type of derivative is normally described in terms of MS, i.e., the average number of molecules of reactant combined per AGU.

Although useful, these parameters do not completely character-ize a cellulose derivative. Wirick and others (1,2) have con-

401

cluded that uniformity of substitution is also an important indicator of the polymer properties. Therefore, a convenient procedure for determining the mole % unsubstituted anhydroglucose
would be valuable as a measure of uniformity of substitution. This
chapter includes results of enzymatic hydrolysis studies on HEC
showing a linear relationship between molecular chain breaks and
concentration of unsubstituted AGU's at a constant enzyme to AGU
ratio; a description of an improved analytical technique for determining the concentration of unsubstituted AGU's in HEC or other
cellulose derivatives; and finally a study of a previously proposed
phthalation method for measuring the DS of HEC.

     A recent description of cellulose chemistry can be found in
Ref. 26.

                   ENZYMATIC HYDROLYSIS STUDIES

     Studies were made on a series of five hydroxyethylcellulose
samples all prepared the same way from the same starting cellulose
and having MS values of 1.5 to 5.15 (see Table I). Other properties of these particular samples have been described previously
(3). In Figure 1, the loss in Brookfield viscosity as a function
of time is given for 1% aqueous solutions of these HEC samples,
containing 400 ppm of a commercial cellulase based on the HEC. In
the MS range, 1.5 to 3, there is very little difference in response
to enzymatic hydrolysis. This is somewhat surprising in view of
Figure 2 which shows that there is a significant decrease in unsubstituted AGU with increase in MS. In going from MS 3 to MS 5.15,
there is a significant improvement in resistance to enzymatic hydrolysis despite the fact that the unsubstituted AGU is still 6
mole % at MS 5.15. These data were obtained at constant weight
ratio of enzyme to polymer. Since the average weight of the repeating unit is increased with increase in MS, the ratio of enzyme
to number of AGU's increases with MS. At higher molar ratios of
enzyme to polymer, the rate of hydrolysis would be greater.

          TABLE I. Hydroxyethylcellulose Used in This Study

Sample	1	2	3	4	5
MS	1.47	1.82	2.44	2.96	5.13
Brookfield viscosity in H_2O (cP)	590	340	360	160	260
$[\eta]$, dl/g in H_2O	4.1	4.05	3.90	3.73	4.58

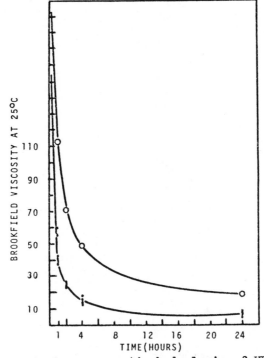

Fig. 1. Effect of MS on enzymatic hydrolysis of HEC.
1% aqueous solution of HEC; 400 ppm enzyme based on solute;
● MS = 1.47, 1.82, 2.44, 2.96; o MS = 5.13.

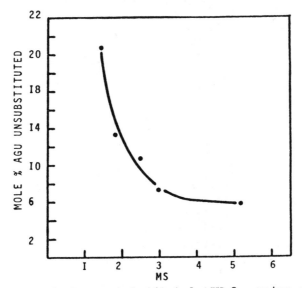

Fig. 2. Effect of MS on unsubstituted AGU for water-soluble HEC.

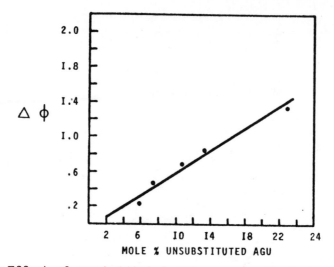

Fig. 3.   Effect of unsubstituted AGU on enzymatic hydrolysis;
          HEC concentration 0.0055 M;   enzyme concentration
          0.0008 g/l;   hydrolysis 24 hr at 25°C.

     Therefore, another series of enzymatic hydrolyses were made
in which account was taken of the change in average weight of the
repeating unit with increase in MS. The solution concentration
was 0.0055 molar with respect to HEC. As a result the ratio of
enzyme to total number of AGU's was kept constant. Also, the data
were obtained in dilute solution and the enzymatic hydrolysis was
expressed as the intrinsic fluidity change, $\Delta\Phi$, which is generally
believed to be proportional to the number of chain breaks (4).
The data (Figure 3) indicate that there is a direct proportionality, within experimental error, between the mole % of unsubstituted
AGU's and the number of chain breaks.

                    DETERMINATION OF ANHYDROGLUCOSE CONTENT

              Hydrolysis of Cellulose and Cellulose Derivatives

     The total acid hydrolysis of cellulose and cellulose deriva-
tives for analytical purposes has received considerable attention.
Various researchers in the field are in agreement as to the general
requirements for hydrolysis, but differ substantially on preferred
procedural details. Usually the sample is treated with concentrated
(72% most common) sulfuric acid at or near room temperature to

effect initial solution. This is followed by dilution and continued hydrolysis at elevated temperatures. Under these conditions it
has been shown (5) that there are three distinct types of reactions
occurring during the hydrolysis: (a) the hydrolysis of cellulose
to intermediate products, and ultimately to glucose; (b) the reversion, or polymerization or condensation of glucose to sugars of
higher molecular weight, probably a variety of polysaccharides;
(c) the destruction (charring) of glucose. Reaction (a) involves
hydrolytic splitting of the glucosidic linkage between the AGU's
of the cellulose chain to yield lower polysaccharides and ultimately glucose. The mechanism of the hydrolysis has been investigated
by a number of workers (6,27). Reaction (b), the reversion or
polymerization of glucose to sugars of higher molecular weight,
has also been extensively studied (5). This reaction is bimolecular and thus should be most pronounced at high glucose concentrations. Reaction (c) is the degradation of glucose to various products. This has been found to be most pronounced at high temperatures. Also, in the analysis of HEC, Samuelson has observed an
intramolecular condensation reaction to give 1,2-O-ethylene-D-glu-
cose (7).

Thus, in devising or modifying hydrolysis conditions for ana-
lyzing cellulose or cellulose derivatives, all three of these possible reactions must be considered. Conditions should be chosen
so as to maximize hydrolytic splitting of the glucosidic bonds and
to minimize both reversion to products of higher molecular weight
and degradation of the glucose to charred products. These competing reactions, coupled with the variations found among types from
different sources, are probably the major factors causing discrepancies in analytical procedures among various workers. For comparison, a few of the more widely used procedures are described
below.

Ritter et al (8) studied the conversion of purified spruce
cellulose to glucose and found that maximum reducing sugar yield
was obtained with a primary hydrolysis of two hours at $35^{\circ}C$ in 72%
H_2SO_4 followed by a 4 hr secondary hydrolysis in 4% H_2SO_4 at $100^{\circ}C$.
They also found that below the optimum temperature, yields were reduced because of insufficient hydrolysis, and above the optimum
temperature, yields were reduced because of charring. Work with
other materials led them to the conclusion that some cellulose materials char more readily than others.

Saeman et al (9) studied the effect of acid hydrolysis conditions on various monosaccharides and concluded that glucose was
the most stable of all the common sugars. Pure glucose carried
through their hydrolysis conditions (1 hr in 72% H_2SO_4 at $30^{\circ}C$,
1 hr in 4% H_2SO_4 in an autoclave at 15 psi steam pressure) was recovered in 96% yield. They also showed that glucose submitted to

simulated total hydrolysis conditions was recovered in lower yield
when the time of secondary hydrolysis was less than 1 hr in an
autoclave. They claimed that this was due to the formation of reversion products in the primary hydrolysis which were hydrolyzed
in the secondary hydrolysis. The time of hydrolysis used by Saeman
was substantially shorter than that used by Hägglund and co-workers
(6). Hägglund's procedure requires 4 hours of primary hydrolysis
with 72% H_2SO_4 at room temperature, followed by intermediate dilution and standing for 6 hours, and finally a 6-hour secondary
hydrolysis at reflux in dilute sulfuric acid. Saeman claimed that
his substantially shorter primary hydrolysis time was effective because of the higher temperature he used in the primary hydrolysis.

Thus, while the hydrolysis of cellulose and cellulose derivatives has been widely investigated, there is little agreement on a
standard analytical procedure. All procedures, however, share the
requirement for a primary hydrolysis in fairly concentrated mineral
acid. No publications have indicated whether the primary hydrolysis
is necessary for the quantitative hydrolysis of water-soluble hydroxyalkylcelluloses.

Methods for Analysis of Hydrolyzates of
Cellulose and Cellulose Derivatives

Procedures for analysis of the total composition of cellulose
hydrolyzates have been described (10-12). All of these procedures,
after the hydrolysis step, involve neutralization, concentration,
and a chromatographic analysis of the hydrolyzate. They all require a considerable expenditure of time and effort. A simpler
approach can be used to determine the unsubstituted AGU content of
hydroxyalkylcelluloses, a measure of the uniformity of substitution.
Determination of the unsubstituted glucose in the hydrolyzate is
possible using the glucose oxidase method (1). The glucose oxidase
reagent consists of the enzymes, glucose oxidase and horseradish
peroxidase, and the chromogen, o-dianisidine (3,3'-dimethoxybenzi-
dine). With these reagents, the quantitative, colorimetric determination of glucose is possible by the following reactions.

$$\text{glucose} + O_2 + H_2O \xrightarrow{\text{glucose oxidase}} H_2O_2 + \text{gluconic acid}$$

$$H_2O_2 + \text{chromogen} \xrightarrow{\text{horseradish peroxidase}} \text{oxidized chromogen} + H_2O$$

The absorbance (13) of the oxidized form of the chromogen is
measured for standards and a calibration curve is prepared from
which the concentration of unknown may be determined. The time for

color development (14), effect of pH on color development, and effect of concentration of reagents have been previously investigated (15).

The advantage of the glucose oxidase method, particularly over reducing sugar methods, is that it is highly specific. It has been shown (16) that the glucose oxidase reagent will react only with glucose, not with a substituted or otherwise modified glucose molecule. The only significant exception to this is 2-deoxy-D-glucose, which has been reported to react at 12% of the rate for glucose. Other possible interferences in the glucose oxidase method have been evaluated (17) and none of them should be encountered in a hydrolysis solution of cellulose or its derivatives.

Recovery of Glucose from Cotton

Since hydroxyalkylcelluloses of known unsubstituted anhydroglucose content are not available, cotton was used as a standard. Cotton should consist almost entirely of β-D-glucose units and thus should serve as a useful indicator of the overall efficiency of both the hydrolysis and the colorimetric procedure. This is not meant to imply that a specific hydrolysis procedure for cotton will proceed to the same degree of completion with hydroxyalkylcelluloses.

The effect of hydrolysis time and hydrolysis conditions on recovery of glucose from cotton is given in Table II. In the notation 1P, 2P, etc., the number indicates the hours of primary hydrolysis and the P specifies that the conditions are 72% H_2SO_4 at 25°C. The notation 5S stands for 5 hours of secondary hydrolysis in 5.5% H_2SO_4 in an oven maintained at 105°C. The notation 7SR indicates that the predissolved sample is refluxed in 5.5% H_2SO_4 for 7 hours. These abbreviations will be used throughout the body of this chapter.

Using hydrolysis conditions of 1P5S, a recovery from cotton of 84.2% of the expected glucose was realized.

$$\% \text{ recovery} = \frac{\text{g anhydroglucose found}}{\text{g cotton taken}} \times 100$$

By extending the primary hydrolysis time to 2 hours and the secondary hydrolysis to 7SR, 95.6% recovery was achieved. Further increase of the primary hydrolysis time to 4 and 6 hours while holding the secondary hydrolysis time constant did not yield a further increase in percent recovery. Thus, the maximum recovery of glucose from cotton under these conditions is 95-96% which is consistent with results found by other workers (18,19). This percent recovery is also consistent with results obtained for destruction of

TABLE II.   Effect of Hydrolysis Conditions on Recovery of Glucose
            from Cotton

Sample	Conditions[a,b,c]	Mean Recovery, %[d]	Rel Std Dev[e]
A	1P5S	84.2	2.5
B	2P7SR	95.6	2.4
C	4P7SR	95.7	-
D	6P7SR	96.0	-

- - - - - -

[a]1P, 2P, etc. = hours of primary hydrolysis in 72% H_2SO_4 at 25°C.

[b]5S = 5 hours of secondary hydrolysis in 5.5% H_2SO_4 at 105°C in
    an oven.

[c]7SR = 7 hours of secondary hydrolysis in 5.5% H_2SO_4 at reflux.

[d]% recovery = [g AGU found/g cotton] x 100.

[e][Absolute standard deviation/mean] x 100.

TABLE III.   Recovery of Pure Glucose[a] Under Acid Hydrolysis
             Conditions

Sample	Conditions[b]	Recovery, %[c]	Mean, %
E	1P5S	97.6	97.6
F	1P5S	97.6	
G	4P7SR	96.9, 96.0	96.0
H	4P7SR	96.0	
I	7SR	96.0	96.0
J	7SR	96.0	
K	1P NOS	56.8	58.8
L	1P NOS	60.8	

- - - - - -

[a]All samples are National Bureau of Standards (NBS) glucose.

[b]Explanation of abbreviations are given in Footnote a of Table II.

[c]All results reported as [g glucose taken/g glucose found] x 100.

glucose under the acid conditions used for hydrolysis. Recoveries
of pure glucose under acid hydrolysis conditions are tabulated in
Table III. From these results, it can be concluded that: (a) the
small loss of glucose (4-5%) occurs mainly in the secondary hydrolysis rather than the primary since samples G and H which were carried through 4P and 7SR were recovered in the same yield as samples
I and J which were subjected only to 7 hours of secondary hydrol-

ysis; (b) for any of these hydrolysis conditions, it would not be possible to recover 100% of the glucose from cotton; (c) the secondary hydrolysis at elevated temperature is necessary not only to complete hydrolysis but also to hydrolyze any reversion products formed in the primary hydrolysis.

Samples K and L represent a 1 $\underline{N}$ solution of glucose in 72% H_2SO_4 which was not diluted and carried through a secondary hydrolysis. Only 50-60% of the glucose was recovered. This is consistent with results found by Saeman et al (6). In interpreting these results, it should be kept in mind that during an actual hydrolysis of cellulose or a cellulose ether, all of the glucose would not be present as such until the end of the hydrolysis and practically none would be present at the beginning. Thus, these experiments represent the most severe conditions possible.

Recovery of Glucose from Cellulose Ethers

The effects of hydrolysis conditions on recovery of glucose from hydroxyethylcellulose (HEC), hydroxypropylcellulose (HPC), and sodium carboxymethylcellulose (NaCMC) were studied. The goal was to obtain the highest recovery of glucose from each of these cellulose ethers by use of a simple procedure. Although the conditions necessary to recover glucose from cotton in 95% yield had been established, this was no guarantee that the same conditions would be best for HEC, HPC, and CMC. Since each of these three cellulose ethers is water soluble, it was not clear that the primary hydrolysis in 72% sulfuric acid was necessary since one of its primary functions is to effect solution of the cellulose or cellulose derivative. Another question was whether the MS or DS of the cellulose ether would affect its degree of hydrolysis under given conditions.

In Table IV are listed results obtained on HEC, HPC, and CMC under various hydrolysis times and conditions. Each of the HEC samples M and N was analyzed using the following three sets of conditions: 1P5S, 1.5P5S, and 7SR. No significant difference was found between the results for either sample under these conditions. This indicates that a primary hydrolysis is not necessary for hydolysis of HEC at this MS level (MS = 2.5-3.0). Also, since an additional 30 min of primary hydrolysis with both samples did not increase the recovery of glucose, it would appear that these conditions are sufficient to effect maximum hydrolysis. This was also found to be the case with HEC samples O, P, and Q, where the results using 1 hr versus 1.5 hr of primary hydrolysis were the same. Also sample R (MS = 5.13) was found to give comparable results when run by the procedure 1P5S (5.73%) versus 7SR (5.83%). Thus, the MS of the HEC does not have an effect on the recovery of glucose under these hydrolysis conditions.

TABLE IV. Effect of Hydrolysis Time and Conditions on Recovery of
           Glucose from HEC, HPC, and NaCMC

Sample	Type	MS or DS	Conditions	Mole % Unsubst Glucose[a]	No. of Detns	Rel Std Dev[b]
M	HEC	2.50	1P5S	6.38	3	1.3
M	HEC	2.50	1.5P5S	5.96	3	6.3
M	HEC	2.50	7SR	6.21	2	2.2
N	HEC	2.59	1P5S	12.1	3	6.1
N	HEC	2.59	1.5P5S	12.4	2	1.5
N	HEC	2.59	7SR	13.1	3	2.0
O	HEC	3.01	1P5S	4.72	2	-
O	HEC	3.01	1.5P5S	4.83	1	-
P	HEC	2.5	1P5S	6.34	1	-
P	HEC	2.5	1.5P5S	6.50	1	-
Q	HEC	2.9	1P5S	3.42		-
Q	HEC	2.9	1.5P5S	3.43		-
R	HEC	5.13	1P5S	5.73	1	-
R	HEC	5.13	7SR	5.82	1	-
S	NaCMC	0.77	7SR	24.2	1	-
T	NaCMC	0.83	7SR	31.6	1	-
U	HPC	3.15	2P7SR	4.88	2	-
U	HPC	3.15	4P7SR	4.72	2	-
U	HPC	3.15	6P7SR	4.66	2	-

- - - - - -

[a]Mole % unsubstituted glucose =

$$\frac{\text{g glucose found}}{\text{g sample taken}} \times \frac{\text{monomer weight}}{\text{mole weight glucose}} \times 100.$$

where monomer weight equals: $162 + (44 \times MS)$ for HEC
and $162 + (58 \times MS)$ for HPC.

[b][Absolute standard deviation/mean] x 100.

Although CMC was not studied in detail, the two samples analyzed (S and T) seemed to hydrolyze under the same conditions.
Thus, based on this evidence, and in view of the results obtained
with cotton and glucose, it can be said with reasonable certainty
that recoveries of 95% of the expected glucose are possible for
water-soluble HEC and NaCMC using only 7SR.

The hydrolysis of HPC by the above procedure was not possible,
however, due to the fact that HPC precipitates at elevated temperatures. Therefore, with HPC it is necessary to carry out a primary
hydrolysis to raise the cloud point sufficiently to avoid precipitation at the elevated temperatures of the secondary hydrolysis.
As can be seen from the results in Table IV, 2 hr of primary hydrolysis is sufficient for this purpose.

Effect of Neutralization Conditions on Recovery of Glucose

After hydrolysis of the cellulose or cellulose ether, the
glucose is in 5.5% sulfuric acid solution. Before the glucose
oxidase method can be applied, it is necessary to neutralize the
solution to a pH between 4 and 5. The favored method has been to
use either $Ba(OH)_2$ or $BaCO_3$. A more recent approach has been to
use a weak-base anion-exchange resin. Use of an anion-exchange
resin is attractive since it obviates the necessity of centrifuging
or filtering to remove the $BaSO_4$. However, we have found that results using an anion-exchange resin (Amberlite IR45-free base form)
are consistently lower by 10% in most cases, and 20-30% relative
in a few cases, compared with results obtained using a barium hydroxide neutralization procedure.

Although no extensive work was done to determine the mechanism
of loss of glucose on the resin, the reaction responsible must be
reversible since continued washing with fresh portions of water resulted in recovery of a portion of the bound glucose. Murphy et al
(20) observed the same effect with xylose and glucose on Amberlite
IR45 (OH^-). Based on kinetic studies, they concluded that the reaction product between the resin and sugar was most likely a
Schiff's base or glycosylamine. They found that glucose reacted
much less readily with the resin than xylose and attributed this to
the smaller concentration of the open-chain form of glucose in solution. Based on this conclusion, the use of a weak-base anion-exchange resin for neutralization could lead to erroneous results
even if only the ratio of the sugars present and not the absolute
results are desired. Therefore, we recommend using either $BaCO_3$
or $Ba(OH)_2$ for neutralization.

DETERMINATION OF DS

Senju (21) first explored the possibility that a DS method for hydroxyethylcellulose could be based on the assumption that under mild conditions phthalic anhydride in pyridine would only esterify the pendant hydroxyethyl hydroxyls. Quinchon (22) reported an improvement of the method, concluding that complete phthalation of pendant hydroxyls occurs in 5 hours at 40°C. Froment (24) has also used the method but at much milder reaction conditions.

Work in our laboratories has shown that water-soluble HEC of MS range from 1.5 to 5.0 is not completely soluble in pyridine (Table V). A sample of MS 1.47 was insoluble. At MS 1.8 to 3.0, the samples were highly swollen but did not dissolve. A product of MS 5.13 was more soluble than the lower DS samples, but was not completely soluble. After addition of the phthalic anhydride and agitation for 16 hr, the samples in the MS range 2.5 to 3 were still not completely dissolved.

The early workers on this method did not indicate whether the HEC used by them dissolved. If it is in solution, the esterification would not be expected to be so selective. On the other hand, if it is not in solution, gelled portions would not be completely accessible and reliable results would not be obtained. In one preliminary experiment in our laboratory with methylcellulose of DS 1.8 reacted at 30°C for 20 hr, a substitution of 0.33 hydroxyethyl hydroxyls was measured; i.e., 0.33 of the 1.2 free hydroxyls reacted with the phthalic anhydride. This result made the phthalation method suspect.

TABLE V.   Solubility of Hydroxyethylcellulose

MS	Solubility in Pyridine	Solubility in Pyridine Plus Phthalic Anhydride
1.47	Insoluble, swollen	Soluble
1.82	Much insoluble material	Few insoluble gel specks
2.44	(Partly soluble ((Many gel specks	(Many swollen ((Insoluble gel specks
5.13	Mostly soluble, few gel specks	Soluble

TABLE VI.   Effect of Reaction Conditions on Apparent DS by Phthalation Technique[a]

Volume of Reagent, ml	Time hr	Temp, $^{\circ}$C	Apparent DS[b]			
			A	B	C	D
10	5	40	1.38	1.34	-	-
20	5	40	1.72	1.51		
10	65-70	23	1.55	1.49	2.19	2.00
20	65-70	23	1.70	1.63	2.41	2.23

- - - - -

[a]Sample weight, 100 mg; 0.34 $\underline{M}$ phthalic anhydride in pyridine.

[b]MS of samples: A = 2.54;   B = 2.46; C = 2.96;   D = 2.80

Limited work carried out HEC samples in the MS range of 2.5-2.9 also raised some serious questions. Starting with conditions specified by Quinchon (22), it was found that by using twice the recommended volume of reagent (Table VI) an apparent DS value significantly higher was obtained. Again, this raises the question of whether the limiting factor is the solubility of the HEC samples in the pyridine. This experiment was repeated using a lower temperature and longer reaction time (Table VI) and the result was found to be the same; a larger volume of reagent gave a higher apparent DS value. Thus, the phthalatiom method does not appear to be a reliable way to determine DS in HEC. An approach for the determination of DS in HPC by use of nuclear magnetic resonance has recently been described (23), but no NMR work has been done on HEC.

CONCLUSIONS

An improved analytical method has been developed for determing mole % unsubstituted AGU in water-soluble hydroxyethylcellulose and sodium carboxymethylcellulose. This method involves hydrolysis in dilute sulfuric acid and eliminates the prehydrolysis step required to hydrolyze cellulose. With hydroxypropycellulose a prehydrolysis step in 72% H_2SO_4 is necessary because HPC is insoluble in hot water. Enzymatic hydrolysis studies on HEC indicate that the number of chain breaks (intrinsic fluidity change) is a linear function of the mole percent of unsubstituted anhydroglucose units.

The previously published phthalation method for determining DS of HEC has been found to give unreliable results.

EXPERIMENTAL METHODS

The HEC samples used in this work were all prepared by the same procedure from the same starting cellulose, the only variable being the ethylene oxide input (see Table I). Glucose used in this work was NBS standard material. Dry surgical cotton (Johnson & Johnson) was used as a standard cellulose.

Enzymatic Degradation. The enzyme used in this work was Cellase 1000 (Wallerstein Laboratories), a commercial enzyme complex derived from Aspergillus Niger. The hydrolyses described in Figure 1 were made in distilled water. Those in Figure 3 were made in water buffered to pH 5.5 with a phosphate buffer. Viscosity data in Figure 3 were obtained with a Ubbelohde viscometer. In fact, the hydrolyses were conducted in the viscometers.

Primary Hydrolyses. A 0.25-0.35 sample was dissolved in 3.0 ml of 72% H_2SO_4 with stirring. The hydrolyses were carried out under nitrogen at $25^{\circ}C$ for the indicated lengths of time.

Secondary Hydrolyses. The solution from the primary hydrolysis was transferred to an Erlenmeyer flask and diluted to give a solution 5.5% in H_2SO_4. The mouth of the flask was covered with aluminum foil, the flask was brought to a boil on a hot plate and was then transferred to an oven maintained at $105^{\circ}C$. The flasks were left in the oven for five hr and were then removed and allowed to cool to room temperature.

For secondary hydrolysis at reflux after a primary hydrolysis, the solution from the primary hydrolysis was transferred to an Erlenmeyer flask, diluted with distilled water to 5.5% H_2SO_4, and heated under reflux for 7 hr.

Hydrolysis in 5.5% H_2SO_4 at Reflux with No Primary Hydrolysis. A 0.25-0.35 g portion of the sample was predissolved in 84 ml of distilled water in an Erlenmeyer flask with stirring. 3.0 ml of 72% H_2SO_4 was then added and the solution heated on a hot plate at reflux for seven hr.

Neutralization by $Ba(OH)_2$ Titration. A saturated solution of $Ba(OH)_2$ was added slowly with stirring to the hydrolysis solution until the pH was 4-5. The precipitate was then allowed to settle for 45-60 min and the pH readjusted to 4-5 with $Ba(OH)_2$ solution. The pH during neutralization was not allowed to exceed 6.0. The $BaSO_4$ was removed by centrifugation and the $BaSO_4$ precipitate was washed with water.

Glucose Oxidase Method. The reagent was prepared from Gluco-stat kits using the manufacturer's directions (24). A 0.20 $\underline{M}$ phosphate buffer (pH = 7.0), rather than distilled water, was used to prepare the reagent. Color development was carried out for 1 hr at 37 C after which the solutions were made 6 $\underline{N}$ in H_2SO_4 by adding sufficient 10 $\underline{N}$ H_2SO_4.

The absorbance of the solutions was measured at 625 nm using either a Beckman Model B or a Beckman DK2A. The amount of glucose in the unknowns was determined from a calibration curve prepared using glucose standards run at the same time as the unknown.

Solubility. A bottle containing 0.2 g HEC and 40 cc pyridine was tumbled 16 hr at room temperature. After the solubility was recorded, 1.0 g phthalic anhydride was added and the bottle again tumbled for 16 hr.

BIBLIOGRAPHY

1. M. G. Wirick, J. Polymer Science, Part A-1, 6, 1705 (1968).

2. S. S. Bhattacharjee and A. S. Perlin, J. Poly. Sci., Part C, No. 36, 509 (1971).

3. E. D. Klug, J. Poly. Sci., Part C, No. 6, 491 (1971).

4. L. F. McBurney in E. Ott, H. Spurlin, and M. Grafflin, Eds., Cellulose and Cellulose Derivatives, Interscience Publishers, New York, 1954, p. 108.

5. W. R. Fetzer, E. K. Crosby, C. E. Engel, and L. C. Kirst, Ind. Eng. Chem., 45, 1075 (1953).

6. E. Hägglund, Chemistry of Wood, Academic Press, New York, pp. 86-103 (1951).

7. O. Ramnas and O. Samuelson, Svensk Papperstidning, 71, 674 (1969).

8. G. Ritter, R. M. Selory, and R. L. Mitchell, Ind. Eng. Chem. Anal. Ed., 4, 202 (1932).

9. J. F. Saeman, J. Bubl, and E. L. Harris, Tappi, 37, 336 (1954).

10. ASTM Standard Method D 1915-63.

11. M. L. Lauer, A. F. Root, F. Shafizadeh, and J. C. Lowe, Tappi, 50, 618 (1967).

12. O. Ramnas and O. Samuelson, Svensk Papperstidning, $\underline{71}$, 829 (1968).

13. R. R. Barton, Analytical Biochemistry, $\underline{14}$, 258-260 (1966).

14. A. Dahlquest, Analytical Biochemistry, $\underline{7}$, 18-25 (1964).

15. A. S. Hester, "Specific Colorimetric Reagents for Glucose" Abstracts of Papers, 129th Meeting ACS, p. 31C, April 1956.

16. R. B. McComb, W. D. Yushok, and W. G. Butt, J. Franklin Inst., $\underline{263}$, 161-165 (1957).

17. O. Hansen, Scandinav. J. Clin. and Lab. Investigations, $\underline{14}$, 651-655 (1956).

18. E. T. Reese et al, J. Bacteriol., $\underline{59}$, 485 (1950).

19. E. T. Reese, Ind. Eng. Chem., $\underline{49}$, 89 (1957).

20. P. T. Murphy, G. N. Richards, and E. Senogles, Carbohyd Res., $\underline{7}$, 460 (1968).

21. R. Senju, J. Ag. Chem. Soc. Japan, $\underline{22}$, 50 (1958).

22. J. Quinchon, Comp. rend., $\underline{248}$, 225 (1959).

23. F. L. Ho, R. Kohler, and G. Ward, Anal. Chem., $\underline{44}$, 178 (1972).

24. G. Froment, Ind. Chem. Belg., $\underline{23}$, 115 (1958).

25. "Glucostat," Worthington Biochemical Corp., Freehold, New Jersey.

26. N. M. Bikales and L. Segal, Eds., Cellulose and Cellulose Derivatives, Parts IV and V, Wiley-Interscience, New York, 1971.

27. Ibid, pp. 991-1006.